CONTRIBUTIONS TO THE HISTORY AND PHILOSOPHY OF SCIENCE

By

Aldemaro Romero Jr., Ph.D.

**Volume I of the series
"Scholarly Contributions by Aldemaro Romero Jr."**

Published in the United States of America

© Aldemaro Romero Jr. All rights reserved.

This publication is in copyright. Subject to statutory exception
and to the provisions of relevant collective licensing agreements,
no reproduction of any part mat take place without the
written permission of the authors.

First published 2020

Library of Congress Cataloging in Publication data

Romero Díaz, Aldemaro, 1951-

Institutional Failure in higher education / Aldemaro Romero Jr.
p. cm.
Includes bibliographic references.
ISBN: 9798559698779
Library of Congress Control Number: 2020921935

1. History of Science 2. Philosophy of Science 3. Science in General 4. Collected Papers by Aldemaro Romero I. Title

Volume I of the collection "Scholarly Contributions by Aldemaro Romero jr."

Cover design by Mayayo, Olga. 1953-

The author has no responsibility for the persistence or accuracy of
URLs for external or third-party internet websites referred to
in this publication, and does not guarantee that any content on such
websites are, or will remain, accurate or appropriate

To my grandchildren Jordi and Laia, who someday will see me as part of "history." As someone who had his own ideas on how we got to where we are today in science.

Other Books by Aldemaro Romero Jr. Available at Amazon

Romero, A. 2020. Romero, A. & J. Pratt. 2020. *Institutional Failure in Higher Education: Why some colleges and universities fail and what we can do to prevent it.* Amazon Books. 236 pp.

Romero, A. 2020. *Letters from Academia. Ideas for Reforming Higher Education and Other Essays.* Amazon Books. 372 pp.

Romero, A. & S.J. Peters (Eds.). 2019. *Leonardo's Children: Stories on Creativity by Fine Arts Leaders that will Blow your Mind.* International Council of Fine Arts Deans (ICFAD). 134pp.

Hentzi, G. & A. Romero (Eds.). 2018. *From Departure to Destination: Reminiscences of the Weissman School of Arts and Sciences.* New York: Weissman School of Arts and Sciences 137pp.

Romero, A. & E.O. Keith (Eds.). 2012. *New Approaches to the Study of Marine Mammals.* Rijeka, Croatia: InTech. 248 pp.

LaFond, L., W.A. Retzlaff & A. Romero (Eds.). 2012. *After the Academy: Memories of Teaching and Learning in the Land of Lincoln.* Edwardsville, Illinois: College of Arts and Sciences, SIUE. 184 pp.

LaFond, L., C. Berger & A. Romero (Eds.). 2010. *Adventures in the Academy: Professors in the Land of Lincoln and Beyond.* Edwardsville, Illinois: College of Arts and Sciences, SIUE. 182 pp.

Noakes, D.L.G., A. Romero, Y. Zhao & Y. Zhou. (Eds.). 2009. *Chinese Fishes.* Developments in Environmental Biology of Fishes, Volume 28. Dordrecht, Netherlands: Springer. 278 pp.

Romero, A. 2009. *Cave Biology: Life in Darkness.* Cambridge, UK: Cambridge University Press. 291 pp.

Trauth, J. & A. Romero. (Eds.). 2008. *Adventures of the Wild: Experiences from Biologists from the Natural State.* Fayetteville, Arkansas: The University of Arkansas Press. 160 pp.

Romero, A. & S. West. (Eds.). 2005. *Environmental Issues in Latin America and the Caribbean.* Dordrecht, Netherlands: Springer. 299 pp.

Romero, A. (Ed.). 2001. *The Biology of Hypogean Fishes.* Developments in Environmental Biology of Fishes, Volume 21. Dordrecht, Netherlands: Kluwer. 370 pp.

Romero, A. 1994. *Vida Verde.* Barcelona, Spain: Apóstrofe, 199 pp.

Romero, A. 1992. *Canaima.* Caracas, Venezuela: Palmaven, S.A., 208 pp.

Romero, A. & A. Mayayo. 1992. *Manual de Ciencias Ambientales.* Caracas, Venezuela: Bioma, x + 212 pp.

TABLE OF CONTENTS

Preface 1

SECTION I: Major Analyses ....<u>5</u>

Chapter 1. The Unending Conflict. Religion vs. Science: Purpose vs. Uncertainty......... 7
Chapter 2. Destined to Fail: Influence of Religion on Science.
    Predestination in Biospeleology..25
Chapter 3. Biospeleologists in Their Labyrinth: A History of Cave Biology................ 47
Chapter 4. Scientists Prefer them Blind: The History of Hypogean Fish Research.......... 95
Chapter 5. When Whales Became Mammals: The Journey of Cetaceans
    from Fish to Mammals and Intellectual Inertia 135

SECTION II: Histories of Discoveries and Demystifications 171

Chapter 6. The Scientific Discovery of The Amazon River Dolphin
    Inia geoffrensis 173
Chapter 7. Jacques Besson, Cave Eels, and Other Alleged European Cavefishes....183
Chapter 8. The Speleologist who Wrote too Much 191
Chapter 9. The Blind Cavefish that Never Was 189
Chapter 10. Humboldt's Alleged Subterranean Fish from Ecuador 203
Chapter 11. Myth and Reality of the Alleged Blind Cavefish from Pennsylvania 207
Chapter 12. The Discovery of The First Cuban Blind Cave Fish: The Untold Story 215

SECTION III: Biographies 225

Chapter 13. On White Fish and Black Men: Stephen Bishop and the Discovery
    of the Blind Cavefish of Mammoth Cave? 227
Chapter 14. Between the First Blind Cavefish and the Last of The Mohicans: The
    Scientific Romanticism of James E. DeKay 237
Chapter 15. The Life and Work of a Little Known Biospeleologist:
    Theodor Tellkampf 245
Chapter 16. Felipe Poey and the myth of the "Isolated Genius" 251
Chapter 17. Cope, Caves, and Skeletons in the Closet 263
Chapter 18. He Wanted to Know Them All. Eigenmann and His
    Blind Vertebrates 273
Chapter 19. The Unsung Heroes of Speleology... 283
Chapter 20. Charles Marcus Breder, Jr. 1897-1983 289
Chapter 21. Herbert L. Needleman (1927-2017) 291
Chapter 22. Columbus, Christopher 297
Chapter 23. Amerigo Vespucci 299

SECTION IV: Book Reviews 301

Chapter 24: Las plantas del mundo en la historia: Illustraciones botánicas
    de cinco siglos 303
Chapter 25: Del diluvio al megaterio: Los orígenes de la
    paleontología en España 305
Chapter 26. From Popular Medicine to Medical Populism:

Doctors, Healers, and Public Power in Costa Rica, 1800–1940 307
Chapter 27. Floating Gold: A Natural (an Unnatural) History of Ambergris 309
Chapter 28. Constituciones De La Real Y Pontificia Universidad De San Geronimo ...311
Chapter 29. The New Celebrity Scientists. Out of the Lab and into the Limelight 313
Chapter 30. The Gene. An Intimate History 317
Chapter 31. The Beautiful Brain: The Drawings of Santiago Ramón y Cajal...............321
Chapter 32. The Glass Universe: How the Ladies of the
    Harvard Observatory Took the Measure of the Stars 323
Chapter 33: Salvador Gilij's Narrative about the Orinoco fauna.....................325

SECTION V: Letters to the Editor 343

Humboldt 335
Erasmus Darwin 345
Humboldt and Darwin 347

About the author 349

Appendix 1. List of Published Contributions by Aldemaro Romero Jr., Ph.D.,
    To the History and Philosophy of Science in Chronological Order 341

*"History of science without philosophy of science is blind.
Philosophy of science without history is empty"*

Norwood Russell Hanson

Preface. Why this Book?

The field of history and philosophy of science has had a complicated development. Initially, most scholars dealing with this subject were only trained either as philosophers or as historians. As the field progressed, researchers in both camps realized that these two fields were very interconnected. That connection was not merely that of science philosophers adding some history or historians adding some philosophy to their narrative, but much more than that (Patton 2014).

Among those pioneering the bridge between history and philosophy of science were Gaston Bachelard[1], Karl Popper[2], Georges Canguilhem[3], Mario Bunge[4], Paul Feyerabend[5], Thomas Kuhn[6], François Dagognet[7], and Jean Gayon[8]. Thanks to them, we can see history and philosophy of science as a more unified discipline. A current example of that was creating an organization named Integrated History and Philosophy of Science (&HPS), established in 2006. Their mission is to integrate the history and philosophy of science into a unified discipline. Since then, this organization has carried out biennial meetings where they discuss topics around this integration[9]. Part of this movement has been incorporating scholars who hold a degree in science and another either in philosophy or history. Therefore, it is not surprising that all of the pioneers of this area (except for Bachelard), held a science degree in addition to their philosophical training.

Although formally trained as a biologist, my interest in the history and philosophy of science dates back to my years in high school. While on my school bus, I began reading an old biology college textbook in Spanish. The title was *Biología Fundamental* ("Fundamental Biology") by Carlos Morales Macedo, a Peruvian university professor, published in 1949. It was an 800+ pages book that I read from the beginning to the end. I still keep a copy of that book in my library.

That comprehensive treatise was massive in factual information. A lot of it was outdated, mainly when it came to molecular biology, of which it had almost nothing. Yet, it also contained some concepts on the philosophy of biology. While I became very familiar with all the terminology and different life sciences branches, more philosophical discussions opened my mind to aspects of science that were more than purely factual. This would prove very useful once I became interested in issues related to the history of science. History always gives you context.

I also devoured the biographies of scientists for the simple reason that I wanted to be like them. I was trying to learn as much as possible about how to become a successful one. One particular book that I read from page one to the end was the Spanish translation of Isaac Asimov's *Asimov's Biographical Encyclopedia of Science and Technology* (1964). This book contained more than 1,000 biographies displayed in chronological order. To me, it was apparent that scientists' paths diverged widely based on personalities, environments in which they developed, and even luck. I found some of them to be remarkable and a beacon of inspiration. The biographical sketch that intrigued me the most was about the Spanish biologist Santiago Ramón y Cajal. The achievements of this Nobel Prize winner were exceptional but even more so given the circumstances in which he developed his scientific career. That career took place in a country, Spain, with not much of a scientific

---

1 *b.* 27 June 1884, Bar-sur-Aube, France; *d.* 16 October 1962, Paris, France.
2 *b.* 28 July 1902, Vienna, Austria; *d.* 17 September 1994, London, UK.
3 *b.* 4 June 1904, Castelnaudary, Aude, France; *d.* 11 September 1995, Marly-le-Roi, Île-de-France.
4 *b.* 21 September 1909, Buenos Aires, Argentina; *d.* Montreal, Quebec, Canada, 24 February 2020.
5 *b.* 13 February 1994, Vienna Austria; *d.* 11 February 1994, Genolier, Vaud, Switzerland.
6 *b.* 18 July 1922, Cincinnati, Ohio, USA; *d.* 17 June 1996, Cambridge Massachusetts, USA.
7 *b.* 24 April 1924, Langres, Haute-Marne, France; *d.* 3 October 2015, Avallon, Yonne, France.
8 *b.* 15 June 1940, Saint-Maur-des-Fossés, France; *d.* 28 April 2018, Paris France.
9 http://integratedhps.org/en/ (accessed 30 October 2020).

tradition and poor scientific infrastructure at that time. I was also impressed by how much he envisioned things proven right many years after he wrote about them. I was also impressed by his propensity to write for the general public, which made him a pioneer as a celebrity scientist. That meant someone popular among the general public because he made an effort to explain things approachable to everyone regardless of their background. I remember that because of those readings, I concluded I had to be myself, be creative, that life is unpredictable, and that the most important thing was to prepare myself for the future. As Ramón y Cajal himself said, "Every man if he so desires, becomes a sculpture of his brain." Something I realized then summarizes that approach, and it was in the form of a quote from another of my most admired scientists: Louis Pasteur: "Only a prepared mind makes discoveries."

In any case, my initial impression of the history of science as a subject was presented as a progressive accumulation of knowledge, in which true theories replaced false beliefs (Golinski 2001). Yet, the history of science has repeatedly demonstrated that errors, fashions, and conceptual inertia have often delayed the development of certain areas of knowledge (Horder 1998).

By the time I got into graduate school at the University of Miami, Florida, I kept reading about the history and philosophy of science even though almost all my working time was being consumed by my field and laboratory studies on the topic of my dissertation: the evolution of behavior in cavefishes.

Once I started publishing in academic journals about my empirical work, I also started doing some research on my scientific subjects' history and philosophy. First were on the history of cave biology and history of marine mammalogy, then on evolution, and later on, topics related to the influence of mysticism, predeterminism, and religion on scientific thinking.

As the reader will see throughout the chapters in this book, I was trying to place scientific development within historical and philosophical contexts. Unlike what most of my biology colleagues believe, my main argument is that scientific thinking is heavily influenced by external factors, from politics to religion. In other words, scientific discoveries are kindled by the human condition. My firm belief in that notion led me to create two courses which I taught while being Chair of the Department of Biological Sciences at Arkansas State University: "History of Biological Ideas" and "History and Philosophy of Science." Because the vast majority of my students were either undergraduate science majors or graduate students in Biology, my approach was to teach them that by understanding the history and philosophy of science, they could be better scientists with a broad view of their specific disciplines. I also overruled the notion of eurocentrism by teaching my students about the scientific achievements in China, the Arab world, sub-Saharan Africa, and pre-Colonial America (as a continent). I proceeded, likewise, when it came to the role of women in science.

I thought that would be a hard sell. For these students whose only past experiences with history had been in high school, where the subject was presented as many dates and names. Others saw philosophy as a cluster of unintelligible jargon. Another challenge I was facing was the fact that when teaching those subjects below the Mason-Dixie line, many of those students had been raised as Southern Baptists. For example, in 1982, the Southern Baptist Convention issued a resolution rejecting the theory of evolution and stating that creation science "can be presented solely in terms of scientific evidence without religious doctrines or concepts."[10]   I even heard of some

---

[10] https://www.sbc.net/resource-library/resolutions/resolution-on-scientific-creationism/ Retrieved 19 November 2020.

Southern Baptist preachers proclaiming from the pulpit that Darwin was "the devil incarnate."

Therefore, I was happily surprised when I saw these students getting enthusiastic about how science developed and scientists' personalities. I even still remember that in one of the teacher's evaluations I received one of the students anonymously wrote in his end-of-the-semester evaluation, "Now I have discovered that everything they taught me in church about science was wrong."

The book in your hands (or your computer screen) contains a collection of my published writings on these issues with several updates. The original formats vary from letters to the editors to articles in popular magazines, book reviews in academic journals, encyclopedia entries, to peer-reviewed articles and book chapters. They are all grouped according to nature and subject.

This book is a collection of 36 peer-reviewed articles, essays, biographies, book reviews, and letters to the editor I have published since 1986 (Appendix 1) after obtaining my doctorate in biology from the University of Miami, Florida. They encompass a wide range of topics and approaches. I have assembled them into five sections. The first one ("Major Analyses") includes in-depth pieces in which I analyze particular areas on the history and philosophy of science ranging from historiography to religion. The second ("Discoveries and Myths") deals with the evolution of specific discoveries and demystifying some alleged findings that were never true. The third ("Biographies") is a set of scientific biographies and obituaries about particular scientists and how they became who they are in science history. The fourth one ("Book Reviews") is a collection of my published opinions on different books dealing with the history and philosophy of science. The fifth and last one ("Letters to the Editor") is a collection of letters published in either highly prestigious peer-reviewed journals such as *Science*, printed by the American Association for the Advancement of Science (AAAS), to well-known newspapers such as *The New York Times*. In those, I either corrected gross errors or clarified some aspects of science publicized by others.

Most of the pieces presented here are modified versions of the original articles or book chapters I wrote. They were updated in terms of new information and expanded in terms of information and/or illustrations available via public domain. In some cases, some of the specific information is repeated in a couple of chapters. Still, I pretty much ignored those repetitions so the reader could go over each chapter individually. I also corrected some typos and maintained the literature cited for each piece, so the reader did not have to be switching back and forth from each chapter to an extensive list of literature placed at the end of the book. Since sources range from scientific writings to writings in history, philosophy, and theology (each one with its terminology and style), I included more precise information in each case so the interested reader can pursue those sources regardless of his/her background. I also used footnotes to mention the time and place on which the individuals mentioned in this book were born and died, so the reader will have a better framework for placing these individuals in time and place. Likewise, I included for each chapter of those individuals, from my co-authors (mostly former students of mine) to colleagues elsewhere who aided me in one way or another in generating the original material, especially in providing with primary source materials when it came to historical work.

I hope the reader will find the chapters of this book both informative and enjoyable. After all, the history and philosophy of science is a magnificent journey in the human endeavor.

NOTE: This is the first in a collection of books titled "Scholarly Contributions by Aldemaro Romero Jr.," aimed at critically assembling my published contributions in different areas of my academic pursuits. In addition to this one on history and philosophy of science, there more are in preparation: Paleontology and Evolution, Marine Mammalogy, and Biospeleology.

Literature Cited

Asimov, I. 1971. *Asimov's Biographical Encyclopedia of Science and Technology.* New York: Doubleday & Company Inc.
Golinski, J. 2001. *Making Natural Knowledge: Constructivism and the History of Science.* Chicago: University of Chicago Press.
Horder, T. J. 1998. Why do scientists need to be historians? *Quarterly Review of Biology* **73**:175-187.
Morales Macedo, C. 1949. *Biología Fundamental.* Barcelona: Salvat Editores.
Patton, L. 2014. *Philosophy, Science, and History: A Guide and Reader.* New York: Routledge.

Reproduction of "The Alchemist Discovering Phosphorous" by Joseph Wright of Derby originally completed in 1771 and reworked in 1795. This painting is at the Derby Museum and Art Gallery in Derby, England. It epitomizes the characterization of scientists as lone pursuers of "the truth" (whatever that may mean) developing their work in isolation from the world. None of that is true nowadays.

Section I. Major Analyses

This section contains five chapters that encompass original analyses I have made on five different topics on the history and philosophy of science.

The first chapter ("The Unending Conflict. Religion vs. Science: Purpose vs. Uncertainty') is about what seems the eternal conflict -particularly in the United States- between science and religion. I used the case of biospeleological ideas to argue that discussions on this general topic cannot obviate the fact that religion has had, and continues to have, an influence on how we practice science even if we do not want to recognize it. The bottom line is that science, despite being a fountain of knowledge, provides uncertainly, and that is what motivates its pursue. On the other hand, religion gives people comfort that there is a purpose in nature that justifies our existence and, therefore, provides us with the belief that we are in the universe as individuals because of some reason.

The second chapter ("Destined to Fail: Influence of Religion on Science. Predestination in Biospeleology") deals with the religious idea of predestination and how it has influenced evolutionary stances in science. Predestination is a concept familiar to the three major monotheistic religions. We use it as a motivation for self-importance and, hence, self-preservation. This concept has tainted and continues to spoil the whole vision of biological evolution as something either dominated or overshadowed that the presumption of it has been foreshadowed "it is written," if not in a revealed book, by our consciousness.

The third one ("Biospeleologists in Their Labyrinth: A History of Cave Biology") has to do with the development of cave biology as an academic pursuit. Although it seems to be a narrow subdiscipline, the fact of the matter is that it lacks verifiable unifying principles because it has developed many in many different cultures through time and space. This has resulted in disputed conceptions of how life has evolved in an environment dominated by the lack of light. Such conceptual divergences have resulted in biological explanations of the evolution of life in caves that contradict well-established principles in modern evolutionary biology. This is a perfect case study of intellectual inertia.

The fourth chapter ("Scientists Prefer them Blind: The History of Hypogean Fish Research") is a historical analysis of the research on cavefishes and how many people worldwide have taken different routes in approaching the study of these vertebrates, especially when it comes to their classification and evolutionary explanations about the phenotypic features they share, such as blindness and depigmentation.

The final chapter ("When Whales Became Mammals: The Journey of Cetaceans from Fish to Mammals and Intellectual Inertia") of this section deals with the difficulties researchers have had for over two thousand years in understanding that cetaceans (whales and dolphins) were mammals. This is another example of intellectual inertia in the history of a human endeavor as science. Scientific approaches should be informed by objectivity, reason, and innovation, not cultural biases.

Chapter 1. The Unending Conflict. Religion vs. Science: Purpose vs. Uncertainty [11]

Summary

The way scientists responded to evolutionary ideas can yield powerful insights into understanding the historical resistance against the idea of evolution by means of natural selection as understood in current neo-Darwinian thinking. From the beginning, evolutionists, including Darwin himself, struggled in trying to find an explanation for the loss of features during evolution, particularly the loss of eyes and pigmentation among many cave organisms. Although Darwin responded to this challenge by embracing neo-Lamarckian ideas, most biologists, at least until the advent of the Modern Synthesis, strongly advocated directional evolution propelled by more or less mystical forces. Currently, many biospeleologists still employ jargon that epitomizes this view of evolution. Today's controversies surrounding the evolution-creation debate are not really about biblical literalism versus scientific evidence but rather on the disgust created in many quarters of viewing evolution as a materialistic, purposeless process.

Introduction

To understand the rejection by many to the idea of evolution by means of natural selection, I have chosen to examine the interpretations by scientists themselves to evolutionary phenomena that defy conventional Darwinian wisdom. To that end, I have chosen a phenomenon that Darwin himself found puzzling and that even late twentieth-century scientists have had trouble dealing with: the explanation for the loss of phenotypic features (e.g., eyes and pigmentations) among cave animals.

Let me state from the onset that when I say "evolution by means of natural selection" I refer to a purely naturalistic explanation of changes in both genotype and phenotype, regardless of the fact that other mechanisms such as genetic drift may play in the process.

When it comes to the explanation of loss of features, particularly among cave organisms, many evolutionists have embraced a number of mechanisms that regardless of their labels (e.g., Lamarckism, neo-Lamarckism, orthogenesis, vitalism), they all share two common foundations: one is progressionism (the belief that evolution moves in a direction aiming at more complexity and/or "perfection"); the other is that the mechanism behind progressionism is mystical in nature. It is particularly interesting, for example, that despite the fact that Lamarck's evolutionary ideas preceded those of Darwin's for more than half century, Lamarckism never encountered the strong resistance from the religious establishment as Darwinism did.

By the end of this chapter, I hope to convince you that what has driven this resistance against materialistic evolution is the deep belief among humans that there is a purpose in nature, ranging from subtle expressions in natural laws along the lines of natural theology to direct divine intervention in everyday life[12].

Early Evolutionary Ideas

---

[11] Based on Romero, A. 2006. (Published in 2007). The big issue between science and religion: purpose vs. uncertainty. *Forum on Public Policy* 2(4):867-881.
[12] I thank Ronald L. Numbers for his comments and suggestions on a previous draft of this article.

Evolutionary ideas and their mystical interpretations predate the writings of the two most famous names in evolutionary biology: Lamarck and Darwin[13]. In fact, the first generation of Greek philosophers (ca. 600-550 BCE) came out with evolutionary interpretations that were quite naturalistic and materialistic. Thales of Miletus[14], Anaximander of Miletus[15], and Empedocles[16] epitomize that first generation of Greek philosophers characterized by operating in an environment in which there was no belief on a single god, revealed truth, or dogmatic book. Probably influenced by Middle Eastern cultures, most of them believed that: (1) the creation of the world was the product of the forces of nature, (2) that there was no design, (3) that what happened was the result of necessity, (4) that the world was eternal, and (5) in general, rejected supranatural explanations in favor of materialistic ones. An exception to that line of thought was Anaxagoras[17] who believed that there was a plan in nature.

The second generation of Greek philosophers (ca. 550-400 BCE) epitomized by Pythagoras of Samos[18], Heraclitus[19], Alcmaeon[20], and Hippocrates[21] continued that materialistic tradition that now included beliefs in the inheritance of acquired characters, the principle of use and disuse, and spontaneous generation.

The third generation of Greek philosophers (ca. 400-322 BCE) had many representatives that took a turn towards idealism and progressionism. Plato[22], for example, believed in creation by supranatural powers while his student Aristotle[23] founded teleology, the doctrine of purposiveness in nature.

The advent of Christianity as the official religion of the Roman Empire meant the erosion of rational thought regarding the workings of nature by using Christian beliefs to explain everything including the belief that all species were designed by God. In Medieval Times (ca. 500-1450) explanations about the natural world developed in closed conjunction with Christian thought which was dominated by the concept of creation, that all the knowledge is in the "revealed" book and the birth of natural theology, i.e., that God exists because of the order and harmony of the world which requires an intelligent being. The eleventh century also saw the rise of scholasticism which further meant: (1) lack of freedom of thought, (2) that truth was determined by logic, not observation, and (3) blind faith in the "Authorities" (such as Aristotle).

At that time other cultures had less of an issue with natural explanations of the world. The Chinese culture, for example, were more interested in the practical applications of science than on speculation. Islam, for its part, was sympathetic to science regardless of its origin: Greece, India or China. Muslim scholars were also more interested in practical pursues and saw scientific discoveries (including evolution) as a confirmation of their religions tenants.

With the Renaissance some major changes took place. Facts like the discovery of species of plants and animals not mentioned in the Bible meant that such a book could not be taken as the sole source of the truth. Advances in observation of nature by using new instruments such as the telescope and the microscope and the defense of

---

13 This summary was based on Mayr 1982 where the sources for this information can be found. Some other ideas on the history of evolutionary in general are from Romero 2001 and from Romero 2009 pp. 1-61.
14 *b*. Miletus, Ionia, Asia Minor, ca. 626-623 BCE; *d*. ?, 548-545 BCE.
15 *b*. Miletus, Ionia, Asia Minor, ca. 610 BCE; *d*. ?, ca. 546 BCE.
16 *b*. Akragas, Magna Graecia, ca. 494 BCE; *d*. ?, ca. 434 BCE.
17 *b*. Clazomenae, Ionia, Persian Empire, ca. 500 BCE; *d*. Lampsacus, Ancient Greece, ca. 428 BCE.
18 *b*. Samos, Graecia, ca. 570 BCE; *d*. either Croton or Metapontum, Magna Graecia, ca. 495 BCE.
19 *b*. Ephesus, Ionia, Persian Empire, ca. 535 BCE; *d*. Ephesus, Ionia, Delian League, ca. 475 BCE.
20 *b*. Croton, Magna Graecia, ca. 510 BCE; *d*. ?
21 *b*. Kos, Ancient Greece, ca. 460 BCE; *d*. Larissa, Ancient Greece, ca. 370 BCE.
22 *b*. Athens, Greece, 428/427 or 424/427 BCE; *d*. Athens, Greece, 348/337 BCE.
23 *b*. Stagira, Chalcidian League, 384 BCE; *d*. Euboena, Macedonian League, 3*22 BCE.*

experimentalism as epitomized by Francis Bacon[24] meant that knowledge could be acquired via personal experience not just by reading books.

With the advent of Modern Science (ca. 1650-1800) we see the how: (1) direct observation replaces scholasticism, (2) the first attempts are made to classify living being based on biological similarities, and (3) how the center of gravity of science moves from the Mediterranean to Northern Europe.

That does not mean that materialism replaced religious mysticism, but rather that teleological explanations using divine intervention were ideal to link religion and the new scientific observations. Everybody believed that species were fixed but, for example, when the nature of fossils was confirmed as extinct species, people asked if they were created by God, what their purpose was. Furthermore, how come some species such as parasites looked "imperfect" and if so, how they could be produced by divine designer?

Divergent Lines of Thought: Lamarckian Mysticism and Darwinian materialism.

To see how scientists confronted a materialistic view of evolution versus a mystical one, we need to look at the way Lamarckism and its allies and the different brands of Darwinism, confronted the issue of evolution of cave animals. This confrontation had numerous overtones, not only religious but also political and sociological ones.

Lamarckism and its Spinoffs

Nineteenth century biology first developed in France. By the time the first evolutionary ideas were articulated by French naturalists, there was a strong mystical view of history and society in that country. For example, the Marquis de Condorcet[25] used the idea of progress into virtually all of his historical interpretations and beliefs that humanity's destiny was progressive perfection. This vision set the foundations for the positivism of the French philosophers such as Auguste Comte[26] and the speleologist and naturalist Marcel de Serres[27] who saw life was a manifestation of progressive perfecting (Fig. 1.1).

Within that intellectual environment, virtually all French naturalists (with the only notable exception of Georges Cuvier[28]) embraced some sort of "transformism" as evolution was known then. Paradigmatic of this point of view was Jean-Baptiste Lamarck[29] who called himself a "naturalist-philosopher". Although a naturalist by training, Lamarck relied heavily on speculations and metaphysics and used the classification of animals, particularly invertebrates, to sustain his view that nature was organized along clear lines of increasing progress toward complexity (Burkhardt 1977, p. 58 & fol.).

Although some of Lamarck's ideas (mostly wrong) were naturalistic such as: (1) "use and disuse", (2) belief in spontaneous generation, and (3) that fossils represented species that had evolved into new ones, many more of his notions of nature, were mystical. They included: (1) the existence of a metaphysical "power of life" as the main

---

24 *b*. The Strand, London, England, 22 January 1561; *d*. Highgate, Middlesex, England, 9 April 1626.
25 Marie-Jean-Antoine-Nicolas de Caritat. *b*. Ribemont, Picardy, France, 17 September 1743; *d*. Bourg-la-Reine, France, 29 March 1794.
26 Isidore Marie Auguste François Xavier Comte. *b*. Montpellier, France, 19 January 1798; *d*. Paris, France, 5 September 1857.
27 Pierre Toussaint Marcel de Serres de Mesplès. *b*. Montpellier, France, 3 November 1780; *d*. 22 July 1862, Montpellier.
28 Jean Léopold Nicolas Frédéric, Baron Cuvier. *b*. Montbéliard, Holly Roman Empire, 23 August 1769; *d*. Paris, France, 13 May 1832.
29 Jean-Baptiste Pierre Antoine de Monet, chevalier de Lamarck. *b*. Bazentin, Picardy, France, 1 August 1744; *d*. Pris, France, 18 December 1829.

mechanism leading toward increasing complexity and (2) that "needs" (*besoins*) created by a changing environment are experienced by a "sentiment interieur," an unconscious reaction to external stimuli found in animals with central nervous system, able to direct "Vital Fluids" that promote changes in parts of the body and that those changes were inherited by the next generation[30].

Figure 1.1. Marquis de Condorcet, Auguste Comte, and Marcel de Serres (from left to right, public domain images).

When trying to explain decrease in complexity as in the case of parasites, Lamarck proposed two possible explanations: (1) they were either a recent product of spontaneous generation (and therefore they did not have time to "progress" towards complexity) or (2) they lacked the "desire" to have such organs. He proposed that the lack of teeth in whales or eyes in subterranean moles were evidence of his ideas.

Although some of these explanations for evolutionary mechanisms were dismissed (and even ridiculed) by some of his contemporaries, he and his followers never faced any significant religious opposition. Somehow the Catholic Church never saw in these mystical ideas the threat that materialistic Darwinism would pose later in Protestant England and the U.S.

Even Georges Cuvier, a creationist and an adversary of Lamarck, believed in "progression" in the succession of the geologic record. Other French naturalists took an even more mystical/religious position: Étienne Geoffroy Saint-Hilaire[31], a protégé of Lamarck, saw nature so logically aimed toward perfection that when he was forced to explain the origin of vestigial organs, he interpreted them as "disgraces" of natural beauty. Similar tenants can be found in *Naturphilosophie* [32] (Fig. 1.2).

---

30 Lamarck's ideas are often confusing, difficult to follow and even contradictory. Most of the ones cited here can be found in Lamarck, J.B. 1809, 1815.
31 *b*. Étampes, France, 15 April 1772; *d*. Paris, France, 19 June 1844.
32 This was a romantic philosophy that sought metaphysical correspondences and interconnections within the natural world. It was generated in early nineteenth century Germany by Friedrich Schelling and G.W.F. Hegel who were essentially idealists. This philosophical current was extremely popular among scientists particularly in Germany which at that time was one of the most important science centers of the world, thus its influence. Contrary to experimentalist and observational science, its tenat was that spirit and/or mind were closely connected to the body.

Figure 1.2. Jean-Baptiste Lamarck (Portrait by J. Pizzetta, 1893), Georges Cuvier, and Étienne Geoffroy Saint-Hilaire (from left to right, public domain images).

Even when Darwin's *Origin* was translated into French, it was colored with dramatic Lamarckian overtones. The translation published in 1862, by the French feminist Clémence-August Royer[33], was done with the explicit intent of using Darwin's work as a confirmation of Lamarck's ideas (Harvey 1999). She chose to translate Darwin's third edition of the *Origin* which was more Lamarckian that previous ones in the explanation of the loss of organs among cave organisms. She added a preface and footnotes along the lines of "I told you so, Lamarck was right" and then changed the word "selection" for "election" giving the impression that in nature things did not occur by chance but by design.

If that was not enough, she changed the title of Darwin's book to *De l'origine des espèces, ou Des lois de progrès chez les êtres organizes* (The origin of species, or the laws of progress among organized beings) giving the impression that Darwin emphasized the idea of progress, something for which he was ambiguous at best (Fig. 1.3). Thus, Darwin's work was presented to the French public[34] as a confirmation of Lamarck's mystical ideas about nature. No wonder French Catholics in particular and Catholics in general -for the most part- did not express a strong, generalized anti-Darwinian sentiment since their version of Darwinism was a mystical one.

But things were not as simple for Lamarckians as it seemed. When Louis Pasteur demonstrated in 1859 that spontaneous generation was a fallacy, such scientific victory was seen as a victory of experimentalism over materialistic simplifications. Further, the political and military humiliation of the French by the Prussians during the 1870-1871 War, was seen by many as a confirmation of the Spencerian notion of "survival of the fittest" and, therefore, was rejected in France and supplanted by mysticism in the belief that France, would continue to progress until achieving national grandeur. Therefore, the implication of natural selection as an ineludible law of nature was dismissed.

Post-Darwinian French biologists in general and biospeleologists in particular, developed French neo-Lamarckism that emphasized two major features: progressionism and mysticism in the form of vitalism. The father of these neo-Lamarckian ideas in France was Henri-Louis Bergson[35] (Bergson 1907). Bergson was a philosopher (among other things) familiar with the American neo-Lamarckism that we will discuss later. He

---

33 *b*. Nantes, France, 21 April 1830; *d*. Neuilly-sur-Seine, France, 6 February 1902.
34 Because French was a major scientific language at that time, many scientists did not read Darwin's *Origin* until it had been translated into French. See Romero, A. 2006c for an example of that in Latin America.
35 *b*. Paris, France, 18 October 1859; *d*. Paris, 4 January 1941.

championed and popularized orthogenesis, the idea that evolution occurred along specific directions aiming at increasing complexity and perfection[36].

Figure 1.3. Clémence-August Royer and the cover of her translation of Darwin's *The Origin of Species* (from left to right, public domain images).

Bergson was also an intense French patriot who dismissed the notion of natural selection not only because his abhorrence of the implication that Prussian victory over the French meant the survival of the fittest but also because it was materialistic. He proposed in 1907 the idea of an *élan vital* or vital impetus[37]. He used this term to refer to a characteristic of life that, according to him, always pushes in the direction of complexity; that, for Bergson, was the mechanism of orthogenesis, which directed evolution from the domain of the divine into the natural world. Since Bergson could not find strong evidence supporting the inheritance of acquired characters, he thought that the *élan vital* was the mechanistic explanation for evolution.

For Bergson evolution was impregnated with finalism[38] and what made it possible was his mystical force, *élan vital*. Catholics found no problem with this interpretation because that mystical force could be synonymized with God's will and above all, it was not materialistic. Bergson, a Jew by birth, felt so close to the Catholic mysticism that he almost became a Catholic.

---

[36] This term was first proposed by Haacke (1893) while others used different terminologies to express the same thing: orthoevolution (Plate, L. 1922, p. 11), nomogenesis (Berg, L.S. 1926, p. 8), aristogenesis (Osborn, H.F. 1933), and the omega principle (Teilhard de Chardin, P. 1955).
[37] This term is so obscure that it is usually left untranslated; yet it is somewhat similar to Lamarck's "power of life".
[38] This idea also known as a teleologism was originated by Aristotle and is based on the idea that things have ends or purposes into themselves and that I why they happen.

Others followed Bergson's path: the biologist Lucien Cuénot[39] expanded Bergson's ideas by arguing that species succeed in a particular environment because they were "preadapted." The term he coined was *préadaptation* (Cuénot 1911, vol. IV, p. 306) and it served perfectly the aims of progressionists: species could succeed in new environments because they had been "programmed" to that end. And who else could have programmed those species but God?

Figure 1.4. Henri-Louis Bergson, Lucien Cuénot, and Édouard-Alfred Martel (from left to right, public domain images).

At the beginning of the twentieth century virtually all speleologists were French or French educated and they all showed the philosophical influence of their compatriots. Such was the case of Édouard-Alfred Martel[40] (Fig. 1.4). He was known for his pioneer work in 1894 on the physiography and accessibility of caves, and was who coined the term speleology (in both French and English) in the 1890s (Martel 1894, 1896). In 1895 he founded the *Société de Spéléologie* in France and later became a professor of subterranean geography at the Sorbonne (the first speleological academic post in the world). He is often called "the father of modern speleology" and his publication record includes more than 1,000 articles and books on the subject. In 1904 Armand Viré[41], another Frenchman, coined the term biospeleology (*biospeleologie*) (Viré 1904).

However, the two figures that would ultimately consolidate biospeleology as a science and gave it many of the distinctive features that it has today, were Emil Racoviță[42] and René Gabriel Jeannel[43]. Racoviță, a Rumanian-born, French-educated naturalist, started exploring caves in the Pyrenees in 1905 together with his protégé Jeannel. In 1920 Racoviță founded in Cluj, Romania, the world's first speleological institute. He was greatly influenced by the American neo-Lamarckians (see below) and had a great deal of distaste for natural selection.

---

39 Lucien Claude Marie Julien Cuénot. *b*. Paris, France, 21 October 1866; *d*. Nancy, France, 7 January 1951.
40 *b*. Pontoise, France, 1 July 1859; *d*. Montbrison, France, 3 June 1938.
41 *b*. Lorrez-le-Bocage-Préaux (Seine-et-Marne), France, 28 January 1869; *d*. Moissac (Tern-et-Garonne, 11 July 1951.
42 Emil Gheorghe Racoviță. *b*. Iași, United Principalities of Moldavia and Wallachia (today Romania), 15 November 1868; *d*. Bucharest, Romania, 17 November 1947.
43 *b*. Paris, France, 23 March 1879; *d*. Paris, 20 February 1965.

Figure 1.5. Armand Viré, Emil Racoviţă', and René Gabriel Jeannel (from left to right, public domain images).

Racoviţă's two main publications dealing with biospeleological theory were his 1907 *Essai sur les probleme Biospeologiques* (*Essays on biospeleological problems*). This book was published at the same time that Bergson was proposing his *élan vital*) and his little known 1929 *Evolutia si problemele ei* (*Evolution and its problems*) book. In those publications he clearly delineated his evolutionary thinking about cave organisms, which can be summarized as follows: (1) all cave organisms were "preadapted" to the cave environment, (2) lack of use made eyes disappear among cave animals, (3) natural selection is of little importance because natural variation is virtually non-existent, and (4) evolution is directional. Similar views were endorsed by his student Jeannel (Jeannel 1950, p. 7.), and these ideas continue to have a tremendous impact on biospeleologists today (Fig. 1.5). That is evidenced by the common usage of the term "regressive evolution" by many biospeleologists when referring to the loss of eyes and pigmentation among cave animals.

Thus, the founders of biospeleology were not only progressionists in their views of evolution but also mystics when it came to explain its mechanisms. Thus, all these ideas were entirely compatible with Catholic mysticism and, therefore, the Catholic Church never had a major problem with evolution as presented by Catholic or neo-Catholic thinkers.

Enter Charles Darwin[44]

To understand Darwin's influence in the debate about the loss of eyes and pigmentation among cave animals, we need first to examine the state of biospeleological research in the U.S. before Darwin's ideas came into play. At the time of the publication of Darwin's first edition of *The Origin* (Darwin 1859), biospeleological research in the U.S. mainly involved taxonomic and morphological descriptions of species being collected at Mammoth Cave, Kentucky. This cave opened to tourism in the 1830s, and some of the wealthy visitors from the east coast took specimens of its fauna back to scientists in New England. Thus, beginning in the 1840's, numerous papers describing the fauna of

---

44 Charles Robert Darwin. *b*. The Mount, Shrewsbury, Shropshire, England, 12 February 1809, *d*. Down House, Downe, Kent, England, 19 April 1882.

Mammoth Cave were published, including the first description of a blind cavefish, *Amblyopsis spelaea*[45] (Fig. 1.6)

Unlike papers describing species being found elsewhere, the reports on species from Mammoth Cave generated a lot of speculation about the origin of such fauna. Most of the discussions concerned the question of why these animals were blind and depigmented in the first place. Jeffries Wyman[46], for example, described *A. spelaea* as with "imperfect" eyes and proposed in a Lamarckian fashion that this "might be owing to a want of stimulus through a series of generations" (Wyman 1854, p. 19).

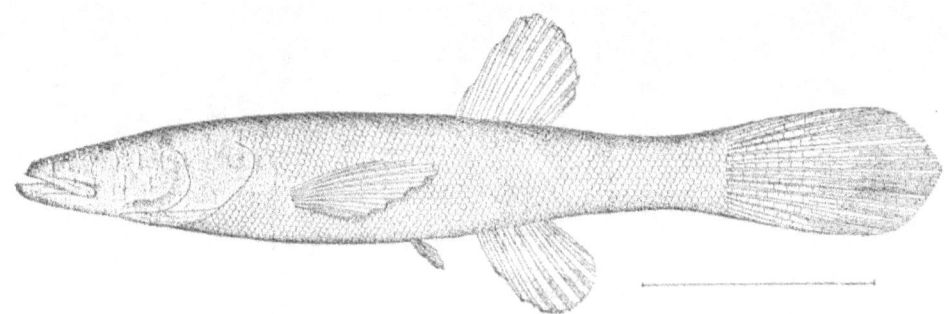

Figure 1.6. "*Amblyopsis spelaeus*" (currently *A. spelaea*). Drawing by David Starr Jordan (public domain image).

Theodor Tellkampf[47] when studied *A. spelaea* concluded that "While it is true, in general, that all animals retain their essential form, and that no species passes over into another by transformation, we know that less material changes of form are produced by external influences such as changes in climate or food, lasting though many generations of the same species". Obviously, he was not embracing evolution at the species level but rather temporal influences on development that led to the loss of eyes and pigmentations (Tellkampf 1844, p. 393; Romero 2002b).

To test this hypothesis Louis Agassiz[48] (Fig. 1.7), America's most famous naturalist of his time and later a strong anti-Darwinian, proposed a "Plan for an investigation of the embryology, anatomy and effect of light on the blind-fish of the Mammoth Cave, *Amblyopsis spelaeus*" which essentially called for raising these fish under both conditions of light and darkness with the hope that his creationist ideas will be vindicated (Agassiz 1847, p. 180; Romero 2001).

Agassiz's ideas originated from his belief in *Naturphilosophie* and its idealistic view that all nature must be deducible from a single first principle which could be equated with the concept of God. Thus, it is not difficult to understand why Agassiz believed in special creation, i.e., the direct intervention of God in the design and destiny of each species. Accordingly, American thought about biospeleology at the time of Darwin's publication of the first edition of the *Origin* in 1859 was a mixture of creationist views and intriguing questions about environmental effects on development.

---

45 This fish was first scientifically described by James DeKay in 1842. For historical essays on this discovery see Romero, A. 2002a and Romero and Woodward, 2005.
46 *b*. Chelsmford, Massachussetts, 11 August 1814; *d*. Bethlehem, New Hampshire, 4 September 1874.
47 August Otto Theodor Tellkampf. *b*. Heinde, Germany, 27 April 1812; *d*. Hannover, Germany, 7 September 1883.
48 Jean Louis Rodolphe Agassiz. *b*. Haut-Vully, Switzerland, 28 May 1807; *d*. Cambridge, Massachusetts, 14 December 1873.

Figure 1.7. Charles Darwin, Jeffries Wyman, and Louis Agassiz (from left to right, public domain images).

Darwin on "rudimentation"

From Darwin's correspondence we know that one of the aspects that interested him the most about cave fauna was the question about the cause of the phenomenon of rudimentation or reduction or loss of organs, i.e., the eyes among cave animals. Darwin's writings, including his notebooks and correspondence, show that he saw this phenomenon as part of a larger compensatory-process issue, i.e., the enlargement of other sensory organs, regardless of whether compensation occurred among cave fauna or not. In other words, he only saw a naturalistic explanation to this phenomenon. He also concluded that cave fauna had derived from eyed and pigmented forms found in areas surrounding caves.

However, in his correspondence with American naturalists sympathetic to his ideas, Darwin was receiving messages related to the idea of progress. In a letter from James Dana[49] dated 8 December 1856 about the cave fauna of Mammoth Cave, Dana told Darwin, that progress was "a law which involves the expression of a type-idea in forms or groups of increasing diversity, and generally of higher elevation; always resulting in a purer & fuller exhibition of the type" and that "it is the simple before the complex" (Burkhardt & Smith 1990, p. 299-300). Here we can see how strong the idea of progressionism was in the minds of American naturalists even before evolutionary ideas became a matter of discussion. Also, Darwin was reading in this account a message of order in nature, not necessarily an evolutionary one, but one confirming the idea of the Great Chain of Being already present in Plato's and Aristotle's writings. According to this account, nature is characterized by plenitude, continuity, and gradation. The universe is "full" of everything that is possible, in other words that the universe is composed of an infinite series of forms, each of which shares with its neighbor at least one attribute, and these elements in nature can be arranged in hierarchical order from the smallest, simplest type of existence, to God himself.

When it came to the explanation of the evolution of cave creatures, Darwin at first considered the mechanisms of both natural selection and disuse to explain blindness and depigmentation as well as the enlargement of some sensory systems and appendages. To Darwin, this meant a "contest (...) between selection enlarging and disuse alone reducing

---

49 James Dwight Dana. *b.* Utica, New York, 12 February 2013; *d.* New Haven, Connecticut, 14 April 1895.

these organs" (Darwin 1859, p. 296). He did not seem to make up his mind about which of the two mechanisms was the real one acting in this case.

By the third edition of *The Origin* Darwin de-emphasized the importance of natural selection by eliminating the speculation of a "contest" between selection and disuse. In fact, in the first two editions, in the paragraphs relative to cave animals and rudimentation, he used the words disuse and selection seven times each; by the third edition, it was five and two, respectively. Despite this use of a Lamarckian mechanism, Darwin never accepted any of the mystical portions of Lamarckism and his explanations were always naturalistic in what it was view as materialism in the more religious quarters, especially when he used the same arguments to explain human evolution.

Thus, most of the attacks on Darwin's ideas were not against evolution as a phenomenon but its mechanism: natural selection which was viewed as a force without any purpose and whose result was the survival of the fittest without any plan or design and, therefore, godless.

Although Darwin was never explicit about progress, there is no question that he held a modified version of the Great Chain of Being of seeing nature ordered in a hierarchical way (Bowler 1983, pp. 55-59). This was a pre-Darwinian idea championed by the Swiss naturalist Charles Bonnet[50] and the French Philosopher Jean-Baptiste Robinet[51] who happened to endorse the idea of organic progress (Burkhardt 1977, p. 80-84) (Fig. 1.8). Bonnet, in particular, articulated the idea of progressive development in 1770. Essentially, he wrote that changes in earth conditions allowed for already existing forms of life to manifest higher levels of complexity. For Bonnet, the term "evolution" meant the unfolding of a providential plan to replenish the earth with life (Richards 1992, 2002).

While Darwin avoided mentioning any purpose or plan, he was considered a heretic. Thus, for his contemporaries the issue was not that he had proposed evolution (many had done so well before him without creating any major controversy) but that he saw the process as devoid of any spiritual connections.

Figure 1.8. James Dana, Charles Bonnet, and Jean-Baptiste Robinet (from left to right, public domain images).

---

[50] *b*. Geneva, Republic of Geneva, 13 March 1720; *d*. Genthod, near Geneva, Republic of Geneva, 20 May 1793.
[51] Jean-Baptiste-René Robinet. *b*. Rennes, France, 23 June 1735; *d*. Rennes, 24 March 1820.

The American neo-Lamarckism

The publication of Darwin's *Origin* in 1859 stimulated American naturalists not only intellectually but also sociologically. With the exception of Agassiz who dismissed the idea of transmutation of species altogether, the rest of the American naturalists participated in one form or another in the development of neo-Lamarckism in the United States.

One of the founders of American neo-Lamarckism was Alpheus Hyatt[52]. Hyatt was a former student of Agassiz, visited Mammoth Cave in September 1859, and collected specimens of its fauna (Bocking 1988). Hyatt's evolutionary ideas were based on three tenants: (1) species have, as do individuals, an inevitable life cycle that includes decline as age advances, (2) for a species the preceding step before extinction is "degeneration" of the species (cave creatures with their lack of eyes and pigmentation epitomized to him this degeneration), and (3) species "transmutation" is the result of the speeding ("acceleration") or slowing ("retardation") of development which, in turn, is caused by use and disuse (Brooks 1909).

In other words, Hyatt was given species a life cycle as if they were individual organisms. Hyatt, of course, had no proof of that and his ideas were influenced by Agassiz himself, through his version of *Naturphilosophie*, based on Lorenz Oken's[53] German idealism and transcendentalism and Ernst Haeckel's[54] (also a German) "Principle of Recapitulation" in the form of progressionism known as "Biogenic or Biogenetic Law" (i.e., "ontogeny recapitulates phylogeny") or the recapitulation theory (Fig. 1.9). Although Haeckel was impressed with Darwin's *Origin*, like most contemporaries he was not very enthusiastic about natural selection and preferred Lamarckian explanations. Its materialism ran against his romantic idea of nature.

The other early champion of progressionism in the U.S. was Edward Drinker Cope[55]. Together with Hyatt he developed what was to be known as the Hyatt-Cope school, which emphasized an alleged parallelism between embryology and phylogeny. Cope was also against natural selection as an important evolutionary force (Cope 1864) and preferred Lamarckian mechanisms. He even amplified Lamarck's ideas by representing evolution as a phenomenon governed by trends: "The method of evolution has apparently been one of successional increment or decrement of parts along definite lines" (Cope 1896, p. 24). This is what was later called orthogenesis, the view that evolution has a life of its own that can take it in certain directions. As Hyatt had also done, Cope proposed evolutionary principles such as the "Law of the Unspecialized" which when applied to cave organisms meant that these cave creatures without eyes and pigmentation were at the end of their phylogenetic life because they were too specialized to evolve into something else; therefore, the next step had to be extinction (Cope 1896, pp. 172-174).

No wonder cave fauna gave to Hyatt, Cope, and their followers bases on which to build the idea that evolution was governed by mystical trends and that natural selection was an unimportant mechanism. In fact, these ideas have remained extremely popular among biospeleologists despite the fact that natural selection has been demonstrated to be a major factor in the evolution of cave organisms (Romero & Green 2005).

---

[52] *b*. Washington, D.C., 5 April 1838; *d*., Washington, C.D., 15 January 1902.
[53] *b*. Bohlsbach, near Offenburg, Baden, Germany, 1 August 1779; *d*. Zurich, Switzerland, 11 August 1851.
[54] Ernst Heinrich Philipp August Haeckel. *b*. Postdam, Prussia, 16 February 1834; *d*. Jena, Weimar Republic, 9 August 1919.
[55] *b*. Philadelphia, Pennsylvania, 28 July 1849; *d*. Philadelphia, Pennsylvania, 12 April 1897.

Figure 1.9. Alpheus Hyatt, Lorenz Oken, and Ernst Haeckel (from left to right, public domain images).

Despite the tremendous popularity of American neo-Lamarckism among cave researchers, some European scientists were not satisfied with the metaphysical explanations for the evolution of cave fauna in particular and the general dismissal of natural selection as the major driving force of evolution. The main opposition came from August Weismann[56] and Edwin Ray Lankester[57], both pro-selectionists with a very skeptical view of the idealism contained in *Naturphilosophie* (Fig. 1.10). For Lankester the loss of eyes among cave animals occurred as follows: some animals are, by chance, born with defective eyes, and occasionally a few of them, some of which have normal eyes and some defective eyes, fall or are swept into caves. Then in each generation, those that have good eyes are able to see the light and escape while only those that are blind will remain in the cave (Lankester 1839).

Figure 1.10. Edward Drinker Cope, August Weismann, and Edward Ray Lankester (from left to right, public domain images).

---

[56] August Friedrich Leopold Weismann. *b*. Frankfurt am Main, Hesse, Germany, 17 January 1834; *d*. Freiburg, Germany, 5 November 1914.
[57] *b*. London, England, 15 May 1847; *d*. London, 13 August 1929.

The Impact of the Modern Synthesis (1936-1947)

By incorporating population genetics to evolutionary ideas, the architects of the Modern Synthesis proved that you did not need to assert metaphysical ideas to explain evolution and by doing so they furthered a neo-Darwinian materialistic agenda. A key element of the development of the Modern Synthesis was the incorporation of Mendelian genetics into Darwinism. However, Mendelian genetics was also rejected or simply ignored by most French biologists during the first half of the twentieth century. This rejection was largely due to the fact that Mendelian genetics was incompatible with the mysticism of neo-Lamarckism (Bonneuil 2006). In fact, one can argue that Mendelian genetics was a purely materialistic explanation of heredity, the weakest area of Darwinism up to that time.

When the non-French biologists of the 1930's and 1940's saw how Mendelian genetics would provide a strong support to evolution by means of natural selection, they rushed to combine both and explicitly disprove any mystic idea of evolution. Theodosius Dobzhansky[58] when dealing with cave fauna made very clear that the rudimentation of loss of eyes and other characters were the direct result of natural selection and mutations. He further emphasized the role of opportunism to explain the ubiquity of life with no plan or design attached to it (Dobzhansky 1970, pp. 405-407). Ernst Mayr[59] also acknowledged that "(the) evolutionary phenomena dealing with regression and the loss of structures (…) are entirely consistent with the synthetic theory of evolution" (Mayr 1960, p. 351) (Fig. 1.11). Despite these clear statements by some of the most influential biologists of the twentieth century, they seemed to have had little impact among biospelologists who consciously or unconsciously were pushing the mystical agenda of neo-Lamarckism. Part of the problem was that biospeleology as a science continued to flourish in France and somehow those ideas found a good reception among neo-Lamarckian Americans.

Among the French writers that supported that view of life were Lucien Cuénot[60], René Jeannel[61], Maurice Caullery[62], Jean Rostand[63], Pierre-Paul Grassé[64], and Albert Vandel[65]. They kept espousing neo-Lamarckian mechanisms for heredity as well as a staunch finalism through orthogenesis. The most prominent figure of these ideologues was the French Jesuit priest and paleontologist Pierre Teilhard de Chardin[66]. He went so far as to propose that evolution was constantly pushing living things toward some sort of point of perfection (the "Omega point"). The fact that he was a priest and a paleoanthropologist who explained evolution in mystical terms, made his evolutionary philosophy not only palatable among many Christians but also a paradigm in the Catholic Church. Thus, it is not surprising that evolution was much better welcomed among Catholics than among evangelical Protestants who tended to view Catholic vision of the world with suspicion.

---

[58] Theodosius Grygorovych Dobzhansky. *b*. Nemyriv, Russian Empire, 25 January 1900; d. San Jacinto, California, 18 December 1975.
[59] Ernst Walter Mayr. *b*. Kempten, Bavaria, Germany, 5 July 1904; *d*. Bedford, Massachusetts, 3 February 2005.
[60] Lucien Marie Julien Claude Cuénot. *b*. Paris, France, 21 October 1866; *d*. Nancy, France, 7 January 1951.
[61] René Gabriel Jeannel. *b*. Paris, France, 23 March 1879; *d*. Paris, 20 February 1965.
[62] Maurice Jules Gaston Corneille Caullery. *b*. Bergues, France, 5 September 1868; *d*. Paris, France, 13 July 1958.
[63] Jean Edmond Cyrus Rostand. *b*. Paris, France, 30 October 1894; *d*. Ville-d'Avray, France, 4 September 1977.
[64] *b*. Périgueux, Dordogne, France, 27 November 1895; *d*. Carlux, France, 9 July 1985.
[65] Pol Marie Albert Vandel. *b*. Besançon, France, 26 December 1894; *d*. Toulouse, France, 11 October 1894.
[66] *b*. Orcines, Auvergne, France, 1 May 1881; *d*. New York City, 10 April 1955.

To be sure, some liberal protestant thinkers also espoused orthogenetic ideas. Samuel Alexander[67], Jan Smuts[68], Alfred North Whitehead[69], and Michael Polanyi[70] are examples of that.

Figure 1.11. Theodosius Dobzhansky and Ernst Mayr (from left to right, public domain images).

Other Influences

In addition to the above-mentioned philosophical currents, we need to explore two other movements on their influence on American neo-Lamarckian ideas. One of those is Romanticism. This intellectual and artistic movement originated in the late Eighteenth-century Europe and had, among other characteristics, a rebellion against the rationalization of nature. This movement had a tremendous influence on the way science was viewed, developed, and utilized during the nineteenth century (Richards 2002, Heringman 2003a, Fulford *et al.* 2004). Heringman, for example, cites passages from English geologist William Smith[71] that are quite revealing (Heringman 2003b). In Smith's publications, that date as far back as 1815-1817, one can read teleology all over the interpretation of the geological strata[72]. As I showed, one of the leading American post-Darwinian naturalists, James DeKay[73], spent long periods of time in contact with the

---

[67] *b*. Sydney, West South Wales, Australia, 6 January 1859; *d*. Manchester, England, 13 September 1938.
[68] Jan Christian Smuts. *b*. Bovenplaats, Cape Colony, 24 May 1870; *d*. Irene, Union of South Africa, 11 September 1950.
[69] *b*. Ramsgate, England, 15 February 1861; *d*. Cambridge, Massachusetts, 30 December 1947.
[70] *b*. Budapest, Austria-Hungary, 11 March 1891; *d*. Northampton, England, 22 February 1976.
[71] *b*. Churchill, Oxfordshire, England, 28 August 1839; *d*. Northampton, Northamptonshire, England, 28 August 1839.
[72] "there seems to have been one grand line of succession, a wonderful series of organization successively proceeding in the same train towards perfection" in Heringman 2003b, p. 63, which is not only a romantic narrative, as Heringman argues, but also almost perfectly consistent with the American neo-Lamarckism wordiness of the late nineteenth century.
[73] James Ellsworth DeKay. *b*. Lisbon, Portugal, 12 October 1792; *d*. Oyster Bay, New York, 21 November 1851.

leading representatives of the American Romantic literary movement (Romero 2002a). It has also been argued that Friedrich von Schelling[74], one of the founders of both *Naturphilosophie* and the Romantic Movement was an evolutionist (Richards 2002, p. 311).

Another angle that still requires exploring is whether both, American neo-Lamarckism and the popular opposition and/or skepticism toward Darwinian evolution is the product of what has been termed "American Exceptionalism" (Lipset 1996).

Conclusions

The examples presented above, the big real issue in today's controversy in the United States is not evolution vs. creationism and all of its versions including "intelligent design." The real issue is between believing whether our existence as humans in the universe is the result of chance or part of someone's elaborated plan, whether that plan is guided minute by minute as the more fundamentalist evangelicals assert or just by a spiritual force laid in nature.

Neither natural selection nor Mendelian genetics provide any of the mystical characteristics that are palatable to the Biblical literalists in particular, nor to most of those who see the history of life on earth as guided by a superior being in general.
That is the real big divide: whether we want to believe that we, humans, are the result of probabilistic events or the desired outcome of someone's wisdom.

Literature Cited

Agassiz, L. 1847 [1848]. [Plan for an Investigation of the Embryology, Anatomy and Effect of Light on the Blind-fish of the Mammoth Cave, *Amblyopsis spelaeus*]. *Proceedings of the American Academy of Arts and Sciences* **1**:1-180.

Berg, L.S. 1926. *Nomogenesis*. London: Constable.

Bergson, H. 1907. *L'évolution créatice*. Paris: Félix Alcan.

Bommeuil, C. 2006. Mendelism, plant breeding and experimental cultures: agriculture and the development of genetics in France. *Journal of the History of Biology* **39**:281-308.

Bowler, P.J. 1983. *The Eclipse of Darwinism. Anti-Darwinian Evolution Theories in the Decades Around 1900*. Baltimore: The John Hopkins University Press.

Brooks, W.K. 1909. Biographical memoir of Alpheus Hyatt (1838-1902). *Biographical Memoires of the National Academy of Sciences* (USA) **6**:311-325.

Burkhardt, F. and S. Smith (Eds.). 1990. *The Correspondence of Charles Darwin. Volume 6. 1856-1857*. Cambridge: Cambridge University Press.

Burkhardt, R.W. 1977. *The Spirit of System. Lamarck and Evolutionary Biology*. Cambridge Harvard University Press.

Cope, E.D. 1864. On a blind silurid from Pennsylvania. *Proceedings of the Academy of Natural Sciences of Philadelphia* **1864**: 231-233.

Cope, E.D. 1896. *The Primary Factors of Organic Evolution*. Chicago: Open Court.

Cuénot, L. 1911. *La Genèsis des espèces animals*. Paris: Librairie Félix Alcan.

Darwin, C. 1859. *On the Origin of the Species by Means of Natural Selection*. London: J. Murray.

DeKay, J. E. 1842. *Zoology of New York or the New-York Fauna, Part IV, Fishes*. Albany: W. & A. White & J. Visscher.

Dobzhansky, T. 1970. *Genetics of the Evolutionary Process*. New York: Columbia University Press.

---

[74] Friedrich Wilhelm Joseph von Schelling. *b*. Leonbnerg, Würtemberg, Holy Roman Empire, 27 January 1775; *d*. Bad Ragaz, Switzwrland, 20 August 1854.

Fulford, T.; D. Lee and P.J. Kitson. 2004. *Literature, Science and Exploration in the Romantic Era. Bodies of Knowledge.* Cambridge: Cambridge University Press.

Haacke, W. 1893. *Gestalt und vererbung; eine Entwicklungsmechanik der Organismen.* Leipzig: Weigel.

Harvey, J. 1999. A focal point for feminism, politics, and science in France. The Clémence Royer Centennial Celebration of 1930. *Osiris* **14**:86-101.

Heringman, N. 2003a. *Romantic Science. The Literary Forms of Natural History.* Albany: State University of New York Press.

Heringman, N. 2003b. The Rock Record and Romantic Narratives of the Earth, pp. 53-84, In: N. Heringman (Ed.) 2003, *Romantic Science. The Literary Forms of Natural History.* Albany: State University of New York Press.

Jeannel, R.G. 1950. *La marche de l'évolution.* Paris: Presses universitaires de France.

Lamarck, J.B.P.A.M. 1809. *Philosophie zoologique, ou exposition des considerations relatives à l'histoire naturelle des animaux.* Paris: Dentu et L'Auteur.

Lamarck, J.B.P.A.M. 1815. *Histoire naturelle des animaux sans vertèbres.* Paris: Verdière.

Lankester, E.R. 1893. Blind animals in caves. *Nature* **47**:389.

Lipset, S.M. 1996. *American Exceptionalism. A Double-Edged Sword.* New York: W.W. Norton & Company.

Martel, E.A. 1894. La spéléologie. *Comptes Rendus de l'Association Française pour le Advancement des Sciences* **22**:60.

Martel, E.A. 1896. p. 721 In: *Report of the Sixth International Geographical Congress: held in London, 1895.* London: J. Murray.

Mayr, E. 1960. The emergence of evolutionary novelties, pp. 349-380, In: S. Tax (Ed.). *The Evolution of Life. Its Origin, History, and Future.* Chicago: The University of Chicago Press.

Mayr, E. 1982. *The Growth of Biological Thought.* Cambridge: The Belknap Press of Harvard University Press.

Osborn, H.F. 1933. Aristogenesis, the observed order of biomechanical evolution. *Proceedings of the National Academy of Sciences USA* **19**:699-703.

Plate, L. 1922. *Allgemeine Zoologie und Abstammungslehre.* Jena: Gustav Fischer Verlag.

Richards, R. 1992. *The Meaning of Evolution. The Morphological Construction and Ideological Reconstruction of Darwin's Theory.* Chicago: The University of Chicago Press.

Richards, R. 2002. *The Romantic Conception of Life. Science and Philosophy in the Age of Goethe.* Chicago: The University of Chicago Press.

Romero, A. 2001. Scientists prefer them blind: the history of hypogean fish research. *Environmental Biology of Fishes* **62**:43-71.

Romero, A. 2002a. Between the first blind cave fish and the last of the Mohicans: The scientific romanticism of James E. DeKay. *Journal of Spelean History* **36**:19-29.

Romero, A. 2002b. The life and work of a little known biospeleologist: Theodore Tellkampf. *Journal of Spelean History* **36**:68-76.

Romero, A. 2006c. The discovery of the first Cuban blind cave fish: the untold story. *Journal of Spelean History* **41**(131):16-22.

Romero, A. & S.M. Green. 2005. The end of regressive evolution: examining and interpreting the evidence from cave fishes. *Journal of Fish Biology* **67**:3-32.

Tellkampf, T. 1844. Uber den blinden Fisch der Mammuthöhle in Kentucky. *(Muller's) Archives fur Anatomie und Physiologie* **1844**:381-395.

Teilhard de Chardin, P. 1955. *Le phénomène Humain.* Paris: Editions du Seuil.

Viré, A. 1904. La biospéologie. *Comptes rendus de l'Académie des Sciences du Paris* **139**:826-828.

Wyman, J. 1854. On the eye and the organ of hearing in the blind fishes (*Amblyopsis spelaeus* DeKay) of the Mammoth Cave. *Proceedings of the Boston Society of Natural History* **4**:395-396.

Chapter 2. Destined to Fail: Influence of Religion on Science. Predestination in Biospeleology [75]

Summary

Evolution has always been considered a battleground between religion and science. Despite that perception, there are some indications that religious beliefs figure 2have influenced and continue to influence some current interpretations in evolutionary biology. To that end I present evidence on how pervasive the theological idea of predestination, which has been long discussed in the Jewish, Christian and Muslim traditions, has influenced some of the interpretations of the nature of biological evolution. I will concentrate on the history of ideas about the evolution of cave organisms to epitomize the strong influence of religion on some evolutionary ideas as shown not only by some of the interpretations but also by the terminology still used today. I conclude that scientists need to understand the historical and philosophical framework of their research if they really want to claim that their work is really value-free.

Introduction

Part of the conventional wisdom in scientific circles dominated by reductionist views of research is that science is or can be both value-free and ahistorical. However, there has been mounting criticism to this position by arguing that ideology has intruded and will continue to intrude into scientific approaches (Kincade *et al.* 2007).

The idea of predestination defined as the doctrine that contends that God predestines from eternity the salvation of certain souls, has been debated for a long time in theological circles[76]. A survey of the *WorldCat* database on books that are catalogued in (mostly) academic libraries around the world up to November 2020 shows that there are more than 8000 entries that deal with this idea. Yet, only recently scholars from non-theological fields have started to take a look of the possible influence of the notion of predestination in their own area of knowledge. Economists (Glaeser & Glendon 1998) have suggested that the differences of biblical interpretations about individual fate may be largely responsible for the way Protestant countries developed economically when compared to Catholic ones. Geographers have argued that the emergence of regional inequality within developing countries and of the emergence of giant urban centers are the result of conflict between "predestination" and "self-organizing" approaches to economic geography (Krugman 1999). In the field of psychology, Goodey (2001) has suggested that within the seventeenth century reformation movement in France, Calvinism and its notion of predestination was challenged by the belief that the mentally disable was free in his/her destiny from natural law.

In the natural sciences, the phrase "biochemical predestination" was coined by Kenyon and Steinman (1969) and reiterated by de Duve (1995). Their basic argument was that since there are strict laws that govern physicochemical phenomena in nature, those very same laws must have made life an "imperative" phenomenon beyond our earth.

---

[75] Based on Romero, A. 2016r. The influence of religion on science: the case of the idea of predestination in

[76] For earlier discussions on this issue see Weizsäcker (1859), Das Dogma von der göttlichen Vorherbestimmung im 9. Jahrhundert in Jahrbücher für deutsche Theologie; Dieckhoff (1883), Zur Lehre von der Bekehrung und von der Prädestination; Dieckhoff (1885), Der missourische Prädestinianismus und die Concordienformel; Scheibe (1897), Calvins Prädestinationslehre; Köstlin (1901), Luthers Theologie; Müller (1903), Die Bekenntnisschriften der reformierten Kirchen, s. v. Erwählung; Jacquin (1904), La question de La prédestination au Ve et VIe siècle in Revue de l'histoire ecclésiastique; van Oppenraaij (1906), La prédestination de l'église réformée des Pays-Bas.

Therefore, the origin of life and its later evolution were irrepressible and we should expect to find life ubiquitous in the universe. These authors' explanations, confined to the biochemical realm, stop at the moment in which the "RNA world" is formed and say nothing of organic evolution once the first organized beings (cells) appear.

In this chapter I will argue that the notion of predestination has had a strong influence in the biological evolutionary ideas developed in the western world, particularly when it comes to explanations relative to the loss of phenotypic (morphological, behavioral, and physiological) features during evolution as epitomized by organisms living in caves and other light-deprived environments. I will further argue that such ideas have hampered and continue to hamper our understanding of the phenomenon of evolutionary loss of features.

My approach will be, first, to show how predestination has had very deep roots in all monotheistic religions since their inception. Then, I will show how that idea was adopted –in some cases very explicitly- by evolutionary biologists as late as the Twentieth Century and continues to dominate the conversation when it comes with the explanation of some evolutionary processes, particularly in the realm of biospeleology. I will conclude by showing that we need to understand intellectual inertia and the influence of those ideas if we really want to assert the scientific process as one that is really objective and free of superfluous influences.

Predestination in History

The Jewish Tradition

There is abundant pre-New Testament material such as select apocalypses[77] wisdom books, and the Qumran (Dead Sea scrolls) documents that attest for a sense of predestination in the understanding of life, destiny, and human relationship with God. The firm belief on predestination of the Jewish faith would cool off from the Deuteronomic (faithfulness to Yahweh and obedience to his commands bring blessings) approach to Israel's salvation to the spiritual wisdom (sapiential) tradition. This may have been as a response to persecution among Jews which may have compelled the wisdom teachers to adopt a new eschatological dualism, according to which salvation was ultimately determined not just on the basis of covenantal election, but also on the basis of fidelity to the law (Eskola 1998). Therefore, predestination although important, is not deterministic.

However, by the Twelfth Century, Judaism would turn into a more mystical conception of life through the Kabbala (Hebrew for tradition) with the publication of *Sefer ha-bahir* or "Book of Brightness." Kabbalism has its roots in First Century Palestine and was a form of esoteric Jewish mysticism whose initiation into its doctrines and practices was conducted by a personal guide that included the knowledge of some "secret wisdom" of the unwritten Torah that was communicated by God to Moses and Adam. Although observance to the law remained a pillar of Judaism, the Kabbala gave means to approach God directly and introduced the notion of transmigration of souls (*gilgul*).

Yet, the major influence on non-Judean thought in terms of predestination would come from two other works: *Sefer ha-temunah* or "Book of the Image" and *Sefer ha-zohar* or "Book of Splendor." They deal with the notion of cosmic cycles and speculations about soul and salvation. They are important not only because they recapture in part the original ideas about predestination rooted in ancient Judaism but also because they were published in late Medieval Spain, a point we will return later.

---

[77] E.g., 1 Enoch, Jubilees.

Figure 2.1. Front page of *Sefer ha-bahir, Sefer ha-temunah,* and *Sefer ha-zohar* (from left to right, public domain images).

The Origins and Christian Tradition

This notion of predestination has its roots in a number of pre-Christian religious documents whose prime example is the Qumran. There we can find the notion of predestination together with eschatological/apocalyptic concepts (Merrill 1975, Lange 1995). A passage that epitomizes these beliefs is:

"In your wisdom you es[tablished] eternal [...]; before creating them you know all their deeds for ever and ever. [...] [Without you] nothing is done and nothing is known without your will" (García Martínez & Tigchelaar 1999, p. 159).

Figure 2.2. Fragment of the Dead Sea Scrolls (public domain image).

The first Christian author to suggest the notion of predestination was Paul the Apostle[78]. Originally a Jew and a fervent antichristian, he converted into the new religion shortly after the death of Jesus and went on to become one of the leading figures of early Christianity. He wrote:

"For those whom he [God] foreknew he also predestined to be conformed to the image of his Son, in order that he might be the first-born among many brethren. And those whom he predestined he also called; and those whom he called he also justified; and those whom he justified he also glorified" (Rom. 8:29–30).

However, the idea of predestination was not fully articulated until the writings of St. Augustine[79]. His influence on Christian thought derives from both his synthesis of Platonism, Roman, and early Christian ideas that developed into a theological system that later made its mark on both Catholicism and Protestantism. St. Augustine's notion of predestination was that human beings could not attain righteousness by their own efforts and were totally dependent upon the grace of God. In other words, the actions of God were the ones that foreordain the future lot and fate of all mankind in this life and after death, including their salvation or perdition[80].

Figure 2.3. St. Augustin. Unknown artist (public domain image).

St. Augustine did not propose these ideas in an intellectual vacuum but was rather responding to the ideas of Pelagius[81]. He and his followers stated that humans were essentially good and that their fate depended entirely on their will. Concerned about the lowering of moral standards among Christians, he hoped that by stressing personal responsibility their moral behavior would improve. St. Augustine attacked these ideas on philosophical grounds while the Christian church felt threatened largely because

---

[78] *b.* Saul of Tarsus in Cilicia [now in Turkey], c. AD 10; *d.* Rome [Italy], AD 67?
[79] *b. Thagaste, Numidia [now Souk Ahras, Algeria], 13 November 354; d. Hippo Regius [now Annaba, Algeria],* 28 August 430.
[80] See De spiritu et littera, De gratia Christi et de peccato originali, and, especially, De praedestinatione sanctorum and De dono perseverantiae.
[81] *b.* Britain, AD 354; *d.* Palestine[?], after AD 418

Pelagius and his followers rejected any claims of original sin by insisting that God created humans free to choose between good and evil, making sin an act of individual responsibility. Therefore, the baptism of the infant was unnecessary. This led to the labeling of Pelagianism as heresy and the excommunication of Pelagius and some of his followers.

Despite St. Augustine's theological attacks and the Church's political actions, the controversy was far from over. Others like Julian of Eclanum[82] continued their support for Pelagianism despite the Church's threats and actions against them. At the end, a new ideology was developed. What was later called Semi-Pelagianism, can be defined as a movement that in some ways tried to reconcile both Pelagian and Augustinian thoughts. On one hand they agreed with St. Augustine that the original sin was a corruptive force among humans and that without God's grace this corruptive force could not be overcome, and they therefore agreed that Baptism of the infant was necessary. On the other hand, they agree with the Pelagians in that humans' will was very powerful. Therefore, they concluded, the innate corruption of humans was not too great as not to be overcome through the powers of individual determination.

The Semi-Pelagians were led by Johannes Cassian[83]. He was an ascetic monk and theologian whose writings gave rise to the Western idea of monasticism as a result of his experiences in the hermits of Egypt. This influenced his beliefs on the importance of individual determination[84]. Because in the final analysis Semi-Pelagians were asserting that there was no need for God's supernatural intervention for the empowering of man's will for saving action, their ideas were also considered heresy, but Cassian and his followers were not personally persecuted by the Church.

During Medieval Europe, the idea of predestination continued to be discussed by Christian theologians. Godescalc or Gottschalk of Orbais[85],[86] was a monk and theologian who believed that Christ's salvation was limited and that his power of redemption extended only to the elect, thus the elect went to eternal glory and the reprobate went to damnation. This was considered heresy and Godescalc was imprisoned.

The continuation of ideas of predestination would be carried well into Medieval Europe by Thomas Aquinas[87] and, particularly, by Gregory of Rimini[88],[89]. For Aquinas, God wills the salvation of all souls although certain souls are granted special grace that in effect foreordains their salvation; thus, the damned are sent to hell only in the sense that God foresees their resistance to the grace given them. Gregory, on the other hand, believed that goodwill was insufficient to acquire the perfect love necessary for the vision of God to which Christians aspire. He reaffirmed the Church teachings on Baptism by stating that children dying without Baptism would suffer eternal punishment[90].

Peter Auriol[91] a philosopher and critical thinker (he was a forerunner to William of Ockham) criticized St. Thomas Aquinas' theory of (scholastic) knowledge by emphasizing the part played by experience in knowledge against that played by reasoning. He wrote on predestination in *Commentariorum in primum librum sententiarum*, *Tractatus de paupertate*, *Tractatus de principiis naturae*, and *Tractatus de conceptione beatae*

---

[82] *b*. Eclanum [Italy], *c*. AD 386; *d*. Sicily, *c*. AD 455.

[83] *b*. Dobruja, Scythia Minor, AD 360; *d*. Marseille [France], AD 435.

[84] See Cassian's Institutes of the Monastic Life and Collations of the Fathers.

[85] *b*. Saxony [Germany], *c*. 803; *d*. Hautvillers, near Reims, France, *c*. 868.

[86] See Godescalc's *De praedestinatione*.

[87] *b*. Roccasecca, near Aquino, Terra di Lavoro, Kingdom of Sicily, 28 January 1225; *d*. Fossanova, near Terracina, Latium, Papal States, 7 March 1274.

[88] *b*. Rimini, near Venice [Italy], *c*. 1300; *d*. Vienna [now in Austria], November 1358.

[89] See his Summa Theologica.

[90] See his Lectura in librum I et II sententiarum.

[91] b. Gourdon, Guyenne [today France], *c*. 1280; *d*. Aix-en-Provence/Avignon, Provence. [today France] 10 January 1322.

Mariae Virginis and proposed that God offers his grace freely to all human beings; therefore, salvation comes to those who passively accept this free offer of grace.

The Islamic Tradition

Shortly after these controversies were taking place among Christians, the birth of Islam would reinforce the notion of predestination and its influence on the western world would be felt strongly all the way to the Renaissance. The Arabic word *islam* literally translate as "surrender" meaning that the believer (Muslim) has to accept "surrender to the will of Allah". These beliefs are articulated in the sacred scriptures known as the Qur'an (Koran) as Allah revealed to his messenger, Muhammad in the Seventh Century.

The idea of subordination to Allah is omnipresent throughout the Qur'an to the point that even those that do not believe in Him are so because it is Allah who wants them to be infidels ("God misleads whom He will and whom He will He guides" [Qur'an xiv. 4]).

The idea of predestination as a result of God omnipotence while the dismissal of man's free will is pervasive throughout the Islamic scriptures despite a few contradictory statements (for an analysis of these see van Ess 1975, Meier 1981). Here are some quotes from the Qur'an:

- All things have been created after fixed decree (liv. 49).
- No one can die except by God's permission according to the book that fixes the term of life (iii. 139).
- The Lord has created and balanced all things and has fixed their destinies and guided them (lxxxvii. 2).
- God killed them, and those shafts were God"s, not yours (viii. 17).
- By no means can anything befall us but what God has destined for us (ix. 51).
- All sovereignty is in the hands of God (xiii. 30).
- The infidels whose eyes were veiled from my warning and had no power to hear (xviii. 101).
- If We had so willed, we could have given every soul its guidance, but now My Word is realized - "I shall fill Hell with jinn and men together." (xxxii. 32).
- Say unto them, O Muhammad: Allah gives life to you, then causes you to die, then gathers you unto the day of resurrection... (xlv. 26).
- No disaster occurs on earth or accident in yourselves which was not already recorded in the Book before we created them (lvii. 22).
   This set of beliefs may have led to certain sense resignation and or fatalism in the Muslim world, as epitomized by the common usage by many Muslims of the phrase "it is/was written" (for more on fatalism and Islam see Cohen-Mohr 2001).

The "Spanish Connection" of Predestination and its Influence in the Western World

Between the twelve and fifteen centuries, Spain lived through a unique convergence of ideas for a single country in Western Europe: *Cristianos* (Christians), *Moros* (Muslims), and *Judíos* (Jews) were all influencing Spanish thought not only religiously in the strictest sense but also in philosophy and literature.

We can find good examples of the preoccupation with predestination among Spanish writers such as Diego de Valencia[92] a Franciscan monk and a *marrano* (originally a Jew converted into Christianity)[93], Ferran Sánchez Talavera or Calavera[94], Juan Alfonso

---

[92] *b.* Valencia [today Spain] *c.* 1350; *d. c.* 1412.
[93] See his Los campanilleros y el Rosario de la Aurora.
[94] *b.* 1370-1385; *d.* before 1443.

de Baena[95] a Calatravan monk and possibly a *marrano*[96], Fray Martin Alfonso de Córdoba[97] an Augustinian monk[98] and, particularly, Ausias March[99]. March was the first major poet to write in Catalan and who would have a great influence on Romance poetry to this day. Of all his books *Cant espiritual* ("Spiritual song") is the one in which he expresses the confusion about the notion of predestination in pre-Renaissance Spain, probably as a result of the somewhat conflicting views generated by Christianity, Judaism, and Islam in that country. These writers provide an insight on the struggle of different views between free will and divine grace, between good and evil, and on the apparent incongruence of how God could create human beings predestine to damnation.

These views about predestination somehow expanded and became more universal due to three historical facts that took place in 1492: (1) The defeat of the moors in Spain imbedded a sense of fatalism in the psyche of Muslims setting the basis for rancor toward the western world that is still present today; (2) The expulsion of the Jews from Spain (or their forced conversion into Christianity), impregnating them with further messianic hopes and eschatology, furthering Kabbalism; and, (3) the notion that if Spain had been the one discovering America, it was because of predestination and that they had to take place their religious fervor under which both atrocities and humanitarian feats would be carried out (see Fuertes Herreros 1992, for more on the concept of predestination and the history of Spain). As we will see later, these concepts would be mirrored later on during the American Revolution all the way to the Romantic era as epitomized by the concept of "manifest destiny."

Enter Double Predestination

The new impetus and discussion on the issue of predestination in the Sixteenth Century comes by the hand of John Calvin[100]. Calvin generated the concept of "double predestination" (*Gemina Predestinatio*) according to which God has actively chosen some people for damnation as well as for salvation confirming God as omniscient and omnipotent which is closely related to the doctrines of divine providence and grace. This is a contrasting view with the Catholic church (God wills the salvation of all souls but that certain souls are granted special grace that in effect foreordains their salvation, so the Roman Catholic Church teaches that predestination is consistent with free will).
Further discussion on how the concept of predestination continued to evolve can be found in Evans (1982), James (1998) and Behringer (1999). In any case we can conclude that the idea of predestination was deeply rooted in all monotheistic religions and, therefore, we cannot be surprised that such a notion would spill over scientific ideas.

Predestination and National Claims

Although national pride has surfaced in many countries at many times, only two western countries can claim that the sense of being a predestined nation has become part of their psyche. They are the United States of America and France.

---

[95] *b*. Baena, Córdoba, [today Spain]; *d. c.* 1404; *d.* Baena 1454.
[96] See his *Cancionero*.
[97] *b.* Córdoba, [today Spain] *c.* 1400; *d.* 1476.
[98] See his Tratado de la Predestinación.
[99] *b.* Beniarjó, Valencia [today Spain], 1400; *d.* Valencia, 3 March 1459.
[100] *b.* Noyon, Picardy, France; 10 July 1509; *d.* Geneva Switzerland, 27 May 1564.

Figure 2.4. John Calvin by an unknown artist (public domain image).

The American Experience

Unlike the colonization process that took place elsewhere in the American continent, the colonization of what is today the Unites States had little to do with the search of riches and rather had political and religious overtones. Ideologically speaking, the main actors were the Puritans. They were made up of an assortment of groups united by some common themes: (1) an ideology of religious reformation that originated within the Church of England during the middle of the Sixteenth Century; (2) their common Calvinist theology; and, (3) the same critical stance toward the Anglican Church in particular and the English society and government in general. After their ascent to power in the person of Oliver Cromwell as a result of the English Civil War (1642-1651), their influence declined steadily as a result of the restoration of the Stuart monarchy in 1660. Because they were identified with radicalism and the autocratic Cromwell and his government, many moved to British North America (a phenomenon that actually started in the 1620s as a result of religious intolerance in England), Scotland, and Northern Ireland. In North America they formed two main communities: The Congregationalists, settled in Massachusetts, Connecticut, and Rhode Island, and The Presbyterians, who settled mostly in New York, New Jersey, and Pennsylvania during the late Seventeenth Century and throughout the Eighteenth Century.

As the Pelagians and Semi-Pelagians, Puritans were concerned with what they considered social and moral corruption and developed a series of rules that governed many aspects of individual behavior, from dress codes to religious observances. As they continue to break away from the Church of England, they also wanted to make sure that there were no vestiges of rituals and practices that may resemble those of the Roman Catholic Church. Thus, their worship services were simple, austere, and centered on long, learned sermons in which their clergy expounded on passages from the Bible. The parishioners were expected to live an exemplary life dominated by temperance and restrain. They were possessed by a sense of predestination and that America was the Promised Land where they could act according to their own beliefs and with very little outside interference.

By the Eighteenth Century most of these conceptions of life had given away to a more competitive, individualistic, and secular society as a result of the growth in commercial capitalism and the intellectual challenges of the Age of Enlightening (Bushman 1967). Yet, many of their philosophical traits remained in place including the sense of predestination (Innes 1996).

The idea of predestination would resurface at least twice as a major component of the American social and political scene. One was during the American Revolution. Although the idea of religious predestination was severely criticized by Thomas Paine[101], politically speaking there is abundant literature that speaks for a development of a sense of national predestination. The next resurgence for the ideology of predestination will be seen during the era of American Romanticism in the mid 1800s when a sense of frontier, "go west," experimentation with new institutions, the idealization of Native-Americans, and the integration of new immigrants (from Anglo-Saxon Europe other than from England) took place. James E. DeKay[102], for example, the author who first Scientifically described a blind cavefish, was strongly influenced by the American romantic authors of his time (Romero 2002). As we will see later, this context will help to explain why biospeleological ideas that espoused a sense of determinism became very popular among American researchers.

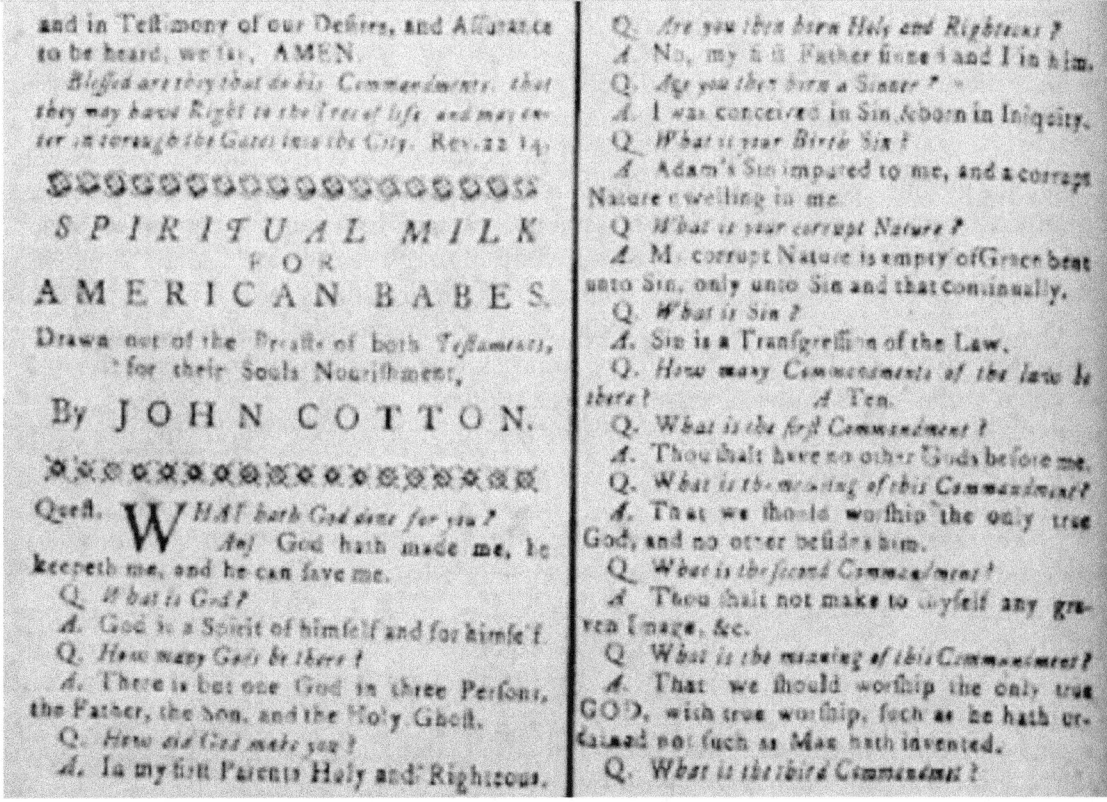

Figure 2.5. Puritan Catechism in *The New-England Primer, 1646.* "I was conceived in Sin & Born in Inequity" (public domain image).

Biospeleological Ideas in France and Elsewhere in Continental Europe

---

[101] See http://www.deism.com/paine_essay_predestination_calvinism.htm. Retrieved 5 January 2016.
[102] *b*. Lisbon, Portugal, 12 October 1792; *d*. Oyster Bay, New York, 21 November 1851.

There is a long history of the idea of predestination in France as a national ideology that includes emphasizing the teaching of nationalistic topics throughout the curriculum in schools[103]. They have had a profound and lasting influence even in today's official policies to the extent that the government even subsidizes articles written in French and by French scientists[104]. The Maintenance of the Purity of the French Language Act enacted in 1975 introduced fines for the use of banned Anglicisms. In 2006 the French subsidiary of General Electric Medical Systems was fined for more than €500,000 simply for issuing software manuals in English only[105].

Within the scientific realm, the lasting influence of the idea of predestination is quite apparent by French or French-based researchers on the intellectual influence on biological evolution in general and biospeleological ideas in particular to the point that their way of thinking and terminology has been pervasive in cave biology since Lamarck until the 1950s and beyond. To understand why this is so, we must (1) review the political and intellectual environment in France previous to the publication of Darwin's *Origin* in 1859; (2) examine how Darwin's book was received and investigate how and (3) why the French developed an evolutionary ideology of their own, particularly when it came to interpreting the nature of cave fauna.

Ideas on evolution (biological and otherwise) in pre-*Origin* France abound, but all have something in common: a strong philosophical rather than an empiric basis. Jean Baptiste Lamarck[106], a physician by training, considered himself a "naturalist-philosopher", and therefore much of his narrative was shaped with speculations and metaphysics rather than facts. In addition, his evolutionary views (mostly expressed in his 1809 *Philosophie Zoologique* and the 1815 supplement to the *Histoire naturelle*) were never very well formulated and even sometimes contradictory. To make things worse, Lamarck's writings were translated into numerous languages, but such translations were not always accurate and some of his statements were reproduced out of context which contributed to the general confusion on to what Lamarck really said (Corsi 2005). But one thing is for sure: he was an early organicist and progressionist who viewed nature as being linearly organized and saw today's organisms as the result of increasing complexity (Burkhardt 1977, p. 58 & fol.). Lamarck was the main (although not the first) advocate of the idea of inheritance of acquired traits concept and of evolution as a goal-oriented process striving towards progressive complexity and perfection. In fact, he did not believe in the extinction of species but rather on the constant transformation into new ones.

He described a metaphysical "power of life" (*puissance de la nature*) leading this process of increasing complexity. That, together with the modifying power of the environment was responsible for the life forms we see on earth. Although he never wrote about cave fauna, the case of parasites with simplified organization amused him; yet, he had a perfect explanation: they appeared primitive because they had been the recent product of spontaneous generation. External circumstances were responsible for deviations from the rule of progression and some contingency (e.g., the disuse of an organ) could alter the path to complexity generating lateral ramifications in his linear view of progression. For him the lack of teeth in whales and eyes in subterranean moles were perfect examples. Lamarck had a great influence on many scientists not only at his time but through the Twentieth Century. The progressionists ideas of Lamarck had also

---

[103] See, for example, http://www.napoleon-series.org/research/society/c_education.html, retrieved 1 January 2016.
[104] See http://www.axl.cefan.ulaval.ca/europe/france-2politik_francais.htm.
[105] See http://www.nytimes.com/2015/04/01/opinion/france-gives-in-to-the-hashtag.html. Retrieved 1 January 2016.
[106] *b*. Bazentin-le- Petit, Picardy, France, 1 August 1744; *d*. Paris, France, 28 December 1829.

a great influence not only in Europe but in America as well where a vigorous neo-Lamarckian school developed. That school was following Lamarck's tenants with the exception of those that were more mystical in nature (Burkhardt 1977). Therefore, we can interpret that progressionism toward complexity as a form of predestination that all forms of live evolve as predestined by his mysterious "power of life."

Two Lamarck contemporaries would also make their own contributions to the notion of increasing complexity in nature. Jean Léopold Nicolas Frédéric (Georges) Cuvier[107], for example, although a creationist, noticed some "progression" in the succession of the geologic record. Geoffroy Saint-Hillaire[108], was a believer in evolution, progressionism, and the Great Chain of Being, always looking for transitional forms (Bourdier 1972b, Appel 1987). He discussed the issue of the origin of vestigial organs from a mystic/religious viewpoint and interpreted them as "disgraces" of natural beauty. Saint-Hillaire, a protégé of Lamarck, was even less materialistic than his mentor and added an aura of mysticism to evolutionary ideas.

At the same time French philosophers were thinking along the same lines. For example, Marie-Jean-Antoine-Nicolas de Caritat, Marquis de Condorcet[109], a brilliant mathematician, philosopher, and political activist, infused the idea of progress into virtually all of his historical interpretations. He adopted the concept of inheritance of acquired characters in constructing his vision for the social and organic progressive improvement of humankind. This idea was also espoused by other philosophers such as Herbert Spencer[110], Friedrich Engels[111], and Lester Ward[112] (Condorcet 1802). They strongly influenced the positivist school founded by the French philosopher Auguste Comte[113] and the ideas of another French philosopher, Marcel de Serres[114]. The latter proposed the view that life was a manifestation of progressive perfecting.

Thus, the intellectual environment in pre-*Origin* France was not anti-evolution as in other parts of Europe and the United States; actually, one can say that no well-educated

---

[107] *b*. Montbéliard, Holy Roman Empire, 23 August 1769; *d*. Paris, France, 13 May 1832.

[108] *b*. Etampes, France, 15 April 1772; *d*. Paris, France, 19 June 1844.

[109] *b*. Ribemont, Picardy, France, 17 September 1743; *d*. Bourg-la-Reine, France, 28 March 1794. His book *Esquisse d'un tableau historique des progrès de l'esprit humain*, published posthumously in 1795, analyzed human history under the view of progressiveness. For him humanity was destined to achieve an evolutionary apex through the education of the masses. His ideas were mirrored later by Theilhard de Chardin (Granger 1971, Leith 1989, Baker 2004).

[110] *b*. Dderby, England, 27 April 1820; *d*. Brighton, England, 8 December 1903. He was a Lamarckian who tried to apply evolutionary ideas to support free market ideologies, he believed that humans were on a natural progressionist route and that the state might create obstacles to economic progress by trying to regulate free society. He rejected the notion of special creation and believed that species were the result of modification of preexistent ones. Interestingly, he was an agnostic who believed that science and religion were trying to answer different questions and that even if you believe in God that was not necessarily incompatible with the idea of evolution. He later embraced Darwinism but still believed in Lamarckian mechanisms to explain the transformation from simple to complex structures in nature. He believed that individuals also evolved as a consequence of learning from good and bad experiences and that at the end the good learner survived. He introduced the concept of 'survival of the fittest' and the rots of social Darwinism can be traced to him.

[111] *b*. Barmen [now part of Wuppertal], Prussian Rhineland, 28 November 1820; *d*. London, England, 5 August 1895. Engels's writings gave the philosophical background to Marxism. His philosophy was based on a materialism that was in accordance with the views of the sciences of the nineteenth century.

[112] *b*. Joliet, Illinois, 18 June 1841; *d*. Washington, D.C., 18 April 1913. One of the founders of American sociology he advocated the intervention of the state at humanizing society by eliminating poverty. He also advocated regulation of competition, the establishment of equal opportunities and cooperation. Ward attacked the very notion of social Darwinism, the *laissez-faire* doctrine and determinism. By doing so he turned against Hebert Spencer who he had admired earlier in his career.

[113] *b*. Montpellier, France, 17 January 1798; *d*. Paris, France, 5 September 1857. He is considered the father of positivism in philosophy.

[114] *b*. Montpellier, France, 3 November 1780; *d*. Montpellier, 22 July 1862. He was a paleontologist and zoologist who believed that the pursuit of truth required violating artificial disciplinary boundaries.

French person at that time harbored any predisposition against evolution (*transformisme*). In fact, in France, the idea of progression could be traced as far back as the development of Modern Science period (1650-1800) at the time of the Enlightenment and the French Encyclopedism. Lamarck contemporaries, with the exception of Cuvier, embraced some sort of transformism: although they were not sympathetic to (and even ridiculed to certain extent) Lamarck's unfounded speculations, particularly the idea that a new organ could be produced by the "desire" of an organism to create it. However, the French were unprepared to view evolution as a materialistic, random process that excluded any metaphysical explanation. And the way Darwin's *Origin* was translated into French made matters worse.

The Origin was translated into French by Clémence-Augustine Royer[115]. This polymath and feminist writer, was not only a great believer in science, but also thought that women should transform it into "female science." Royer probably first heard of Darwin's new work on evolution through a review of *The Origin* by the Geneva-based Swiss entomologist and paleontologist Françoise Jules Pictet de la Rive[116] while lecturing on Lamarck in Geneva in 1860. Pictet was one of the first to receive a copy of *The Origin of Species* directly from Darwin. As soon as Royer read *The Origin*, she convinced her publisher, Guillaumin, to print the first translation of Darwin's work into French. According to Royer, "It was then [after lecturing in Geneva] that I translated the *Origin of Species* of Ch. Darwin, which had appeared in England, during the same winter in which I had affirmed in my course the doctrine of Lamarck. If I translated Darwin, it was because he had brought new proofs to the support of my thesis." (Harvey 1999). In other words, her interest in translating Darwin was not so much to spread the Britton's gospel, but rather to prove how important Lamarck was as the father of evolution as an idea. And it showed.

With the advice of the French zoologist and early Darwinian enthusiast René-Edouard Claparède[117], who had also enthusiastically reviewed Darwin's book, she translated the third edition of *The Origin* (which was, in terms of explanations on rudimentation, more Lamarckian than the first two editions) adding not only numerous footnotes, but also a lengthy prologue in which she espoused eugenics, being probably the first author to do so by applying Darwin's ideas. Darwin, who had authorized to have his book translated into French, was not happy with Royer's preface and footnotes. She not only changed the title of the book, but more significantly, Royer used the word "election" instead of "selection" giving, thus, the impression that nature had a mind of its own directing on purpose evolutionary events.

The title of Darwin's book in French was *De l'origine des espèces, ou Des lois de progrès chez les êtres organizes* ("The origin of species, or the laws of progress among organized beings") giving the impression that Darwin emphasized the idea of progress, a principle for which he was ambiguous at best. Darwin himself, in his correspondence to several of his colleagues such as Jean Louis Armand de Quatrefages[118], Charles Lyell[119],

---

[115] *b*. Nantes, Brittany, France, 21 April 1830; *d*. Paris, France; 6 February 1902.

[116] *b*. Geneva, Switzerland, 27 September 1809; *d*. Geneva, 15 March 1872.

[117] b. Berthezène, near Valleraugue (Gard), France, 12 February 1810; d. Paris, France, 1892. He specialized in *invertebrates and was particularly interested in the degeneration (dégradation) of structures among organisms, although most of his ideas in this matter were wrong. He opposed Darwin's evolutionary (believed in the fixity of species) ideas but maintained very cordial relationships with him.*

[118] b. Berthezène, near Valleraugue (Gard), France, 12 February 1810; d. Paris, France, 1892. He specialized in *invertebrates and was particularly interested in the degeneration (dégradation) of structures among organisms, although most of his ideas in this matter were wrong. He opposed Darwin's evolutionary (believed in the fixity of species) ideas but maintained very cordial relationships with him.*

[119] *b*. Kinnordy, Angus, Scotland, 14 November 1797; *d*. London, England, 22 February 1875. He was the most influential geologist of the 19th. century. His ideas set the stage for Darwin's thinking that life must have been evolving on earth as the geology of our planet had also been changing though long periods of times. He was a

and Asa Gray[120], made it known that he was extremely unhappy with the French translation. Despite this version of *The Origin* being closer to the French state of mind, Darwin sensed that the book had a cold reception in France. In a letter to Quatrefages, a French naturalist who opposed Darwin's ideas on evolution but yet respected him, Darwin wrote, "A week hardly passes without my hearing of some naturalist in Germany who supports my view, & often puts an exaggerated value on my works; whilst in France I have not heard of a single zoologist except M. Gaudry [Albert Jean Gaudry[121]] (and he only partially) who supports my views" (F. Darwin 1896, vol. 2, p. 299). Darwin may have not been happy with this translation; yet, he might not have any other alternatives since he had trouble finding a publisher in France for his book anyway (Herbert 2005).

For years to come, Royer continued publishing and lecturing about Lamarck, her personal hero. She, who was probably the first European woman recognized as a professional anthropologist, was also an enthusiastic caver.

Royer's translation of *The Origin* was very much celebrated by Étienne Rabaud[122]. Rabaud had been a student of Alfred Girard, the first holder of the Chair of Evolution at the Sorbonne and a rabid Lamarckian. Rabaud became such a fanatical supporter of Lamarck's ideas that by the 1930's he even questioned the value of Darwinism (see, for example, Rabaud 1941). When commenting on Royer's preface, Rabaud was enthusiastic because she had restored Lamarck into public attention.

Were this improper translation and the current intellectual climate the only reasons for the poor reception of Darwin's ideas in France? Not really. Just before the publication of *The Origin*, France had witnessed one of the most public and passionate scientific controversies in history. Between 1858 and 1859 French society was inundated with the tales of the dispute between Félix Archimède Pouchet[123] and Louis Pasteur[124], that is, between the belief in spontaneous generation and the belief that the ability to beget life is an exclusive and continual property of living beings. Although Pasteur won the argument and his was a triumph for science as a method of inquiry, Pouchet's sympathizers also supported agnosticism whereas Pasteur's were more comfortable with religious and metaphysical ideas. Thus, despite the fact that the French were not opposed to evolution as an idea *per se*, the mechanism championed by Darwin -natural selection- reminded them of the agnosticism and materialism attached to spontaneous generation. Thus, the land that had given birth to precursors of evolutionary ideas such as Georges-Louis Buffon[125], Lamarck, and Geoffroy Saint-Hillaire, gave Darwin a cold shoulder, and little public controversy of the book took place.

Other political and social events further cemented the French view of evolution as a mystical idea. One experience that generated a nationwide feeling of disgrace was the

---

close friend of Darwin and accepted Darwin's evolutionary ideas being one of the few who immediately accepted the notion of natural selection as a major force of evolution (Wilson 1973).

[120] *b*. Sauquoit (Paris), Oneida County, New York, 18 November 1810; *d*. Cambridge, Massachusetts, 30 January 1888. A physician who became the prominent American botanist of the 19th. century. He embraced Darwinian evolution corresponding extensively with Darwin but was not enthusiastic about natural selection as its mechanism, to say the least. He tried to reconcile Darwinism with religion through a sort of theistic evolutionism.

[121] *b*. St.-Germain-en-Laye, France, 15 September 1827; *d*. Paris, France, 27 November 1908. Affected by the death of his mother when he was very young, he developed a strong mysticism during his entire life. He worked at the *Muséum national d'Histoire naturelle* in Paris and was a great believer of the Great Chain of Being.

[122] *b*. 1868; *d*. 1956. An anti-Darwinian who taught Pierre-Paul Grassé.

[123] *b*. Rouen, France, 26 August 1800; *d*. Rouen, 6 December 1872. A physician, became the Director of the *Muséum d'histoire naturelle* in Rouen. He was a prolific author who gained notoriety because of his dispute with Pasteur over spontaneous generation.

[124] *b*. Dole, Jura, France, 27 December 1822, *d*. Chateau Villeneuve-l'Étang, near Paris, France, 28 September

[125] *b*. Montbard, Bourgogne [Burgundy], France, 7 September 1707; *d*. Paris, France, 16 April 1788. Well known for his 36-volume *Histoire naturelle (Natural History)* (1749–88). He maintained very advanced evolutionary ideas for his time (Roger 1973, 1997, Farber 1974, Sloan 1976, Eddy 1994).

political and military humiliation of the French by the Prussians during the 1870-1871 War (Howard 1981). And as in any nation that has been defeated, their people found consolation in mystical nationalistic ideas. The ideas of national destiny and historical progress became strongly rooted in the French psyche and were reinforced through revisions of school curricula. The Spencerian interpretation of "survival of the fittest" became very unpopular: Prussia had developed into an imperialistic and invincible neighbor and looked like "the fittest" to French psyche. Now French intellectuals threw themselves fully into the arms of mysticism to explain their grand views of nature, and evolution was at the center of all this.

It was in this intellectual atmosphere that the seeds for French Neo-Lamarckism were planted, and these seeds were sown in abundance by French biospeleologists. The father of these neo-Lamarckian ideas in France was Henri Louis Bergson[126]. Bergson was a philosopher and a mathematician whose ideas on evolution were largely anti-materialistic and sustained that organic evolution was just part of a larger, universal cosmic evolution. A Lamarckian follower regarding the canon of use and disuse and principle that evolution was directed by an internal force which he called *élan vital*. He was fiercely patriotic and opposed Darwinism because he did not accept the notion of an undirected mechanism such as natural selection as the major force of evolution. Part of his popularity was due to the fact that by using the notion of an *élan vital*, he was allowing for a role to be played by religion in evolutionary processes (Goudge 1973).

Bergson was familiar with the ideas of Cope and Theodor Gustav Heinrich Eimer[127], a disciple of Rudolf Albert Kölliker[128], who championed the idea of, and popularized the term orthogenesis (Eimer 1887-1888). The term orthogenesis was first proposed by the zoologist Johann Wilhelm Haacke[129] (1893). Others used different terms for essentially the same concept: orthoevolution (Plate 1913), nomogenesis (Berg 1926), aristogenesis (Osborn 1934), and the omega principle (Theilard de Chardin 1955). Bergson, an intense French patriot, proposed in 1907 the idea of the *élan vital* or vital impetus (the term is so obscure that it is usually left untranslated, but reminds that of Lamarck's expression of the "power of life"). He used this term to refer to a characteristic of life that, according to him, always pushes life in the direction of complexity; that, for Bergson, was the mechanism of orthogenesis, which moved evolution from the domain of the divine into the natural world. Given that Bergson did not like natural selection as an idea because of its materialistic implications, and at the same time he could not find strong evidence supporting the inheritance of acquired characters, thus, *élan vital* was for him the answer. Of course, and unlike natural selection or the inheritance of acquired characters, since this idea could not be tested, it could not be disproved either.

According to Bergson, both Darwinian evolution and finalism (the idea that evolution has a sense of directedness toward an end and that such a path has already been laid) could coexist. And what is the unifying force behind such a possibility? It cannot be natural selection, of course, since that is based on apparent randomness, but rather it must be a mystical force, *élan vital*. These ideas may have been interpreted as Lamarckian with a religious twist, but that is also unclear: Bergson, a man profoundly concerned about the fate of his fellow Jews, almost became a catholic. It is evident,

---

[126] *b*. Paris, France, 18 October 1859; *d*. Paris, 4 January 1941.

[127] *b*. Stäfa, near Zurich, Switzerland, 22 September 1843; *d*. Tübingen, Germany, 29 May 1898. In 1875, he became a professor of zoology and comparative anatomy at the University of Tübingen. He described orthogenesis as an intrinsic drive by life towards perfection, a form of directed evolution (Churchill 1990). He dismissed natural selection as a major force in evolution while rejecting vitalism at the same time.

[128] *b*. Zurich, Switzerland, 6 July 1817; d. Würzburg, Germany, 2 November 1905. He studied under Lorenz Oken, Johannes Müller, and F.G.J. Henle and was greatly influenced by *Naturphilosophie* being a close associate of Nägeli. He embraced evolution but opposed the role of natural selection (Hintzsche 1973).

[129] *b*. Clenze, Germany, 23 August 1855; *d*. Lunebourg, Germany, 6 December 1912.

therefore, that his religious views were also complex. Bergson's ideas became extremely popular, and other philosophers such as the French Lucien Cuénot[130] expanded them by arguing that species succeed in a particular environment because they were "preadapted." The term he coined was *préadaptation* (Cuénot 1911, vol. IV, p. 306), and it became an extremely popular idea among biospeleologists, many of whom still firmly believe in it today. Needless to say, Cuénot espoused linear evolution, only that in the new era of experimental genetics of early Twentieth Century, he believed that mutation (*sensu stricto*) was the cause of it.

In summary, Bergson was a progressionist but he did not believe that there was a necessarily pre-designed goal; rather that final progression would lead to a less predictable result trying, thus, to taint Darwinism with the very popular idea of progression.

All of these new philosophies of life were developed at the time when speleology in general and biospeleology in particular were becoming sciences in their own right. All their foundations were being laid by French or France-based naturalists. Such was the case of the French lawyer Édouard-Alfred Martel[131]. Martel was not only a lawyer but also a geographer by training. He was known for his pioneer work in 1894 on the physiography and accessibility of caves, and he coined the term speleology (in both French and English) in the 1890s. He explored the limestone caves of Cévennes and, with others, made descents into previously unknown caves of Europe, Asia, and America. In 1895 he founded the *Société de Spéléologie* in France. Martel was the judge of the tribunal of commerce in Paris from 1886 until 1899, when he became a professor of subterranean geography at the Sorbonne (the first speleological academic post in the world); he was appointed a member of the staff of the Department of Geological Maps of France in 1901. He is often called "the father of modern speleology" and his publication record includes more than 1,000 articles and books on the subject. In 1904 Armand Viré[132], another Frenchman, coined the term biospeleology (*biospeleologie*). Viré had written his doctoral thesis on cave fauna in 1899 and thereafter established an underground laboratory in the catacombs of Paris.

However, the two figures that would ultimately consolidate biospeleology as a science and give it many of the distinctive features that it has today were Emil G. Racoviţă[133] and René Gabriel Jeannel[134]. Racoviţă', a Romanian-born, French-educated naturalist, started exploring caves in the Pyrenees in 1905 together with his protégé Jeannel. Racoviţă initiated an extensive international research program under the umbrella of *Biospeologica* (a supplement to the scientific French journal *Archives de Zoologie Experimentale et Generale*), primarily intending to document and collect cave fauna. In 1920 he founded in Cluj, Romania, the world's first speleological institute. He explored 1200 caves in Europe and Africa, collected about 50,000 specimens of cave animals, and published 66 papers on subterranean fauna totaling almost 6000 pages (Motas 1962). He read and was greatly influenced by, Eimer and Cope (on orthogenesis), Packard (on Neo-

---

130 *b*. Paris, France, 21 October 1866; *d*. Nancy, France, 7 January 1951. Cuénot was a brilliant scientist being the first French biologist to accept Darwinism ideas in Lamarckian France although he was not fully convinced of the all-powerful role played by natural selection. He also pioneered genetics studies in France aimed to prove Mendelian inheritance. Yet, he refused to completely accept neo-Darwinism because he maintained a finalistic view of evolution. One of his most lasting influences in biospeleology was the development of his notion of preadaptation according to which new ecological niches were occupied by mutants that already had some characteristics that favor them to colonize such environments (Tétry 1971b).

131 *b*. Pontoise, France, 1 July 1859, *d*. Cháteau de la Garde, near de Montbrison, France, 3 June 1938.

132 *b*. Lorrez-le-Bocage-Préaux, Saine-et-Marne, France, 28 January 1869; d. Moissac, France, 15 July 1951.

[133] *b*. Iasi, Romania, 15 November 1868; *d*. Bucarest, Romania, 17 November 1947.

[134] *b*. Toulouse, France, 22 March 1879; *d*. Paris, France, 20 February 1965.

Lamarckism), and Louis Dollo[135] (on general evolutionary ideas). He had a great deal of distaste for the selectionist Weisman (Motas 1962).

Figure 2.6. Front page of the first issue of *Spelunca, Bulletin de la Société de Spéléologie* and cover of an earlier issue of *Biospeleologica* (from left to right, public domain images).

Racoviță's two main publications dealing with biospeleological theory were his 1907 *Essai sur les probleme Biospeologiques* ("Essays on biospeleological problems," published at the same time that Bergson was proposing his *élan vital* and considered to be the birth certificate of biospeleology as a science) and his little known 1929 *Evolutia si problemele ei* ("Evolution and its problems") book. In those publications he clearly delineated his evolutionary thought about cave organisms, which can be summarized as follows: (1) all cave organisms were "preadapted" to the cave environment; (2) function (or lack thereof) creates the organ (or generates its disappearance). He was a strong supporter of the use vs. disuse concept; (3) natural selection is of little importance because natural variation is virtually non-existent (he was a staunch typologist); (4) evolution is directional as evidenced by "phyletic lines."

Similar views were endorsed by his student Jeannel (Jeannel 1950, p. 7) who studied subterranean beetles from Europe and Africa. With Racoviță he founded in 1907

---

[135] *b*. Lille, France, 7 December 1857; *d*. Bruxelles, Belgium, 19 April 1931. An engineer turned into biologist, Dollo became famous for the reconstruction of *Iguanodon* fossils in Belgium and for stating the 'Dollo's Law of Irreversibility' according to which organisms never return to their original state, particularly when losing complex structures.

the journal *Biospeleologica* and in 1926 published *Faune cavernicole de la France*. He considered many of the organisms found in caves as "living fossils", and these ideas continue to have a tremendous impact on biospeleologists all over the world.

Although all this can be presented as a great accomplishment for the French in terms of initiating and developing the systematic study of caves, none of these figures ever embraced any form of Darwinism, but rather different shades of neo-Lamarckism first and different forms of finalism such as orthogenesis and organicism later. Thus, the French biologists who embraced transformism beginning in 1880 did so via neo-Lamarckism while strongly opposing the idea of natural selection (Grimoult 1998, p. 150). This philosophy extended well into the Twentieth Century with Lucien Cuénot, Maurice Caullery[136] and Jean Rostand[137].

Therefore, the utilization of cave organisms as perfect examples for demonstrating the legitimacy of the French version of neo-Lamarckism seemed to be inevitable, and this is exactly what happened. The main points in common of these French intellectuals were: (1) acceptance of evolution as a linear phenomenon (orthogenesis) leading to a perfecting complexity in nature; (2) rejection of natural selection as a phenomenon of any relevance; (3) development of finalism, vitalism, organicism, and other expressions of essentialism in biology; (4) utilization of cave organisms as "perfect" examples of these views of life; (5) mutual reinforcement of ideas concerning biospeleological paradigms (blind, depigmented animals) and philosophical notions of progress within the same country: France.

These ideas were very much espoused by American biospeleologists who followed early directional and deterministic views of evolution (Romero 2009, pp. 59-61).

Conclusions

Since the advent of Modern Synthesis, we have a pretty consistent set of evidence that evolution is not linear, that there is not such a thing as direction for evolutionary processes, and that nothing is predetermined since natural selection, the main evolutionary mechanism, is a process that is not motivated by any mystical force nor directs beings toward a particular end. Yet, biospeleologists continue seeing "preadaptations" and "regressive evolution" (which implies direction) anywhere when it comes to cave fauna (Romero 1985, Romero & Green 2005). Therefore, this paper demonstrates that the imprint of the idea of predestination still casts a shadow in modern evolutionary biology of cave organisms. I am not saying that modern biospeleologists do science under some sort of religious fervor but what many of them seem to neglect is that words matter and that words can hide a lot of the philosophical baggage that sooner or later may influence their ultimate conclusion. This, in turn, has created a sociological intellectual inertia that impedes the progress of science in its larger framework.

Therefore, I hope this paper serves as a warning to scientists that regardless of their reductionist views, if they do not understand the historical roots and the philosophical framework of their research, they are doomed at presenting only a very partial (and many times biased) view of nature.

Acknowledgements

---

136 *b*. Bergues, France, 5 September 1868; *d*. Paris, France, 13 July 1958. He lectured at the University of Paris (1903) where he taught evolution from a neo-Lamarckian perspective (Tétry 1971a).

137 *b*. Paris, France, 30 October 1894; *d*. Ville d'Avray, France, 4 September 1977. A biologist and philosopher who worked on developmental biology and maintained neo-Lamarckian views of evolution.

The idea of exploring the influence of predestination in science originated in a conversation on this topic I had with Phil Regan of the University of Minnesota. I thank Brian J. O'Neill who read an earlier version of this paper and made valuable suggestions.

Literature Cited

Appel, T.A. 1987. *The Cuvier-Geoffroy debate: French biology in the decades before Darwin.* New York: Oxford University Press.
Archer, R. 1993. Ausias March and the *Baena* debate on predestination. *Medium Aevum* **62**:35-50.
Baker, K.M. 2004. On Condorcet's "sketch". *Daedalus* **133**:56-64.
Behringer, W. 1999. Climatic change and witch-hunting: The impact of the Little Ice Age on mentalities. *Climatic Change* **43**:335-351.
Berg, L.S. 1926. *Nomogenesis; or, evolution determined by law.* London: Constable.
Berry, S. 1997. "Biochemical predestination" as Heuristic principle for understanding the origin of life. *Journal of Chemical Education* **74**:950-951.
Bockmuehl, M. 1998. Wisdom and predestination: Sapiential themes and predestination in the 'Dead Sea Scrolls' of Qumran by A. Lange. *Vetus Testamemtum* **48**:126-127.
Bonnet, C. 1770. *Palingénésie philosophique, ou idées sur l'état passé et sur l'état futur des êtres vivans.* Genève: C. Philibert et B. Chirol.
Bourdier, F. 1972a. Gaudry, Albert Jean, pp. 295-297, In: C.C. Gilliespie (Ed.). *Dictionary of Scientific Biography*, Vol. 5. New York: Scribner.
Bourdier, F. 1972b. Geoffroy Saint-Hilaire, Étienne, pp. 355-358, In: C.C. Gilliespie (Ed.). *Dictionary of Scientific Biography*, Vol. 5. New York: Scribner.
Brooks, E.S. 1992. Rrhyme, reason, and absence in Calavera, Ferran, Sanchez debate on predestination. *Romance Notes* **33**:161-168.
Buffon. 1954. *Oeuvres Philosophiques de Buffon.* Paris: Jean Piveteau.
Burkhardt, R.W. 1977. *The Spirit of System. Lamarck and Evolutionary Biology.* Cambridge: Harvard University Press.
Bushman, R. 1967. *From Puritan to Yankee: character and the social order in Connecticut, 1690-1765.* Cambridge, MA: Harvard University Press.
Cameron, E. 2001. Peter Martyr Vermigli and predestination. The Augustinian inheritance of an Italian reformer by F.A. James. *American Historical Review* **106**:675-676.
Churchill, F.B. 1990. Eimer, Theodore Gustav Heinrich, pp. 261-264, In: C.C. Gillispie (ed). *Dictionary of Scientific Biography*, Suppl. 2, Vol. 17. New York: Charles Scribner's Sons.
Clarke, F.S. 1976. Lost and found – Athanasius' doctrine of predestination. Scottish. *Journal of Theology* **29**:435-450.
Cohen-Mohr, D. 2001. *A Matter of Fate: The Concept of Fate in the Arab World as Reflected in Modern Arabic literature.* Oxford, UK: Oxford University Press.
Coleman, W. 1964. *Georges Cuvier, Zoologist: a study in the history of evolution theory.* Cambridge: Harvard University Press.
Collins, J.J. 1997. Wisdom and predestination: Elemental traces of wisdom and prophecy from the Qumran corpus, A. Lange. *Journal of Religion* **77**:283-285.
Condorcet, J.-A. N. de C. 1802. *Outlines of an historical view of the progress of the human mind being a posthumous work of the late M. de Condorcet.* Baltimore: G. Fryer, for J. Frank.
Corsi, P. 2005. Before Darwin: transformist concepts in European natural history. *Journal of the History of Biology* **38**:67-83.
Cuénot, L. 1911. *La Genèsis des espèces animals* Paris: Librairie Félix Alcan.
Darwin, F. 1896. *The Life and Letters of Charles Darwin.* New York: D. Appleton and Company, 2 vols.

de Chardin, T. 1955. *Le phénomè humain*. Paris: Du Seuill.

de Duve, C. 1995. *Vital dust: life as a cosmic imperative*. New York: Basic Books.

de Valencia, D. 1984. Isaac Vazquez Janeiro, *Tratados castellanos sobre la predestinación y sobre la Trinidad y la Encarnación, del Maestro Fray Diego de Valencia OFM (siglo XV). Identificación de su autoría y edición crítica* (coll. *Bibliotheca Theologica Hispana - Textos*, 2). Madrid: Consejo Superior de Investigaciones Científicas.

Delcor, M. 1976. Merrill, E.H. Qumran and predestination. *Biblica* **57**:257-258.

Desmazières de Séchelles, R. 1967. *Essai sur la prédestination de la France*. Paris: Fischbacher.

Eddy, J.H. 1994. Buffon's *Histoire naturelle*. History? A critique of recent interpretations. *Isis* **85**:644-661.

Eimer, T. 1887-1888. *Die Entstehung der Arten auf Grund von Vererben Erworbener Eigenschaften nach den Gesetzen Organischen Waschsens*. Jena: G. Fischer.

Eskola, T. 1998. *Theodicy and Predestination in Pauline Soteriology*. Tübingen: Mohr-Siebeck, 1998.

Evans, G.R. 1982. The grammar of predestination in the 9th. century. *Journal of Theological Studies*. **33**(APR):134-145.

Farber, P.L. 1975. Buffon and Daubenton: divergent traditions within the *Histoire naturelle*. *Isis* **66**:63-74.

Flint, R. 1988. Predestination. *Triquarterly* (73):129-130. (Poetry).

Fuertes Herreros, J.L. 1992. Knowledge, fortune and predestination concerning the discovery of America. *Arbor: Ciencia, Pensamiento y Cultura*. **143**(561):133-160.

García Martínez, F. & E.J.C. Tigchelaar. 1999. *The Dead Sea Scrolls. Study Edition*. Vol. 1. Leiden: Brill.

Ginther, J. 2002. Peter Martyr Vermigli and predestination: The Augustinian inheritance of an Italian reformer by F.A. James. Speculum. *A Journal of Medieval Studies* **77**:196-197.

Glaeser, E.L. and S. Glendon. 1998. Incentives, predestination and free will. *Economic Inquiry* **36**:429-443.

Goodey, C.F. 2001. From natural disability to the moral man: Calvinism and the history of psychology. *History of the Human Sciences* **14**(3):1-29.

Grassie, W. 1997. Postmodernism: What One Needs to Know, *Zygon: Journal of Religion and Science* **32**:83-94.

Granger, G. 1971. Condorcet, Marie-Jean-Antoine-Nicolas, Marquis de, pp. 383-399, *In*: C.C. Gillispie (Ed.). *Dictionary of Scientific Biography*, Vol. 3. New York: Scribner.

Grimoult, C. 1998. *Évolutionnisme et fixisme en France. Histoire d'un combat 1800-1882*. Paris: CNRS Éditions.

Haacke, W. 1893. *Gestaltung und Vererbung. Eine Entwicklungsmechanik des Organismus*. Leipzig: Verlag, T.O. Weigel.

Halverson, J.L. 1998. *Peter Aureol on Predestination: A Challenge to Late Medieval Thought*. Leiden: Brill.

Harvey, J. 1999. A focal point for feminism, politics, and science in France: the Clémence Royer centennial celebration of 1930. *Osiris, 2nd Series*, **14**:86–101

Hazlett, W.I.P. 2000. Peter Martyr Vermigli and predestination. The Augustinian inheritance of an Italian reformer by F.A. James. *Journal of Ecclesiastical History* **51**:626-627.

Herbert, S. 2005. The Darwinian revolution revisited. *Journal of the History of Biology* **38**:51-66.

Hill, P. 2001. *Fate, Predestination and Human Action in the Mahabharat: A Study in the History of Ideas*. New Delhi: Munshiram Manoharlal Publishers.

Hintzsche, E. 1973. Koellikeer, Rudolf Albert von, pp. 437-440, *In*: C.C. Gillispie (Ed.). *Dictionary of Scientific Biography*, Vol. 7. New York: Scribner.

Howard, M. 1981. *The Franco-Prussian War: The German invasion of France, 1870-1871.* London: Routledge.

Innes, S. 1996. *Creating the Commonwealth: The Economic Culture of Puritan New England.* New York: W.W. Norton & Company.

James III, F.A. 1998. *Peter Martyr Vermigli and Predestination.* Oxford: Clarendon Press.

Jeannel, R.G. 1950. *La marche de l'évolution.* Paris: Presses universitaires de France.

Jenson, G.R. 1983. *The problem of determinism: with reference to the Qumran Scrolls & the First Epistle of John.* Calimesa, CA: The Author.

Kenyon, D.H. & G. Steinman. 1969. *Biochemical Predestination.* New York: McGraw-Hill.

Kingdon, R.M. 2000. Peter Martyr Vermigli and predestination: The Augustinian inheritance of an Italian reformer by F.A. James. *Church History* **69**:187-188.

Kincaid, H., J. Dupré & A. Wylie (eds.). 2007. *Value-Free Science? Ideals and Illusions.* Oxford: Oxford University Press

Kraege, J.D. 1998. Fresh look at the doctrine of predestination. *Études Théologiques et Religieuses* **73**:349-369.

Krugman, P. 1999. The role of geography in development. *International Regional Science Review* **22**:142-161.

Kugler, R.A. 1998. 'Wisdom' and predestination: Sapiential primeval order and predestination in Qumranic sources by A. Lange. *Journal of Biblical Literature* **117**:735-736.

Lange, A. 1995. *Weisheit und Prädestination: Weisheitliche Unordnung und Prädestination in den Textfunden von Qumran.* New York: E.J. Brill.

Leith, J.A. 1989. L'Évolution de l'idée de progrès a travers l'histoire. *Transactions of the Royal Society of Canada* **4**:3-8.

Louise, E. 1980. On predestination. *Obsidian-black Literature in Review* **6**:169-169. (Poetry).

Macaghobhainn, I. 1977. 'Predestination' *Poetry Australia* (63):91-91.

Mayr, E. 1982. *The Growth of Biological Thought: Diversity, Evolution, and Inheritance.* Cambridge: Belknap Press.

Meier, F. 1981. Predestination and the doctrines of Ibn-Taimiyya. *Saeculum Ann* **32**:74-89.

Maury, P. 1960. *Predestination, and other papers.* London: SCM Press.

Merrill, E.H. 1975. *Qumran and predestination: A theological study of the thanksgiving hymns.* Leiden: Brill.

Motas, C. 1962. Emil G. Racovitza: Founder of biospeleology. *National Speleological Society Bulletin* **24**: 3-8.

Osborn, H.F. 1934. *Aristogenesis, the creative principle in the origin of species.* New York (pamphlet).

Plate, L. 1913. *Selektionsprinzip und Probleme der Artbildung: Ein Handbuch des Darwinismus.* Leipzig und Berlin: Verlag von Wilhelm Engelmann.

Prince, J.S. 1997. Possible sources of ambiguity in Robert Herrick's "Predestination" sequence. *English Language Notes* **35**:16-22.

Rabaud, E. 1941. *Introduction aux sciences biologiques.* Paris: Armand Colin.

Rainbow, J.H. 1990. *The will of God and the cross: an historical and theological study of John Calvin's doctrine of limited redemption.* Allison Park, Pa.: Pickwick Publications.

Robson, M. 1999. Peter Aureol on predestination. A challenge to late medieval thought by J.L. Halverson. *Revue D'Histoire Ecclesiastique.* **94**:953-956.

Roger, J. 1973. Buffon, Georges-Louis Leclerc, Compte de, pp. 576-582, In: C.C. Gilliespie (Ed.). *Dictionary of Scientific Biography*, Vol. 2. New York: Scribner.

Roger, J. 1997. *Buffon. A Life in Natural History.* Ithaca: Cornell University Press.

Romero, A. 1985. Can evolution regress? *National Speleological Society Bulletin* **47**:86-88.

Romero, A. 2002. Between the first blind cave fish and the last of the Mohicans: the scientific romanticism of James E. DeKay. *Journal of Spelean History* **36**:19-29.

Romero, A. 2009. *Cave biology: life in darkness*. Cambridge, UK: Cambridge University Press.

Romero, A. & S.M. Green. 2005. The end of regressive evolution: examining and interpreting the evidence from cave fishes. *Journal of Fish Biology* **67**:3-32.

Sloan, P.R. 1976. The Buffon-Linnaeus controversy. *Isis* **67**:356-375.

Tétry, A. 1971a. Caullery, Maurice, pp. 148-149, *In*: C.C. Gillispie (Ed.). *Dictionary of Scientific Biography*, Vol. 3. New York: Scribner.

Tétry, A. 1971b. Cuénot, Lucien, pp. 492-494149, *In*: C.C. Gillispie (Ed.). *Dictionary of Scientific Biography*, Vol. 3. New York: Scribner.

Thompson, J.L. 2000. Peter Martyr Vermigli and predestination: The Augustinian inheritance of an Italian reformer by F.A. James. *Sixteenth Century Journal* **31**:509-511.

Vajda, G. 1976. Between Hadith and theology - origins of tradition of predestination. *Revue de l'Histoire des Religions* **189**:105-107.

van Ess, J. 1975. *Zwischen Hadit und Theologie: Studien zum Entstehen prädestinatianischer Überlieferung*. Berlin. New York: DeGruyter.

Watt, W.M. 1948. *Free will and predestination in early Islam*. London: Luzac.

Weiss, J. 1987. Spanish treatises on predestination and on the trinity and the incarnation by Diego de Valencia - an identification of his authorship and a critical edition. *Bulletin of Hispanic Studies* **64**:145-146.

Wensinck, A.J. 1932. *The Muslim creed: Its genesis and historical development*. London: Frank Cass.

Wilson, L.G. 1973. Lyell, Charles, pp. 563-577, *In*: C.C. Gillispie (Ed.). *Dictionary of Scientific Biography*, Vol. 8. New York: Scribner.

Wright, D.F. 2000. Peter Martyr Vermigli and predestination. The Augustinian inheritance of an Italian reformer by F.A. James. *Journal of Theological Studies* **51**:374-377.

Chapter 3. Biospeleologists in Their Labyrinth: A History of Cave Biology [138]

Summary

The history of cave biology is a chapter in the history of science that demonstrate how numerous factors such as religious beliefs, nationalisms, and intellectual inertia can play a lasting role in the development of a particular branch of science. That is particularly true when it comes to evolutionary ideas. I argue that this is so because many biospeleologists have failed to understand the historical framework in which this science has developed. This has led many to uncritically accept both concepts and lexicons that are inconsistent with current biological thought. Thus, this piece provides a historical explanation of the ideological framework surrounding most of biospeleological research. I conclude that biospeleology is a branch of science whose pioneers and major practitioners have been very resistant to any neo-Darwinian ideas and have espoused, rather, neo-Lamarckian views in conjunction with the related concepts of orthogenesis, organicism, and other forms of finalisms.

Conceptual Issues

The understanding of the history of one's area of scientific inquiry is essential in order to really appreciate the significance of the current knowledge and the voids that need to be filled. Most scientists are not particularly interested in pursuing such a task because the history of science is influenced by philosophy, politics, religion, and other expressions of human activities whose comprehension requires interdisciplinary approaches that go beyond what they have been normally trained for at universities. Yet, the history of science has demonstrated again and again that errors, fashion, and intellectual inertia have often delayed the development of certain areas of knowledge (Horder 1998). In this chapter I intend to demonstrate that biospeleology is a perfect example of that.

There are two major ways to present the history of a particular branch of science: one is simply an uncritical chronological narrative, and the other is a critical examination of ideas that influenced the development of that area of study. I will take the second route because an appreciation of the historical background in which biospeleological ideas developed is necessary to understand why the full incorporation of modern biological thought in biospeleology has been delayed. As Ernst Mayr put it "When scientists concentrate on the study of isolated objects and processes, they seem to operate within an intellectual vacuum." (Mayr 1982, p. 66-67). I will also emphasize the notion that geography and religion have played a major role in how this science evolved, first in protestant U.S., and later in catholic France.

To this date there is not a comprehensive published history of biospeleology: Vandel (1965, pp. xxiii-xxiv) outlined a few historical facts; Barr (1966) wrote a brief history of biospeleology in the United States; Bellés (1991) wrote a largely chronological and anecdotal narrative on the subject; Shaw (1992) in his treatise on the history of speleology provided little information on biospeleology *per se*, and Romero's (2001b) article on hypogean fish research dealt mostly with the history of evolutionary ideas.

In this chapter I will demonstrate that the most important issue regarding cave biology has been and continues to be the origin and evolution of cave fauna. Because of

---

[138] Based on Chapter 1 of Romero 2009.

that, Darwinism is a central topic of discussion in the development of biospeleology as a science but not because of what most people would assume. In fact, we can say that the history of biospeleology can hardly be depicted as another triumph of Darwinism as an idea. As I will show, Charles Darwin, who was the first scientist who really tried to provide a scientific explanation about the origin of cave fauna and the phenomenon of what he called "rudimentation" in the form of reduction and/or loss of the visual apparatus, espoused a rather neo-Lamarckian stance about the topic to the point that his explanations were not fully Darwinian in the modern sense of the word. Furthermore, because of this and the fact that biospeleology developed mostly in France where Lamarckism and its philosophical allies were very strong, not even The Modern Synthesis seriously changed the interpretation that most biospeleologists of that country had of biological phenomena in caves.

Therefore, the following is a history of biospeleology focusing on five particular, historic, intellectual, and/or geographic theaters, each characterized by the dominance of a particular idea or set of ideas and that most of the times they overlapped chronologically speaking. These are: (1) pre-Darwinian thought (before 1859), (2) Darwinism and American neo-Lamarckism (1859-1919), (3) European selectionism and the death of the controversies (1880-1921); (4) biospeleological ideas in France and elsewhere in continental Europe (1809-1950); and, (5) the impacts of The Modern Synthesis (1936-1947). I will then discuss the roots of current intellectual inertia. This outline does not follow a strict chronological order but rather expresses the influence of culture on the development of ideas as delineated by geographic and intellectual boundaries. In order to make the narrative more fluid, this chapter contains a number of footnotes aimed at providing some biographical information of the major actors mentioned in the main text as well as explanation of philosophical terms so the reader can better appreciate the context in which many of these developments took place.

As I explain the major ideas that have influenced biospeleology, I will argue that many of those ideas by themselves do not represent paradigms in the Kuhnenian sense of the word (Kuhn 1970, p. 10). They were never original ideas, but borrowed in both form and substance from neo-Lamarckism, and at the same time they ignored the pre-eminence of natural selection as an effective mechanism for the explanation of the evolution of cave organisms. Because of their restrictive nature and their incompatibility with the neo-Darwinian framework, these neo-Lamarckian ideas provide little opportunity for further elaboration and development.

Pre-Darwinian Thought (before 1859)

From Prehistory to Mythology

Caves have been of human interest since prehistoric times, serving as both shelter and as a source of artistic expressions (Morgan 1943, Shaw 1992, Cigna 1993, Romero 2001b). The earliest known human representation of cave fauna dates back to ca. 22,000 YBP (years before the present) (Upper Paleolithic). It is a carved drawing of a wingless cave cricket, *Troglophilus* sp., on a bison (*Bison bonasus*) bone found in the *Grotte des Trois Frères* (Three Brothers Cave) in the central Pyrénées, France (Chopard 1928) (Fig. 3.1.).

From the beginning of history, humans have developed a close mythic-religious association with caves, the underworld, and death. Burials in caves have been common among many cultures (e.g., Watson 1974, Stone 1995, Clottes 2003). The underworld or Hades (Ἀδης) in Greek mythology was believed to be the "Kingdom of the Dead" to which one could gain access via caves (Mystakidou *et al.* 2004). Not surprisingly, ideas about cave creatures were, from the beginning, a mixture of myth and reality. Dragons and other imaginary beasts were described by many authors since before the invention of the printing press.

Such views of cave life survived up to the Seventeenth Century. For example, in 1665 a polymath Jesuit priest Athanasius Kircher[139], published what might be considered as the first book whose title gave the impression of being devoted solely to caves: *Mundus Subterraneus* (Kircher 1665). This was a gigantic, two-volume folio size tome, totaling 892 pages whose second edition, published in Amsterdam in 1678, contained lengthy additions about caves in Switzerland, Austria, Italy, and of the Greek Islands. This latter edition would be the one that achieved more popularity and became the standard geology text in the Seventeenth Century. Despite its title, Kircher dealt with many more topics than just caves such as alchemy, chemistry, and metallurgy, among others.

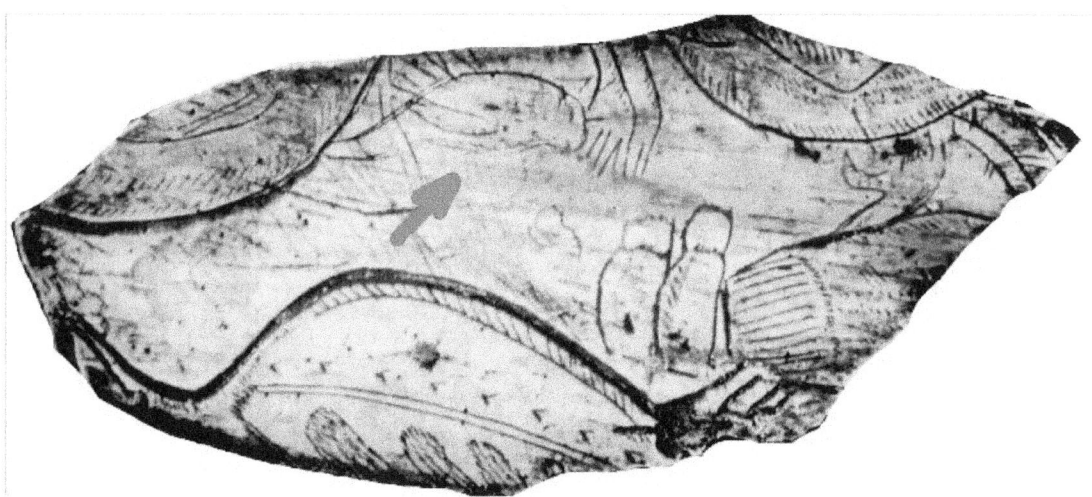

Figure 3.1. A carved drawing of a wingless cave cricket, *Troglophilus* sp., on a bison (*Bison bonasus*) bone found in the *Grotte des Trois Frères* (Three Brothers Cave) in the central Pyrénées, France (Chopard 1928) (From Romero 2009).

Unfortunately, this was an extremely uncritical book full of inaccuracies and odd explanations of how water circulated in the underground (Fig. 3.2). It also contained descriptions of supposed cave fauna that included dragons, unicorns, and giants (he even provided illustrations of such alleged creatures). Yet, no blind and/or depigmented creature was included. Kircher was an uncritical repeater of other people's tales. However, he was a very popular author because of his position as professor of the *Collegio Romano* (the Vatican's University), his reputation for being able to read 16 languages, having published 44 books (most of them huge in size, in large print and with impressive illustrations) on a great array of topics and having written more than 2000 manuscripts and letters (that have survived) (Romero 2000).

Both the Renaissance (ca. 1450-1650) in Europe and the Ming Dynasty (1368-1644) in China were characterized as eras of exploration. They provided the first significant contributions to our knowledge of the world's fauna since Antiquity. In the case of Europe that was particularly true for animals that were new to the ones mentioned in the Bible or by ancient Greek and Roman authors.

The first written record of a true cave organism was in the form of a letter dated in 1537 written by the Venetian poet and philologist Giovanni Giorgio (GianGiorgio)

---

139 *b*. Geisa, Germany, 2 May 1602; *d*. Rome, 28 November 1680.

Trissino[140]. In that letter he mentioned a cave amphipod ("*gamberetti picciolini*", probably *Niphargus costozzae*) from Monti Berici, Veneto, northern Italy. That letter was later reported by the Dominican fray and historian Leandro Alberti[141] in his most famous book, *Descrittione di tutta Italia* (1550, pp. 471-472) in which he portrayed numerous Italian caves in detail (Hill 1974).

Figure 3.2. The underground world according to Kircher (public domain image).
European Renaissance and Ming Dynasty

The first known written account of a cavefish came from China just three years after Trissino's letter. It was a travel report written in 1540 by Yi Jing Xie[142], a local government official. This never published report was found in the records of Luxi County in 1905 by Ying Huang the local governor who had it engraved as an inscription on a stele (Y. Zhao, pers. comm.). In this document, Xie referred to the hyaline fish (*Sinocyclocheilus hyalinus*) from the Alu caves, Yunnan, China (Figs. 3.3. and 3.4.). This fish was not collected for scientific purposes until 1991 and was not scientifically described until 1994 (Chen *et al.* 1994).

That these two discoveries took place almost simultaneously in Europe and China is not totally surprising since, as mentioned above, both cultures were experiencing their golden age of geographic discoveries. For China the Sixteenth Century, which coincided with the first half of the Ming Dynasty, was a century after the Chinese had embarked on impressive maritime explorations. However, by that time, the Yang Ming system of thought established by Shouren Wang[143] had replaced that of Xi Zhu[144]. While Zhu, the most significant Confucian rationalist, insisted in the importance of observation and that

---

b. Bologna, Italy, 1479; *d*. Bologna, 1552?
b. Bologna, Italy, 1479; *d*. Bologna, 1552?
Xie, Yi Jing (*b*. ?; *d*. ?).
[143] Wang, Shouren (Yangming) (*b*. Yuyao, Zhejian Province, China 1472; *d*. Nan'an, Jiangxi, China, 1528).
[144] Zhu, Xi (*b*. Yuxi, Fujian Province, China, 18 October 1130; *d*. China, 23 April 1200).

learning should be based on reason and the "investigation of things" (see his *Four Books*), Wang believed in the "learning of the mind," through intuition. This was, unfortunately, a change in thought opposite to the one that occurred in Ancient Greece when the idealism of Plato[145] based on the recognition of tangible objects via individual perceptions, was replaced by the logic based on observation by his student Aristotle[146].

Aristotle's legacy would have a tremendous importance because it helped to establish one of the fundamental tenants of Western science (particularly in biology) after the Renaissance: knowledge via observation, not pure speculation. On the other hand, Chinese civilization declined due to internal factors and invasions by Mongols and Westerners.

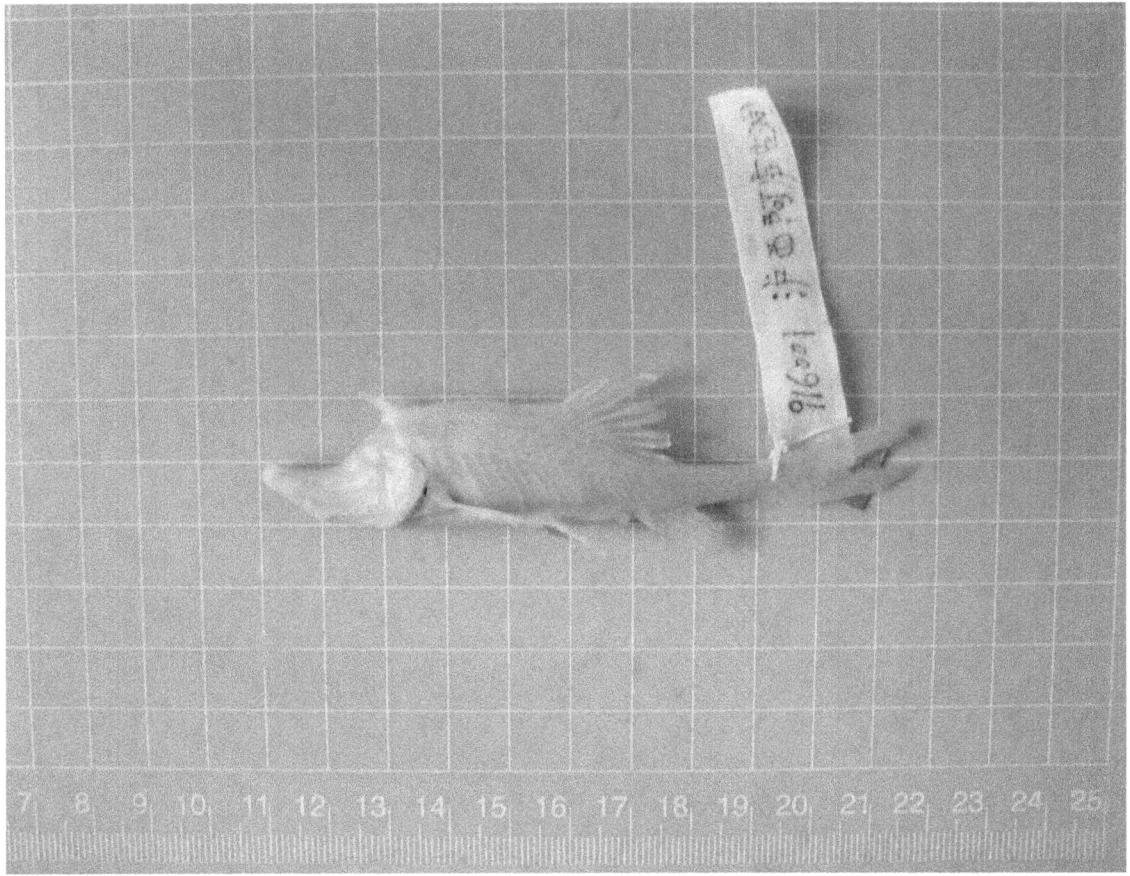

Figure 3.3. The hyaline fish, *Sinocyclocheilus hyalinus*, from the Alu caves, Yunnan, China. Photo by Y. Zhao in Romero *et al.* (2009).

Thus, new scientific discoveries would continue to take place mostly in Europe instead of elsewhere, even when some of those discoveries represented unconfirmed findings and false starts. That was the case of the French engineer and inventor Jacques Besson[147] who reported alleged underground little eels (*petites anguilles*) somewhere in Europe. In his book, Besson (1569) did not indicate the locality nor gave a description of the fish in question. He did not mention the fish as being blind and/or depigmented

---

[145] *b*. Athens [?], 427 B.C.; *d*. Athens, 348/347 B.C.
[146] *b*. Stagira, Macedonia, [in today's northern Greece] 384 B.C.; *d*. Chalcis, Greece, 322 B.C.
[147] *b*. Colombières, France, 1530?; *d*. Orléans, France, 1573.

(what would have been extraordinary characteristics to even the casual observer). Thus, it is unclear whether Besson observed true hypogean fish, actual eels (*Anguilla anguilla*), or European freshwater fishes with eel-like bodies that are sympatric with the areas in which he traveled (France and Switzerland). Those possible fish families include Petromizonidae, Cobitidae, Siluridae, and Clariidae (Blanc *et al.* 1971). Therefore, this description remains unconfirmed (Romero & Lomax 2000).

Figure 3.4. Locality (dot) of the location of the Alu Caves, China. Map provided by Y. Zhao in Romero *et al.* (2009).

Another example of an unconfirmed report of underground fauna was that of Marc-René Marquis de Montalembert[148] a French general and military engineer famous for devising simplified polygonal designs for fortresses that became the standard blueprint for European fortifications until Nineteenth Century. Montalembert reported a blind, subterranean fish in a spring at Gabard, Angoumois, near one of his estates in southwestern France (Montalembert 1748). Yet no specimen was preserved and his description remains unconfirmed (Romero 1999a).

---

148 *b*. Angoulême, Charente, France, 16 July 1714; *d*. Paris, France, 29 March 1800.

These series of casual reports (whether they were confirmed or not) were typical of the natural history of the Renaissance epitomized by "bestiaries" and were later replaced by a more rigorous view of science.

Modern Science (ca. 1650-1800)

Unconfirmed reports and mythical tales typical of the Renaissance were followed in the Seventeenth Century by the flourishing of what has been termed "Modern Science" characterized by direct observation and experimentation. During that time in which precision in description and illustration of the natural world improved considerably, we see some good examples of new accounts of underground biota.

The first of those contributions was the earliest published reference to an underground fungus by the physician and naturalist Martin Lister[149] (Lister 1674, Carr 1973). Lister received samples of this fungus from a Mr. Jessops and he called it "*Fungus subterraneus*", which was found in a mine known as "Old man" in Castleton, Darbyshire, central England. Lister was part of the first generation of English naturalists extremely keened in describing and illustrating natural objects, particularly animals and rocks (Unwin 1995).

The next important contribution came from the Spanish Capuchin fray and missionary Francisco de Tauste[150]. He was the first to publish a reference of a cave bird, the oilbird (*Steatornis caripensis*). Tauste wrote his report based on his study of costumes and languages of the Chaimas, an ethnic group of Native Americans of northeastern Venezuela where these birds inhabit the *Cueva del Guácharo* (Oilbird Cave; Tauste 1678, Longrás Otín, 2002). This species, which had been exploited for many years by the Chaimas for its oil (Anonymous 1833), was not scientifically described until 1817 by the German polymath, explorer and, above all, holistic naturalist Alexander von Humboldt[151] based on a specimen he collected in 1799 (Humboldt 1817). Humboldt's other contributions to our knowledge of hypogean biota includes the first description of underground plants in the mines of Freiberg (Humboldt 1793) and a description of a freshwater species of catfish which he claimed originated from an underground volcano in Ecuador (Humboldt 1805) (Fig. 3.5.), yet this claim remains unsubstantiated (Romero 2001a, Romero & Paulson 2001).

The earliest species of cave animal that underwent intense and continuous scientific study was the first species of a cave salamander ever described: *Proteus anguinus* from a region known then as Carniola, in today Slovenia. This blind amphibian was originally identified as a "dragon's larva" by the traveler and naturalist Janez Vajkard Valvasor[152] (Valvasor 1689). *P. anguinus* was later described scientifically by the Austrian naturalist Josephi Nicolai Laurenti[153] (Laurenti 1768) in the first post-Linnean description of a cave organism.

---

149 *b*. Radclive, Buckinghamshire, England, April 1639; *d*. Epsom, Surrey, England, 2 February 1712.
150 *née* Miguel Torralba de Rada; *b*. Tauste, Zaragoza, Spain, 1626; *d*. Santa María de Los Ángeles del Guácharo, Venezuela, 11 April 1685.
151 *b*. Berlin, Germany, 14 September 1769; *d*. Berlin, 6 May 1859.
152 *b*. Ljubljana, Carniola (today Slovenia), 28 May 1641; *d*. Ljubljana, 19 September 1693.
153 *b*. Vienna, Austria, 4 December 1735; *d*. Austria, 17 February 1805.

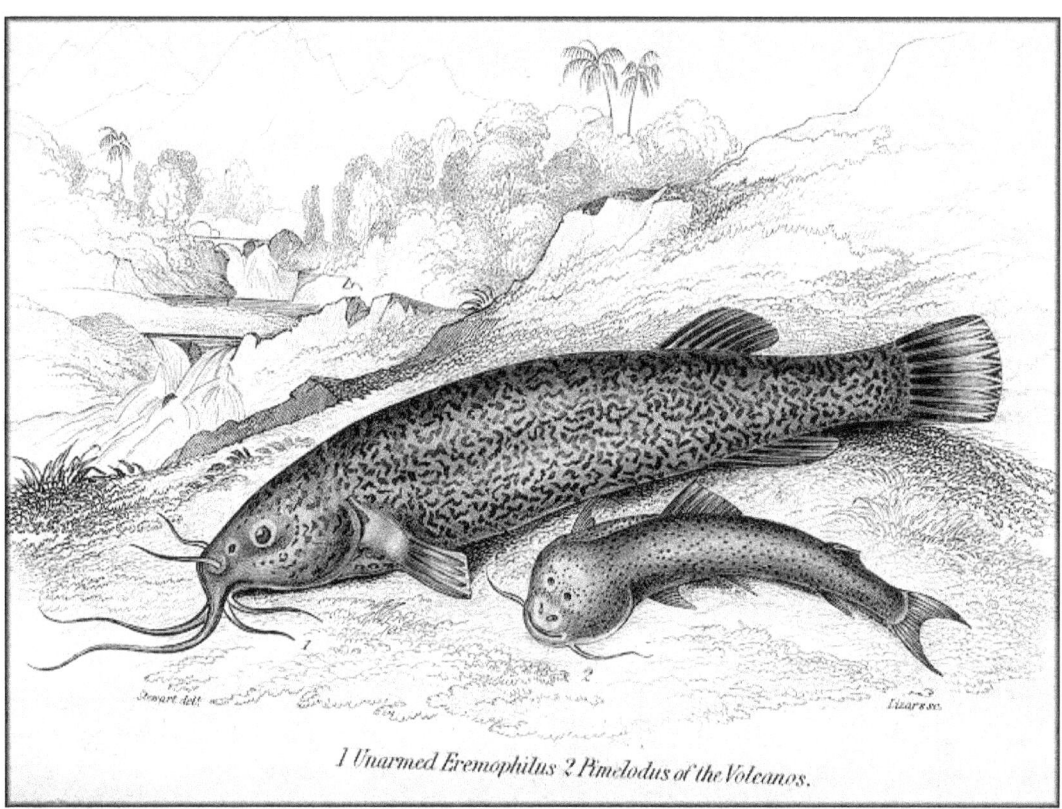

Figure 3.5. The alleged "volcano" fish (right) by Humboldt (public domain image).

First Professional Studies Before Darwin (1800-1859)

The period between 1800 and 1859 is characterized by three major events. The first two were circumstantial in nature: one was the beginning of biology as a formal discipline and for which we find the first generation of professional biologists; in fact, the term "biology" began to be used around 1800 (McLaughlin 2002). The second was the discussions on evolution including the loss or rudimentation of organs such as eyes and pigmentation, a phenomenon common (but not unique) among many cave organisms. The third one was the scientific exploration of two of the most important cave systems in the world: the one occupying the southeastern regions of the United States epitomized by Mammoth Cave, and the other in what is now known as Slovenia. These factors combined themselves to make the discussion on the biology of cave organisms an important aspect of the biological dialectic from Darwin until the end of the Nineteenth Century.

All began with the first scientific description of the cave salamander, *Eurycea lucifuga* (Rafinesque 1822). This description was made by the French-American Constantine Samuel Rafinesque[154] (Fig. 3.6) and took place at a time when he was professor of botany and natural history at Transylvania University in Lexington, Kentucky, between 1819 and 1826. Rafinesque had been exploring the caves of that state since 1818 (Rafinesque 1832) which makes him probably the first professional scientist who studied them. He encountered a salamander that the locals called "cave puppet" in 1821 in caves near Lexington. Kentucky encompasses a great deal of karst formations including large and complex cave systems. Although he did not provide too many details

---

154 *b*. Galata near Constantinople, Turkey, 22 October 1783; *d*. Philadelphia, Pennsylvania, 18 September 1840.

about the exact location not only of the cave but in what portion of it where he found this amphibian, this is one of the cave organisms easier to come across because they are usually seen nearby cave entrances. Thus, finding that salamander does not require in-depth exploration of cave systems. This discovery in itself was not particularly striking to the scientific community at that time because of two reasons: first, the cave salamander is neither blind nor depigmented, so it was not particularly remarkable to the casual observer; the second was Rafinesque's poor reputation as a scientist due to his lack of critical thinking and his almost compulsive behavior at naming species (more than 6,700) many of them previously described by others or just varieties of the same one. Yet, his discovery of the cave salamander was the first indication that the biota of caves in that part of the United States was worth looking at (Ewan 1975, Warren 2004).

Figure 3.6. Constantine Samuel Rafinesque (public domain image).

Rafinesque explorations included Mammoth Cave that since the 1830's had been rapidly becoming a great tourist attraction. Used by Native Americans for about 4,000 YBP, this cave was first reported by people of European descent in 1797 (Goode 1986). Mammoth Cave and its fauna became famous thanks, mainly, to the exploratory work performed by Stephen Bishop[155]. Bishop was born and died into slavery. Stephen was acquired when he was about 13 years old by the lawyer Franklin Gorin[156]. Gorin purchased Mammoth Cave in 1838. Bishop soon became a guide and explorer of the cave. Although by that time the most accessible parts of Mammoth Cave had been visited, explored, and mapped, a major obstacle remained for the expansion of its exploration: Bottomless Pit (Fig. 3.7). Bishop is consistently credited with having suspended either a pole or a log pole ladder across Bottomless Pit and thus, was able to significantly expand the known area of the cave (Anonymous 1981, Barr 1986, Anonymous 1992) collecting its fauna which was given as curiosities to visitors (mostly tourists from the northeastern United States) who, in turn, took those specimens to the museums in New England where they were studied by local biologists (Romero & Woodward 2005).

Figure 3.7. Bottomless Pit at Mammoth Cave (public domain image).

---

155 *b*. Kentucky?, 1821?; *d*. Kentucky, 1857.
156 *b*. Barren County, Kentucky, 3 May 1798, *d*. Glasgow, Kentucky, 10 December 1877.

About the same time the Postojna Cave in Carniola (today Slovenia) was being explored. This and other caves in the area had been known for a long time. Ancient Greek and Roman authors such as Strabo, Virgil, and Plinius, had mentioned them and so did Kircher (1665) and Valvasor (1687). In 1818 Luka Čeč[157] (Lukas Tschesch in the German spelling), an assistant to a lamplighter, discovered new passages which notably expanded the known areas of the cave. With about 20 kilometers of passages this cave is the largest in Europe. This was a feat similar to that of Bishop in Mammoth Cave. Just as Bishop, Čeč was able to collect new species of cave fauna in the newly discovered areas and in September 1831 found the first species of a blind cave beetle ever reported. Čeč donated the specimen to the Earl Franz Josef Hannibal Graf von Hohenwart[158], the district governor, who in turn presented it for study to the Austrian businessman and amateur entomologist, Ferdinand Jožef Schmidt[159]. In 1832 Schmidt published its description as a new genus and species, *Leptodirus hohenwarti*, with the Carniolian name *drobnovratnik* (the "narrow-necked-one"). Since the holotype had been damaged, Schmidt offered 25 florins for a new one and it took many years until other explorers found more specimens, as well as other fauna which included springtails, pseudoscorpions, and crustaceans (Mader 2003, Novak *et al.* 2003, Južnič 2006).

The crustacean *Gammarus puteanus* (today *Niphargus puteanus*) was described by C.L. Koch in either 1835 or 1836 (the publication was not dated) from wells in Regensburg, Germany. The description of "Gammarus puteanus Koch" published by Panzer in Faunae *Insectorum Germanicae Initia* in 1836 is the first unambiguous record of a *Niphargus*. A year later Koch (the describer of the species) published the identical description and drawings in *Deutschlands Crustaceen, Myriapoden und Arachniden*. The genus *Niphargus* was established 13 years later when Schiödte re-examined his own description of *Gammarus stygius* from the Postojnska jama (Postojnska Cave) in Slovenia.

Another example of a species discovery in the region was that of the science teacher and political activist Emil Adolf Rossmässler[160]. He described the first species of cave mollusk: a cave snail under the name of *Carychium spelaeum* (today *Zospeum spelaeum*) from the Postojnska Cave (Rossmässler 1838-1844). Rossmässler, a great admirer of Humboldt who believed in the importance of popularizing science, discovered the cave mollusk in one of his many scientific trips in the 1830's.

These discoveries elicited a great deal of interest for cave fauna and the caves of the karst system of Carniola. Although initially most of the published works from both Carniola and the U.S. were descriptions of new species, European and American scientists had divergent interests beyond purely taxonomic ones. While the European researchers showed more consideration in grouping cave organisms based on ecology and habits, their American counterparts (some of whom were European-born) tried to explain the archetypical or unique morphology (blindness and depigmentation) of many cave organisms via the influence of environmental factors on development.

In Europe, besides Schmidt, other pioneer researchers of the cave system in Carniola included the Danish entomologist Jørgen Matthias Christian Schiödte[161] and the Austrian entomologist Ignaz Rudolph Schiner[162]. Schiödte was particularly interested in the correlations between anatomical characters and the biological conditions under which organisms live. However, he went further by providing the first classification of cave animals (shade animals, twilight animals, animals in the dark zone, and animals living on stalactites) (Schiödte 1849). This categorization, however, was abandoned and

---

[157] *b*. Postojna, Carniola (today Slovenia), 11 October 1785; *d*. Postojna, 31 July 1836.
[158] *b*. Laibach (today Ljubljana, Slovenia), 24 May 1771, *d*. Laibach, 2 August 1844.
[159] *b*. Ödenburg, Sorpon, Hungary 20 February 1791, *d*. Ljubljana, Carniola, (today Slovenia) 16 February 1878.
[160] *b*. Leipzig, Germany, 3 March 1806; *d*. Leipzig, 9 April 1867.
[161] *b*. Slaegt, Denmark, 20 April 1815; *d*. Copenhagen, Denmark, 12 January 1884
[162] *b*. Fronsburg, Horn, Austria, 1813; *d*. Vienna, Austria, 6 July 1873.

replaced by the one proposed by Schiner (1854) who classified cave organisms according to their degree of dependence toward the underground environment as troglobites, troglophiles, and "occasional cavernicoles" (trogloxenes), terms still in use today.

In pre-Darwinian times, American researchers were not only involved in species descriptions as their European counterparts but seemed to be fixated on issues of functional morphology and how that morphology could be used to explain the "types" and development of those organisms. Virtually all of the research was being conducted in Mammoth Cave and began in the 1840's. Of all the organisms discovered in that cave none attracted more attention than the first description of a blind cavefishes, *Amblyopsis spelaea*, by James DeKay[163] (DeKay 1842).

DeKay, originally trained as a physician, was an amateur natural historian and friend of writers that represented the American Romantic Movement. He was hired by The Geological Survey of New York and put in charge of the "zoological productions." Applying his own Romantic ideas, he included in his relation of the fauna of New York species of fauna that were far away from that state such as the northern cavefish from Mammoth Cave (*Amblyopsis spelaea*) becoming the first blind cavefishes described in the scientific literature in the post-Linnean era. Although his scientific work was criticized by many of his fellow naturalists as being shallow, his description of this cavefishes attracted a great deal of attention and generated a lot of research and speculation until the American Civil War interrupted the efforts from scientists in the North to visit and collect at Mammoth Cave which was in the South (Romero 2002a).

Unlike papers describing species being found elsewhere, the reports on species from Mammoth Cave generated a lot of speculation about the origin of such fauna. That is surprising because the authors of those papers published before Darwin's *Origin*, were still in the creationist mood, and were not inclined to believe in evolution or "transformism" as it was called then. Most of the discussions concerned the question of why these animals were blind and depigmented in the first place. The first to engage in this type of debates was Jeffries Wyman[164]. Wyman studied under George Cuvier[165] and Richard Owen[166] (Gifford 1967), both staunch creationists. Because of that influence and also because he was a very modest man who like Cuvier avoided sweeping generalizations, Wyman essentially stuck to purely anatomical studies. For example, in his first paper on the blind cavefishes *A. spelaea*, he reported, "On the most careful dissection no traces of eyes were found" (Wyman 1843, p. 96). Later he wrote that "The optic lobes existed (sic); according to the general rules of physiology these should not exist; as they bare strict relation to the sense of sight, which receives its nerve from them (...) Here the optic lobes were not so large as the allied fishes, but yet they were of good size, and nearly as large as the cerebral lobes" (Wyman 1851, p. 349).

He later re-examined three specimens of this fish species and found "imperfect" eyes covered by tissue, which explained to him why the fish were unable to see. He proposed that this "imperfection" of the eyes "might be owing to a want of stimulus through a series of generations", and though the organ of vision was "imperfect", "it is more like the eyes of other vertebrates" (Wyman 1854a p. 19). The phrase "want of

---

163 *b*. Lisbon, Portugal, 12 October 1792; *d*. Oyster Bay, Long Island, New York, 21 November 1851.
164 *b*. Chelmsford, Middlesex, Massachusetts, 11 August 1814; *d*. Bethlehem, New Hampshire, 4 September 1874.
165 *b*. Montbéliard, France, 23 August 1769; *d*. Paris, France, 13 May 1832. He was one of the most influential biologists of his time being a brilliant comparative anatomist who believed in the Great Chain of Being and that the only changes that had occurred on earth were due to natural catastrophes after all species has been created by God (Coleman 1964, Bourdier 1971, Rudwick 1997).
166 *b*. Lankaster, England, 20 July 1804; *d*. Richmond Park, London, England, 18 December 1892. He was a comparative anatomist and one of the early critics of Darwin's evolutionary ideas. He criticized Putnam's interpretation of the optic lobes in blind cave fishes that their optic lobes are not reduced because they serve other function beyond sight (based on Wyman's ideas). Owen believed that the lobes were atrophied due to lack of light.

stimulus," is a Lamarckian term (see below) that probably had a developmental meaning for Wyman, i.e., the organ did not develop because the environmental stimulus was not there. In any case, he wondered about numerous structures without obvious functions, organs that were of morphological rather than physiological value (Wyman 1854b). He later produced very detailed drawings of the internal anatomy of *A. spelaea* (Wyman 1872), but by that time he had embraced evolution as a natural phenomenon (see below) (Romero 2001a).

Another example of a naturalist interested in taxonomy and morphology of the Mammoth Cave fauna was the German physician August Otto Theodor Tellkampf[167] (Romero 2001b). He apparently developed an interest in cave fauna from visiting Mammoth Cave in October 1842 (Tellkampf 1844a,b), after which he described several species of invertebrates. He also made contributions to the study of the cavefishes *A. spelaea* and concluded that its eyes and those of blind cave crayfishes had become rudimentary as a result of disuse: "While it is true, in general, that all animals retain their essential form, and that no species passes over into another by transformation, we know that less material changes of form are produced by external influences such as changes in climate or food, lasting though many generations of the same species". In other words, he had the idea that disuse led to rudimentation while negating the possibility of evolution above the species level, despite the fact that he could not find the unmodified form that gave rise to the blind and depigmented one. For him, the relationship of the blind fauna to unmodified species could not be settled until "such species, corresponding with them in all essential points, are found" (Tellkampf 1844b, p. 393).

To elucidate this issue, Jean Louis Rodolphe Agassiz[168], America's most famous naturalist of his time, intervened. The son of a minister, Agassiz studied medicine in universities of Switzerland and Germany. His teachers included Lorenz Oken[169], Ignaz von Döllinger[170], and Georges Cuvier. The first two were followers of *Naturphilosophie*. This was a German Romantic philosophy that sought metaphysical correspondences and interconnections within the world of living things. This philosophy was developed in early Nineteenth Century Germany by Friedrich Schelling and G.W.F. Hegel who followed Plato's idealism. Despite its apparent scientific mantra *Naturphilosophie* ideals inundated philosophical postures and the literary movement while opposing the materialistic and mechanist views of modern science. *Naturphilosophie* viewed both mind and body as designed by God and as equally important. Many naturalists that opposed Darwin were followers of *Naturphilosophie*.

Agassiz studied comparative anatomy under Cuvier and developed his ideas along the lines of natural theology, that is, to prove the existence of God through the study of nature. Agassiz became professor of natural history at Harvard (1847-1873) where he established the Museum of Comparative Zoology in 1859 combining research, teaching, and public outreach while securing large amounts of funds both public and private to support such endeavors.

During the 5 October 1847 meeting of the American Academy of Arts and Sciences, Agassiz proposed a "Plan for an investigation of the embryology, anatomy and effect of

---

167 *b*. Heinde, Germany, 27 April 1812; *d*. Hannover, Germany, 7 September 1883.
168 *b*. Motier-en-Vuly, Switzerland, 28 May 1807; *d*. Cambridge, Massachusetts, 14 December 1873.
169 *b*. Bohlsbach bei Offenburg, Baden, Germany, 1 August 1779; *d*. Zurich, Switzerland, 11 August 1851. Although a physician by training, Oken championed *Naturphilosophie* with metaphysical abstractions and mystical speculations about science (particularly biology) and Romanticism despite his scientific background and his rigor as a comparative anatomist. He believed that imagination and feeling should play a part in scientific understanding and in progressive complexity with humans at the zenith.
170 *b*. Bamberg, Germany, 27 May 1770; *d*. Munich, Germany, 14 January 1841. A professor for physiology and general pathology and one of his students was Lorenz Oken. He went beyond the typical *Naturphilosophie* approach to natural sciences by insisting in the importance of observation and experimentation (Risse 1971).

light on the blind-fish of the Mammoth Cave, *Amblyopsis spelaeus*" (Agassiz 1847, p. 180). In this plan he suggested that by studying this fish "there was an opportunity to settle, by actual experiment, the extent of physical influences in causing organized beings to assume their peculiar and distinctive characteristics in relation to the media in which they live." Agassiz, the unrepentant creationist, was not proposing to study the effects of the environment on evolution, but rather the effects of the environment on development. He proposed to raise individuals of *A. spelaea* under different light conditions (darkness, moderate, and intense light) and see if "there is an eye formed in the dark to ascertain when and how (the pigmentation) disappears, as it is entirely wanting in the full-grown individuals, and again notice the differences in this respect between specimens growing under the influence of light" (Agassiz 1847, p. 180).

Although Agassiz never carried out those experiments, he kept insisting on the importance of *A. spelaea* in biological research:

"You asked me to give my opinion, respecting the primitive state of the eyeless animals of the Mammoth Cave. This is one of the most important questions to settle in natural history, and I have several years ago, proposed a plan for its investigation which, if well conducted would lead to as important results, for it might settle, once forever, the question, in what condition and where the animals now living on the earth, were first called into existence. But the investigation would involve such long and laborious researches, that I doubt it will ever be undertaken. (...) If physical circumstances ever modified organized beings, it should be easily ascertained here. (...) Whoever would settle the question by direct experiment might be sure to earn the everlasting gratitude of men of science, and here is a great aim for the young American naturalist who would not shrink from the idea of devoting his life to the solution of one great question" (Agassiz 1851, p. 255).

Agassiz's words leave no doubt about that since he considered *A. spelaea* to be an "aberrant cyprinodont (...) created under the circumstances in which they now live" (Agassiz 1851, p. 256) while "The (rudimentary) organ remains, not for the performance of a function, but with reference to a plan" (Agassiz 1859, p. 11). The latter statement shows that Agassiz (together with Wyman) embraced so-called philosophical or transcendental anatomy, i.e., the search for ideal patterns of structure in nature (Appel 1987). That is why for both Wyman and Agassiz, *A. spelaea* was an excellent subject of study in their quest for evidence of a common plan underlying the differences caused by immediate adaptation through modifications during the developmental process.

Despite these challenging ideas, Agassiz's proposals were not undertaken because of a variety of reasons. One was the scientists' inability then and now to breed amblyopsid fishes in captivity. The second was Agassiz's personality: his refusal to accept ideas other than his own diminished his status among his colleagues as time went by. Agassiz never acknowledged the transmutability of species and fiercely opposed Darwin's theory of evolution. Several of his students, including his son Alexander, showed a great deal of interest in cave fauna, but left Harvard, and accepted the idea of evolution (see below) (L. Agassiz 1847, E.C. Agassiz 1890, Dexter 1965, 1979, Lurie 1960, 1970, Morris 1997, Smith & Brown 2000, Romero 2001b).

In many ways, Agassiz's interpretation of *Naturphilosophie* was a derivation of the idea of *Scala Naturae* also known as the ladder of life or "Great Chain of Being" with man at the top of the pyramid. This is a concept that originated with Aristotle and the stoics and was closely tied to Plato's essentialism, i.e., objects (in this case individuals/organisms/species) have ideal, eternal, unchanging "essence" (*eidos*). These ideas in turn gave rise to the typological notion that all "true" cave animals must be blind and depigmented. Agassiz passed these ideas on to his students at Harvard.

This was pretty much the state of things among American scientists at the time of Darwin's publication of the first edition of the *Origin* in 1859: a mixture of creationist views, intriguing questions about environmental effects on development, and an explicit endorsement of the typological (essentialist) view of life.

Table 1. A summary of major events in the development of biospeleology before Darwin's publication of the first edition of the *Origin of Species by Means of Natural Selection* (1859). This shows that the development of biospeleology was characterized by chance or random discoveries, false starts, and uncritical tales about cave fauna, and ultimately most of these early events were largely inconsequential to the development of later ideas on cave biology.

Date	Fact	Source(s)
20,000 YBP	Knowledge of the cave cricket, *Troglophilus* sp.	Chopard 1928
1537	First written report of a cave organism: a cave amphipod (probably *Niphargus*)	Trissino 1537
1540	First written (but unpublished) reference of a hypogean fish *Sinocyclocheilus hyalinus*. This species was not scientifically described until 1994	Xie 1540, Chen *et al.* 1994.
1550	The first confirmed published record of a cave creature: a cave amphipod (probably *Niphargus*) in Northern Italy	Alberti 1550
1569	Obscure mention of underground little eels (*petites anguilles*) somewhere in Europe	Besson 1569, Romero & Lomax 2000
1665	Publication of *Mundus Subterraneus*, the first speleological treatise	Kircher 1665, Romero 2000
1674	First published reference of underground fungi	Lister 1674
1678	First published reference of a cave bird	Tauste 1678, Humboldt 1871
1689	First published reference of a cave salamander, *Proteus anguinus*	Valvasor 1689, Laurenti 1768
1748	An unconfirmed report of an hypogean fish (a "pike") in France	Montalembert 1748, Romero 1999a
1793	First description of underground plants in the mines of Freiberg	Humboldt 1793
1805	Description of a freshwater species of catfish allegedly originated from an underground volcano in Ecuador	Humboldt 1805
1817	Scientific description of the guacharo or oil bird *Steatornis capipensis*	Humboldt 1817
1822	First scientific description of a cave salamander, *Eurycea lucifuga*, for the American continent	Rafinesque 1822
1832	First scientific description in print of a cave insect, *Leptodirus hohenwarti,* a blind beetle	Schmidt 1832
1835-1836	First scientific description of a hypogean crustacean: *Niphargus puteanus*	Koch 1835-1836
1839	Published mention of a second cavefishes for China	Zheng 1839

1839	First description of a cave mollusk, *Zospeum spelaeum* from Slovenia	Rossmässler 1838-1844
1842	First scientific description of a blind cavefishes, *Amblyopsis spelaeus*, from Mammoth Cave, Kentucky	DeKay 1842, Romero 2002b
1849	First cave species survey for an entire region (Slovenia) in *Specimen Faunae subterraneae*. The author provides first ecological classification of cave organisms (shade animals, twilight animals, animals in the dark zone, and animals living on stalactites)	Schiödte 1849
1854	The terms troglobites, troglophiles, and accidentals are introduced	Schiner 1854

Darwinism and American neo-Lamarckism (1859-1919)

It may be surprising that Darwin is mentioned before Lamarck in this analysis of biospeleological ideas since the latter preceded him by more than half a century. The reasons are two-fold: first, Lamarck never mentioned cave fauna in his evolutionary writings; second, Lamarckism or neo-Lamarckian ideas for explaining biological phenomena in caves did not become popular until after Darwin's publication of his *Origin*. Furthermore, I will argue that Darwin was largely responsible for some (but not all) of the neo-Lamarckian views on cave biota that have survived in both perception and substance to this day.

Although Charles Darwin[171] never studied cave fauna himself, he was interested in the topic, particularly as it related to two issues. The first concerned cave colonization and the similarities between cave fauna and their presumed ancestors in the surrounding areas. The second issue, and the more attractive to him, was the cause of the phenomenon of rudimentation or the loss of organs, i.e., eyes. He saw this trend as part of a larger compensatory-process issue, i.e., the enlargement of other sensory organs, regardless of whether compensation occurred among cave fauna or not. An analysis of Darwin's writings, including his notebooks and correspondence, provides insight on how intrigued he was by these topics and how, also, he changed his opinions on these matters, sometimes as a response to criticism and sometimes as he received new information.

The first written documentation of Darwin's interest in these topics is in his notebook and is dated 8 December 1844. He wrote a few notes after a conversation he had with Joseph Dalton Hooker[172], a distinguished botanist and a close friend of his. Darwin wrote that "I see our cow, which has two abortive mammae, then these two are uniquely developed" adding later "Believe part, which is normally in a species <u>abortive</u> appears often as a <u>rudiment</u>- [Hooker] Has lately seen and describe this in case of pistil of dioecious *Umbellieferous plant*: does not know anything on Bentham's law of variability of abortive parts." George Bentham[173]'s "law of abortive parts" was worded by Darwin himself as follows: "where parts of flower are reduced from normal number, they are apt to vary in number in individuals of same species" (Burkhardt & Smith 1987, pp. 400-403).

However, Darwin did not develop an interest in cave fauna until early 1852, shortly after he read an article on Mammoth Cave. Beginning in the early 1840s a number of blind and depigmented species of both vertebrates and invertebrates had been

---

171 *b*. The Mount, Shrewsbury, England, 12 February 1809; *d*. Downe, Kent, England, 19 April 1882.
172 *b*. Halesworth, Suffolk, England, 30 June 1817; *d*. Sunningdale, Berkshire, England, 10 December 1911. Hooker's initial reaction to Darwin's theory of transmutation of species was not very enthusiastic, but he would later change his mind. He provided Darwin with many botanical facts, particularly in the areas of taxonomy and biogeography (Desmond 1972, Colp 1986, Bellon 2001).
173 *b*. Stoke, Devon, England, 22 September 1800; *d*. London, England, 10 September 1884. A polyglot and polymath very interested in botany who was a friend and supporter of Darwin's (Taylor 1970).

described for that locality and in 1851 Benjamin Silliman Jr.[174] published an article summarizing the current knowledge about those cave creatures (Silliman 1851). In that article, Silliman made a number of statements that, without question, intrigued Darwin. Silliman described several cave species of animals that were not only blind and depigmented but that also displayed elongated antennae. He made a special mention of the "cave rat," of which he had heard that had large but apparently non-functional eyes, and according to Silliman, "By keeping them however in captivity and diffuse light they gradually appeared to attain some power of vision." These presumed facts defied the basic explanation Darwin had already formulated in his mind regarding natural selection's role in determining morphological features of cave organisms. After all, if the alleged cave rat had larger non-functional eyes that would defy the logic of natural selection.

Part of the problem was that Silliman had given a faulty account of this organism. First of all, the alleged "cave rat" (*Neotoma* sp.) was not an obligatory cave organism but rather a nocturnal creature found both in and outside caves. The reason it had such large eyes was that like many other nocturnal vertebrates its eyes were enlarged for better night vision. No wonder Darwin had problems trying to understand the phenomenon.

Yet, intrigued by all this, Darwin, in a letter dated 8 May 1852, asked his friend and colleague the American naturalist James Dwight Dana[175] if he could receive a specimen of the "cave rat" (Burkhardt & Smith 1989, p. 92). The reason that Darwin wrote Dana (one of the most notable American naturalists of the time) and not Silliman was two-fold: Darwin was acquainted with Dana, not Silliman, and Dana was Silliman's bother-in-law, so Darwin probably figured this was the best way to obtain information on the subject.

Another letter to Dana dated 14 July 1856 Darwin wrote that he was "extremely much interested in regard to the blind cave animals, described one time since in your Journal by Prof. Silliman Jun[r.] as the subject is connected with a work of somewhat general nature, which I am endevouring to draw up on variation & the origin of species, classification & c." (Burkhardt & Smith 1990, p. 180). In another letter to Dana dated 8 September 1856 he confirms that most of the species found in Mammoth Cave are "American in type" (Burkhardt & Smith 1990, p. 215-217) meaning that they must be related to other fauna found in adjacent areas. This was an important point for Darwin since he had by then developed the idea that all species were derived from those found in neighboring ecosystems. Darwin makes this point again in a letter to Hooker dated 23 November 1856 (Burkhardt & Smith 1990, p. 281-284).

Apparently, Darwin never obtained the "cave rat" but that did not deter him from asking more questions. Darwin received a letter from Dana dated 8 December 1856 in which the American naturalist told Darwin that he had confirmed with Agassiz that the blind rat of Mammoth Cave is "American in type". For Darwin this confirmed his own hypothesis that cave animals were derived from species of the surrounding areas. In the same letter Dana goes into a long discourse about the idea of progress. He described progress as "a law which involves the expression of a type-idea in forms or groups of increasing diversity, and generally of higher elevation; always resulting in a purer & fuller exhibition of the type" and that "it is the simple before the complex" (Burkhardt &

---

174 *b.* New Haven, Connecticut, 4 December 1816; *d.* New Haven, 14 January 1885. Silliman taught at Yale and helped his father (the founder in 1818 and first editor of the *American Journal of Science and Arts*) to edit that journal where a number of articles on cave animals were published in the nineteenth century.

175 *b.* Utica, New York, 12 February 1813; *d.* New Haven, Connecticut, 14 April 1895. Dana studied at Yale under his future father-in-law Benjamin Silliman. Dana believed that the earth was a changing place through catastrophes (volcanism, erosion and subsidence) which, in turn, generated changes in forms of life toward higher levels of complexity, an explanation he tried to reconcile with his strong religious beliefs by saying that this was the result of God's design (Sanford 1965, Stanton 1971).

Smith 1990, p. 299-300). Here we can see how strong the idea of progressionism was in the minds of naturalists even before evolutionary ideas became a matter of discussion. As Bowler (2005) wrote, Darwin's "theory was sucked into a wave of enthusiasm for progressionist evolutionism (…) which reached its climax in later nineteenth century." Also, Darwin is reading in this account a message of order in nature, not necessarily an evolutionary one, but one confirming the idea of the Great Chain of Being already present in Plato's and Aristotle's writings. According to this account, nature is characterized by richness in forms which show continuity in the form of gradation. Therefore, the universe is stuffed with everything that is possible which sharing characteristics between the neighboring forms and because of that they can be arranged in hierarchical order from the smallest, simplest type of existence to God himself.

Correspondence between Darwin and Dana on the subject of the cave fauna continued, and thus in a letter dated 14 July 1856, Darwin asked Dana if he could get more anatomical information about North American cave fauna, particularly arthropods (Burkhardt & Smith 1990, p. 180). Dana replied on 8 September 1856 with mostly systematic information (Burkhardt & Smith 1990, pp. 215-217). On 29 September 1856, Darwin wrote back to Dana thanking him for the information and inquiring for additional facts about the "blind rat" (Burkhardt & Smith 1990, p. 235-237).

Darwin was evidently unsatisfied by the mostly taxonomic and philosophical answers from Dana and wrote one other scientist for more information on the subject. Case in point was John Obadiah Westwood[176], a British entomologist, who replied to Darwin in a letter dated 23 November 1856 giving him some taxonomical and distributional information about cave insects from both North America and Europe (Burkhardt & Smith 1990, p. 283-284). By 1856 Darwin was already keeping a portfolio on abortive organs (Burkhardt & Smith 1990, pp.253-254).

Darwin's information about cave fauna, however, came not only from what he read about Mammoth Cave and from his correspondence with Dana and Westwood, but also from an article written by the Danish naturalist Schiödte. As mentioned earlier, Schiödte was particularly interested in the correlations between anatomical characters and the biological conditions under which organisms live and also provided the first classification of cave animals (shade animals, twilight animals, animals in the dark zone, and animals living on stalactites) (Schiödte 1849). Although Schiödte's article was originally written in Danish, Darwin (who always struggled with foreign languages, which is the reason he almost exclusively referred to literature that was originally written in or translated into English) had access to the paper because it had been translated and published in English by the also Danish naturalist Nathaniel Wallich[177], who read it at the meeting of the Entomological Society of London on 6 January 1851 (Wallich 1851, Burkhardt & Smith 1990, p. 283-284).

Armed with this information, Darwin speculated in the first edition of his *Origin* (1859, pp. 137-138) that "in the case of the cave-rat natural selection seems to have struggled with the loss of light and to have increased the size of the eyes; whereas with all the other inhabitants of the caves, disuse by itself seems to have done its work." Darwin cited Schiödte's paper as a reference (p. 138). At first Darwin considered the mechanisms of both natural selection and disuse to explain blindness and depigmentation as well as the enlargement of some sensory systems and appendages. To Darwin, this meant a "contest (…) between selection enlarging and disuse alone reducing these organs" (Darwin 1859, p. 296). Later Darwin noted that cave fauna was more

---

176 *b*. Sheffield, England, 22 December 1805; *d*. Oxford, England, 1 January 1893. Entomologist, archaeologist and a superb illustrator with strong religious beliefs who disagreed with Darwin regarding his theory of evolution while respecting him as a scientist.
177 *b*. Copenhagen, Denmark, 28 January 1786, *d*. England, 28 April 1854.

closely related to the fauna of the surrounding regions than elsewhere, as is the case for fauna of other more or less isolated habitats such as islands. Thus, he argued that the cave fauna descended from the fauna of the surrounding region, "the colonists having been subsequently modified and better fitted to their new homes" (Darwin 1859, p. 403).

Although the second edition (1860) of *The Origin* contains very few substantial changes from the first, beginning with the third edition (1861), Darwin makes major changes not only in the book as a whole but on the explanation of the phenomenon of rudimentation of organs among cave animals in particular. Darwin's critics, who had a hard time accepting the role of natural selection in general and its effect on cave animals in particular, found in the latter ammunition for their anti-selectionist criticisms; after all, it seemed that on the surface that Darwin himself was providing the best argument favoring disuse over random selection to explain the reduction and/or disappearance of organs.

Thus, by the third edition of *The Origin*, Darwin de-emphasized the importance of natural selection by eliminating his discussion of a "contest" between selection and disuse. In fact, in the first two editions, in the paragraphs relative to cave animals and rudimentation, he used the words disuse and selection seven times each; by the third edition, it was five and two, respectively.

Yet, criticism mounted. In 1865 Carl Wilhelm von Nägeli[178], a botanist and one of the rediscoverers of Mendel's work, made the point that characters considered useless could not have arisen via natural selection or even Lamarckism for that matter. Darwin respected von Nägeli's opinions very much and, thus, by the sixth edition, he responded to this criticism by significantly expanding his discussion on morphological reductions and natural selection, although suggesting that there were mechanisms yet to be discovered to explain this phenomenon.

Another criticism was expressed by George John Douglas Campbell, Duke of Argyll[179]. With a sort of finalistic ideology, this politician and prolific writer attacked Darwin's explanation of rudimentation by saying that rudimentary organs were not remnants of useful structures but rather incipient structures being prepared for some future use (Argyll 1867, p. 213). Campbell's expression may foreshadow the concept of "preadaptation", so popular among classical biospeleologists. Further, the Duke of Argyll did not understand how Charles Darwin could have proposed natural selection without a "selector", just as animal breeders make selective choices. The Duke of Argyll and Richard Owen were staunch creationists who tried to prevent the influence of Darwin's ideas on British society.

Finally, in 1871 George Mivart[180] (who studied under Owen among others) published his *Genesis of Species*, a work foretold by an article in the *Quarterly Review* of the same year. Mivart articulated many criticisms of Darwin's ideas, criticisms to which Darwin responded in full in the sixth edition (1872) of his *Origin*. One of these criticisms was that "Natural selection utterly fails to account for the conservation and development of the minute and rudimentary beginnings, the slight and infinitesimal commencements of structures, however useful those structures may afterwards become" (Mivart 1871, p. 23).

By the sixth edition (the first of his famous book to use the word "evolution") Darwin remained cautious about the role that natural selection may have played in the reduction of morphological characters by saying that this process may have been "aided

---

178 *b*. Kilchberg, Switzerland, 27 March 1817; *d*. Munich, Germany, 10 May 1891.

[179] *b*. London, England, 30 November 1827; *d*. London, 1 April 1900. A morphologist with a finalistic argument for design by a higher intelligence and a believer of neo-Lamarckism. His books *On the Genesis of Species* (1871) and *Man and Apes* (1873) were sour and personal criticisms of Darwin's ideas on natural selection and human evolution while disparaging the 'bad' influence of Darwinism on British society (Artigas *et al.* 2006).

perhaps by natural selection." By now the number of times he uses the word "disuse" (when discussing cave organisms and rudimentation) has risen to nine and "selection" to ten. He also added the idea that animals subjected to darkness may develop "inflammations of the eyes" and that the covering of those organs by tissue can "be an advantage" (that is where selection might play a role). He then made the statement that rudimentary organs are very common among many organisms (a fact usually overlooked by the practitioners of the "regressive evolution" concept today). He supported the statement by providing numerous examples from plants to whales. Yet, he remained firmed in his idea that "It appears probable that disuse has been the main agent in rendering organs rudimentary." Although he mentioned the benefit derived by an organism in reducing organs that are no longer utilized for the sake of "economy," he had no explanation for why those organs totally disappear of those organs occur[181]

So, which one is the real Darwin when it comes to the evolution of cave faunas and the phenomenon of rudimentation? Was he a selectionist or a Lamarckian? Essentially Darwin's views were neo-Lamarckian in relation to loss or rudimentation of organs; therefore, to say that later neo-Lamarckism was "anti-Darwinian" concerning cave fauna is a misinterpretation of the facts since Darwin himself held neo-Lamarckian ideas. This despite the fact that Darwin wrote to Lyell on 11 October 1859, that "I do not know what you think about it [Lamarck's work], but it appeared to me extremely poor; I got not a fact or idea from it" (F. Darwin 1896, vol. 2, p. 10).

My proposition that Darwin held neo-Lamarckian ideas when dealing with cave fauna is consistent with the interpretation that in many ways Darwin held a modified version of the Great Chain of Being (Bowler 1983, pp. 55-59). This was a position also championed by the Swiss entomologist Charles Bonnet[182] and the French philosopher-naturalist Jean-Baptiste-René Robinet[183]. They both endorsed the idea of organic progress (Burkhardt 1977, p. 8-84).

For Bonnet, God had a plan but his divine role only took place at the beginning of the universe. For Bonnet there were always intermediate forms between species. He was a proponent of the theory of preformation, i.e., that all organisms have a preformed germ in the female germ cell. For him, all these preformed germs were there at the time of the beginning of the universe. He believed that the earth had been affected by cataclysms (similar to Cuvier's catastrophes) that had destroyed life several times over but then every time the "germs" were reborn into better and more perfected (and complex) forms of life culminating in a "paligenesis" or resurrection as interpreted by the Christian gospel. Bonnet's beliefs were very popular, particularly in France (Pilet 1973, Anderson 1976, Rigotti 1986). Thus, Bonnet equated the idea of progressive development with the term "evolution" meaning the unfolding of a providential plan to replenish the earth with life (Richards 1992, 2002).

Both Bonnet and Robinet were strict Lamarckians. Lamarck argued that organisms experience "needs" (*besoins*) that are brought about by the environment and that trigger fluids (including electricity) which, when circulated in the body, enlarge or develop the appropriate organ. According to Lamarck, a crucial causal factor in "higher" animals, is the "inner consciousness" (*sentiment interieur*), which causes body parts to respond and develop. This line of thought resulted in the idea of the inheritance of acquired characters. Although Darwin was less inclined to metaphysical interpretations than his French-speaking colleagues, he favored the idea of inheritance of acquired characters, and his selectionist explanations regarding cave animals were at best weak and at worst confusing.

---

[181] For a word-by-word comparison of these texts, see http://www.clt.astate.edu/aromero/new_page_29.htm).
[182] *b*. Geneva, Switzerland, 13 March 1720; *d*. Geneva, 20 May 1793.
[183] *b*. Rennes, France, 23 June 1735; *d*. Rennes, 24 March 1820.

Given this muddled state of science, the void created by Darwin himself by his lack of a rational explanation for the phenomenon of rudimentation, was filled by orthogenesis and its related conceptions, first in the United States and later in continental Europe.

The publication of Darwin's *Origin* in 1859 stimulated American naturalists not only intellectually but also sociologically. On one hand Louis Agassiz completely dismissed the idea of transmutation of species and Darwin's book as a whole, which he attacked unrelentingly, particularly in academic circles (he qualified Darwin's books as "poor, very poor"; F. Darwin 1896, **2**:63); on the other Wyman (Agassiz's colleague at Harvard) and the students of Agassiz himself including his son Alexander), eventually embraced the idea of evolution though they dismissed natural selection as its main mechanism. These, together with Alpheus Hyatt[184] and Edward Drinker Cope[185], were the founders of the American neo-Lamarckian school, and they saw cave fauna as the perfect example to support their ideas. Since Darwin himself adhered to the explanation of disuse as a mechanism for change and had not articulated a strong argument in favor of selection acting on cave organisms, they did not feel they were contradicting in any significant measure the tenants of the later editions of *The Origin* on this matter.

Three historical factors influenced biospeleological research in the United States after the appearance of Darwin's *Origin*: (1) the American Civil War (1861-1865), (2) the emergence of the Hyatt-Cope Progressionist School (from 1864 on), and (3) the "Salem Secession" of 1864.

1. The American Civil War

The onset of this conflict meant, essentially, that any field and laboratory study of Mammoth Cave fauna stalled. The reason was very simple: the scientists interested in the topic were in the North whereas the Mammoth Cave was in the South. In fact, there would not be a renewal of interest in these fauna until 1871, when after the Indianapolis meeting of the American Association for the Advancement of Science, many of the participants visited Mammoth Cave and collected new specimens. Thus, for more than a decade, American naturalists had to be content engaging in speculation about the cave fauna and their origin without the benefit of direct observation. In at least one instance, a person most interested in the issue was kept away forever. This was the case of Charles Frédéric Girard[186]. He had been brought to the United States by Louis Agassiz in 1847 and had worked at the Smithsonian Institution until 1860. While there, he was given some specimens collected by a J.E. Younglove "from a well near Bowling Green, Ky". He bestowed on those specimens a new species status, *Typhlichthys subterraneus*, which he included in the family Amblyopsidae (Girard 1859). This new species seemed to have "characters apparently transitory" between *A. spelaea* and the other species of the amblyopsid family known at that time: *Chologaster cornutus*, an epigean species. *A. spelaea* lacked eyes but had ventral fins; *T. subterraneus* lacked both eyes and ventral fins, whereas *C. cornutus* had eyes and lacked ventral fins. Obviously, the discovery of this sort of "intermediate" species should have fueled much further discussion on the issue of evolution. However, its discovery occurred right before the beginning of the Civil War, so no more information could be obtained. It is quite possible that Girard would have continued working on it, but hostilities broke out while Girard was in Paris, and he spent most of the rest of his life there practicing medicine. His support for the Confederate cause (by sending drugs, medical supplies, and arms) and the animosity that Agassiz had

---

184 *b*. Washington, D.C., 5 April 1838; *d*. Cambridge, Massachusetts, 15 January 1902.
185 *b*. Philadelphia, Pennsylvania, 28 July 1840; *d*. Philadelphia, 12 April 1897.
186 *b*. Mulhouse, France, 8 March 1822; *d*. Neulilly-sur-Seine, France, 29 March 1895.

developed toward him (Jackson & Kimler 1999) probably led Girard to believe that he would not be welcomed back in the United States, especially since the vast majority of his colleagues had supported the Union. Yet, he must have maintained some interest in cavefishes because later in life he published a number of popular articles on this topic (e.g., Girard 1888) (Romero 2001a).

2. The Emergence of the Hyatt-Cope Progressionist School

Aplheus Hyatt, a former student of Agassiz with whom he broke up, after the "Salem Secession," (see below) and who, to the dismay of Agassiz, embraced evolution, visited Mammoth Cave in September 1859, much earlier than his contemporary colleagues, and collected specimens of its fauna (Bocking 1988, Romero 2001a). Hyatt's evolutionary ideas were based on three tenants: (1) species have, as do individuals, an inevitable life cycle that includes decline as age advances; (2) for a species the preceding step before extinction is "degeneration" of the species (cave creatures with their lack of eyes and pigmentation epitomized to him this degeneration); and (3) species "transmutation" is the result of the speeding ("acceleration") or slowing ("retardation") of development which, in turn, is caused by use and disuse (for a summary of Hyatt's ideas see Brooks 1909).

Hyatt's ideas were influenced by two currents of thought: (1) the Americanized version of *Naturphilosophie* that was based on Oken's German idealism and trascendentalism, which Agassiz had championed and passed on to his students including Hyatt himself and (2) progressionist ideas popularized by Ernst Haeckel's[187] "Principle of Recapitulation" which Hyatt's used to formulate his "Law of Acceleration." The best compendium of Haeckel's philosophy of progressive evolution can be found in his 1891 *Evolution of Man* (*Anthropogenie; oder, Entwicklungsgeschichte des menschen. Keimes- und stammesgeschichte*), which was based on an analytical comparison of embryonic development and evolution, better known today as the "Biogenic or Biogenetic Law" (i.e., "ontogeny recapitulates phylogeny") or the recapitulation theory.

Interestingly enough, although Haeckel was very impressed with Darwin's *Origin*, he was not very enthusiastic about natural selection as the primordial mechanism and tended instead to emphasize Lamarckian mechanisms. This may explain, at least in part, why Haeckel did not like Darwin's natural selection explanation; he was a German idealist and transcendentalist while Darwin represented the best of the British natural theology.

In other words, since Darwin himself advocated disuse as the mechanism to explain the loss of phenotypic features among cave animals, for the American neo-Lamarckians that point was not in dispute. What they disputed was the impression of randomness and lack of direction implied in Darwin's ideas; therefore, their point of contention with Darwin was not a disagreement over evolution as a fact or disuse as a mechanism, but their philosophical view of directionality in nature. This is still the major

---

187 *b*. Postdam, Prussia, Germany, 16 February 1834; *d*. Jena, Germany, 9 August 1919. Haeckel studied medicine and was still practicing when he read Darwin's *Origin of Species* in 1859. That made him abandoned his medical profession to study natural history at the University of Jena where he later became the professor of comparative anatomy and the leading German Darwinist. For him all living organisms were plasmatic bodies differing only in degrees of organization. This was a metaphysical position according to which all living matter is made of the same essence. Another notable influence on him was that by the vitalist and comparative anatomist and physiologist, Johannes Müller. Haeckel was also influenced by *Naturphilosophie* and he ultimately developed pantheistic ideas. He was also a progressionist who believed that advances in science would allow humankind to reach new heights in rationality and morality, while civilizations that did not adhere to these concepts were considered 'degenerative.' He inspired many students, Anton Dohrn among them (Uschmann 1972, Oppenheimer 1982).

philosophical contention that creationists have today. No wonder the triumph of Darwinism has been called "The triumph of chance and change" (Greene 1959).

The other proponent of progressionism in the U.S. was Edward Drinker Cope. He was a highly prolific naturalist who became a very influential, although controversial, figure in his time. His first dealing with alleged cave fauna took place with the description of what he thought to be a new genus and species of troglomorphic fish, "*Gronias nigrilabris*", from Pennsylvania (Cope 1864, p. 231). Although he did not present any evidence that such fish had been captured in the hypogean environment, he was quick to suggest that such fish "is supposed to issue from a subterranean stream, said to traverse the Silurian limestone in that part of the (sic) Lancaster county, and discharge into the Conestoga". Cope was known for his hasty conclusions and the superficiality of some of his work (Romero & Romero 1999). Further studies have shown that the specimens on which he based this description were specimens of *Ictalurus nebulosus* that had eyes present that were asymmetrically developed, probably as a result of a teratological condition. Unfortunately, his assertion concerning this fish continued to be repeated in the literature until recently (see Romero 1999b for full history of this misconception).

But more important than this alleged discovery was the position that Cope himself took about evolution in general and how that influenced biospeleological thinking. Cope and Hyatt developed what was to be known as the Hyatt-Cope position or school, which was based on parallels drawn between embryology and phylogeny. Early in his career Cope took a stand against natural selection, never acknowledging it as an important evolutionary force (e.g., Cope 1864), and like the rest of his contemporaries became a strong supporter of Lamarckism. However, he extended Lamarck's ideas by representing evolution as a phenomenon governed by trends: "The method of evolution has apparently been one of successional increment or decrement of parts along definite lines" (Cope 1896, p. 24). This is what was later called orthogenesis, the view that evolution has a life of its own that can take it in certain directions. As Hyatt had also done, Cope proposed evolutionary principles such as the "Law of the Unspecialized" which when applied to cave organisms meant that these cave creatures without eyes and pigmentation were at the end of their phylogenetic life because they were too specialized to evolve into something else; therefore, the next step had to be extinction (see Cope 1896, pp. 172-174). As we shall see, these ideas became the distinguishing feature of American neo-Lamarckism, and much of the discussion on the evolution of cave species (even today) was heavily influenced by such views as those epitomized by the use of terminology such as "regressive evolution" (Cope 1864, 1872, 1896, Davidson 1997, Romero & Romero 1999, Wallace 1999).

3. The "Salem Secession"

This was the breakup of professional relationships between Agassiz and many of his students at Harvard because of a combination of disputes over labor issues, the ownership of collected material, the freedom of the students to publish, economic issues created by the Civil War, and philosophical differences. Since most of these students went to the Peabody Academy of Sciences in Salem, Massachusetts, their departure from Agassiz was termed the "Salem Secession" by Dexter (1965). The fact that most, if not all, of the most notable of Agassiz's students used cave organisms as either the subjects of their research or for purposes of philosophical disquisitions shows how influential Agassiz was in planting interesting questions in his students' minds. Yet, because Agassiz was an unrepentant creationist, they distanced themselves from their former master sometimes more in form that in substance. Wyman, for example, quickly converted to evolutionism (but without accepting natural selection as its main mechanism) and regarded Agassiz as backwards for his refusal to accept evolution (Appel 1987) while other students of his (Hyatt, Alpheus Packard[188], Edward Morse[189] and to a lesser extent Frederic Putnam[190] and Nathaniel Shaler[191]) went on to contribute to the popularity of neo-Lamarckism in America.

Packard, after breaking with Agassiz in the "Salem Secession," went on to become a leading figure of American neo-Lamarckism, which he championed from his positions at the Boston Society of Natural History, the Peabody Academy of Sciences at Salem, Massachusetts, and at Brown University (Dexter 1965, Bocking 1988). It was Packard who coined the term "neo-Lamarckism" and called Lamarck "the real founder of organic evolution" (Packard 1901, p. v). In 1867, together with Morse and Hyatt, Packard founded *The American Naturalist*, the journal that published the most articles on American cave fauna during the Nineteenth Century. He first examined Mammoth Cave specimens after the Indianapolis meeting of the American Association for the Advancement of Science in 1871 when many of the participants visited that cave and published an account of the fauna that same year. Just as his former teacher had been, he was enthusiastic about the possibilities that cave animals offered to scientists interested in evolutionary studies: "We trust naturalists the world over will be led to explore caves with new zeal". Packard saw the study of Mammoth Cave fauna as the means of fulfilling his higher interest in the issue of evolution, i.e., the knowledge derived from their study could impact broader evolutionary issues (Packard 1871, p. 761). The Mammoth Cave fauna convinced him of their usefulness as a demonstration of evolution (Packard, 1871). For him and Putnam "The comparatively sudden creation of these cave animals affords, it seems to us, a very strong argument for the theory of Cope and Hyatt of creation by acceleration and retardation which has been fully set forth in this journal" (*American Naturalist*) (Packard & Putnam 1872). He thought that cave fauna was of very recent origin and that the loss of certain organs was compensated by the hypertrophy of others.

In 1874, after Packard became associated with the Kentucky Geological Survey, his interest on the fauna of Mammoth Cave and other caverns in the Midsouth U.S. intensified though he never abandoned his neo-Lamarckian views concerning cave faunas (e.g., Packard 1888).

The other leading figure of this time was Frederic Ward Putnam. Like Packard, Putnam studied under Agassiz and was his assistant until the Salem Secession of 1864. He worked either as an ichthyologist or as a vertebrate biologist for the Boston Society of Natural History, the Essex Institute, the Peabody Academy of Science, and Harvard's Museum of Comparative Zoology. He was the one of all of Agassiz's students who took the longest in accepting evolution as a fact, and when he did (between 1872 and 1874), he appeared to do so reluctantly (Dexter 1979). In fact, there is no clear indication that he expressively espoused the neo-Lamarckian ideas of the Hyatt-Cope school, but there is no proof of the contrary either, and he worked very closely with both Packard and Shaler whose sympathies for the American neo-Lamarckian school are indisputable.

Of all of Putnam's experiences, it was his position in 1874 as assistant of the Kentucky Geological Survey that brought him into direct contact with hypogean fauna, particularly fishes. Putnam also first visited Mammoth Cave to collect fishes, crayfish, leeches, beetles, and crickets in 1871 after the adjournment of the meeting of the American Association for the Advancement of Science. He returned in 1874 following an invitation

---

188 *b*. Brunswick, Maine, 19 February 1839; *d*. Providence, Rhode Island, 14 February 1905.
189 *b*. Portland, Maine, 18 June 1838; *d*. Salem, Massachusetts, 20 December 1925). A naturalist, writer, and later director of the Peabody Museum of Archaeology and Ethnology (1880-1914).
190 *b*. Salem, Massachusetts, 16 April 1839; *d*. Cambridge, Massachusetts, 14 August 1915.
191 *b*. Newport, Kentucky, 22 February 1841; *d*. Cambridge, Massachusetts, 10 April 1906. A geologist and paleontologist initially opposed to evolution out of deference to his teacher Agassiz but once he secured his position as Dean at Harvard, accepted Darwin's although maintaining a neo-Lamarckian interpretation of it.

from Nathaniel Southgate Shaler, another of Agassiz's students who, as director of the Kentucky State Geological Survey, appointed Putnam as special assistant to the Survey that year. Although less well known than Packard because he was not very much inclined to provide grandiose generalizations or engage in much speculation, Putnam was very critical of hasty conclusions by others, particularly Cope.

Putnam carried out experiments that suggested that blind cave crayfishes would not take food, unlike the eyed ones and did not acquire pigmentation in subsequent molts even when kept in sunlight (Putnam 1875). He also described a new species of amblyopsid, *Chologaster agassizi* (Putnam, 1872). He always had problems in accepting evolution as an idea: "I think that we have as good reasons for the belief in the immutability and early origin of the species (…) as we have for their mutability and late development, and, to one of my, perhaps, too deeply rooted ideas, a far more satisfactory theory; for, with our present knowledge, it is but theory on either side" (Packard and Putnam 1872, p. 52).

However, when dealing with specifics, Putnam's arguments always seemed to be to the point. For example, he criticized Cope's interpretation that *A. spelaea* was able to survive in hypogean waters because its "projecting under jaw and upward direction of the mouth renders it easy for the fish to feed at the surface of the water (...) This structure also probably explains the fact of its being the sole representative of the fishes of subterranean waters. No doubt many other forms were carried into the caverns since the waters first found their way there, but most of them were like those of our present rivers, deep waters or bottom feeders. Such fishes would starve in a cave river, where much of the food is carried to them on the surface of the stream..." Putnam then asked: where are the surface forms of the "surface feeders"? Why are other surface feeders not found in caves? He did not understand how Cope could justify the above statement when he himself had described an alleged "subterranean" fish ("*Gronias nigrilabris*") from Pennsylvania that was a bottom feeder, and the blind cavefishes from Cuba (discovered by Felipe Poey[192] in 1858) were bottom feeders as well. Putnam noted that studies of stomach contents in *A. spelaea* had shown that they eat mostly crayfish and other fishes. He asked that if blindness is the direct result of darkness, as some contended, "how is it that *Chologaster* from the well in Tennessee or the 'mud fish' at Mammoth Cave are found with eyes?" (Putnam 1872, p. 24).

Putnam (1872, p. 6) also stated that "the blind fish of the Mammoth Cave has from its discovery been regarded with curiosity by all who have heard of its existence, while anatomists and physiologists have considered it as one of those singular animals whose special anatomy must be studied in order to understand correctly facts that have been demonstrated from other sources; and, in these days of the Darwinian and development theories, the little blind fish is called forth to give its testimony, pro or con." He viewed the amblyopsids as former marine and saltwater estuary fishes that were slowly trapped in that geographical area. He substantiated this hypothesis by pointing out that the eyed amblyopsid *C. cornuta* was "now living in the ditches of the rice fields of South Carolina, under very similar conditions to those under which others of the family may have lived in long preceding geological time; and to prove that the development of the family was not brought about by the subterranean conditions under which some of the species now live, we have the ones with eyes living with the one without, and the South Carolina species to show that a subterranean life is not essential to the development of the singular characters which the family possess." He further supported this hypothesis by mentioning that the Cuban blind cavefishes belonged to the genera "with their nearest

---

192 *b.* La Habana, Cuba, 26 May 1799; *d.* La Habana, 28 January 1891. A lawyer by training, he became the foremost Cuban naturalist and discovered two species of cave fishes in that island, the first ones scientifically recognized outside the United States (Romero 2007).

representative in the family a marine form, and with the whole family of cods and their allies, to which group they belong, essentially marine".

How can we summarize, thus, the views of this generation of American naturalists regarding the evolution and ecology of cave fauna? Although generalizations are always dangerous (particularly when dealing with the ideas of people like Cope who kept modifying his), here are some of their views of which we are certain:

1. Disuse, not natural selection, was the major (if not the only) evolutionary force behind the morphological "oddities" (blindness and depigmentation) of cave fauna. Thus, cave fauna provided excellent evidence of the effect of the environment upon the evolution of organisms (Packard attributed more evolutionary importance to the direct effect of the environment than to the effect of changes in habits).
2. The maintenance of rudimentary (but "useless") organs among the cave fauna was explained within the concepts of Bauplane (blueprints or archetypes), homologies, and parallelisms between embryology and phylogeny. These ideas originated from Agassiz as result of his own *Naturphilosophie*.Hypertrophy of certain organs appears as compensation for the rudimentation of others.
4. Cave fauna represented one of the best examples of progressionism or orthogenetic ideas. With orthogenesis comes the idea of progress, with loss of characters the idea of regression. Bowler (1983, p. 57) suggested that Lamarckism and orthogenesis were allies in their war against Darwinism. That was particularly true among American naturalists.
5. All cave fauna species were of recent origin.

European Selectionism and the Death of the Controversies (1880-1921)

Despite the tremendous popularity of American neo-Lamarckism, some European researchers were not satisfied with the metaphysical explanations for the evolution of cave fauna in particular and the general dismissal of natural selection as the major driving force of evolution. The main opposition came from August Weismann[193] and Edward Ray Lankester[194]. Although the first did not specifically study cave organisms, he adopted a pro-selectionist position in part because his teacher Jacob Henle[195] (a very keen observer) had encouraged him to be suspicious of any ideas based on the idealistic *Naturphilosophie*. More explicit regarding cave fauna was Lankester, a comparative anatomist influenced by the German biologist Anton Dohrn[196] (a student of Haeckel's). Lankester wrote that a special kind of natural selection was responsible for blindness among cave animals. His ideas can be summarized as follows: (1) within any population some animal individuals are, by chance, born with defective eyes, and occasionally a sample of both those born with normal eyes and some born with defective eyes fall or are swept into caves; (2) in each generation, those that have good eyes are able to see the light and escape, thus, eventually, only those that are blind will remain in the cave (Lankester

---

193 *b*. Frankfort am Main, Germany, 17 January 1834; *d*. Freiburg im Breisgau, Germany, 5 November 1914. He conducted a series of experiments in which he cut off the tails of mice for 22 generations, disproving the notion that acquired characteristics could be inherited since the mice kept being born with tails.

194 *b*. London, England, 15 May 1847; *d*. London, 15 August 1929.

195 *b*. Fürth, near Nuremberg, Bavaria, Germany, 19 July 1809; *d*. Göttingen, Germany, 13 May 1885. Henle was originally trained in medicine, studied under Johannes Müller and taught biology to August Weismann (Hintzsche 1972).

196 *b*. Stettin, Germany [now Szczecin, Poland], 29 September 1840; *d*. Munich, Germany, 26 September 1909. Dohrn became very enthusiastic about natural history after reading Darwin's *Origin*. His major area of interest was comparative anatomy and the use of embryos to establish phylogenetic relationships. He maintained active correspondence with Darwin (Heuss 1991).

1893); (3) one can find organisms degenerating ontogenetically and phylogenetically; "degeneration" or "a loss of organization making the descendent far simpler or lower in structure than its ancestor," is a phenomenon that is found widespread (Lankester 1880, De Beer 1973). He did not synonymize evolution with progress. "Any new set of conditions occurring to an animal which render its food and safety very easily attained, seem to lead as a Rule of Degeneration; just as an active healthy man sometimes degenerates when he becomes suddenly possessed of a fortune. (…) Let the parasitic life once be secured, and away go legs, jaws, eyes, and ears" (Lankester 1880, p. 33). Note here the influence of the "Hyatt-Cope school" and the fact that he was trying to draw a parallelism between parasitology and the complacency of the British Empire at its zenith.

Despite these controversies and the rediscovery of Mendel's work, little more than speculation was added to the discussion. Hugo De Vries[197], for example, one of the rediscoverers of Mendel's laws, believed that there were two types of mutations: retrogressive mutations (leading toward the loss of characters) and progressive mutations (leading toward complexity). But these were theoretical considerations without a solid experimental backing. Notice that De Vries, despite his distaste for orthogenesis and teleological explanations, was using the jargon of progressionists when referring to these mutations. Lankester had also proposed that each useless character was correlated with a useful one, an idea that found tangential support from Thomas Hunt Morgan[198], who discovered that one gene may have multiple effects (pleitropism). Morgan, the experimentalist, was also speculative about cave fauna: when observing the appearance of eyeless *Drosophila* in the laboratory, he proposed that blind cave animals could be the result of a single mutation, an assertion he never tested.

The last major biologist working on cave fauna who operated more or less under the influence of the American neo-Lamarckian school was Carl H. Eigenmann[199]. Influenced by David Starr Jordan[200] (a student of Louis Agassiz's son, Alexander), Eigenmann became a biologist particularly interested in fishes. His first experience with blind cavefishes took place in 1886 while at Indiana University when he received a living blind fish taken from a well in Corydon, Indiana. The next year he married Rosa Smith[201], an ichthyologist in her own right, and who introduced him to the blind goby *Othonops eos* (formerly *Typhologobius californiensis)* found among the rocks of the California coast, (see Eigenmann 1890 for a historical account of this encounter and how much it impressed him). In 1891 he was appointed Professor of Zoology at Indiana University, a perfect location from which to study the blind vertebrates of the caves in the nearby areas. This motivated him to devote a substantial part of his scientific career to the study of blind vertebrates, most of them from caves (Romero 1986b).

Between 1887 and 1909, much of his work was devoted to understand the process by which cave vertebrates lost their visual structures. He also described two new species of cavefish: *Amblyopsis rosae* from Missouri (Eigenmann 1898) and *Trogloglanis pattersoni*, (Eigenmann, 1919) from the artesian waters of Texas. Eigenmann frequently visited the caves of Indiana, Kentucky, Texas, and Missouri in search of specimens for his work, and in March 1902, he visited Cuba for the first time and secured cave specimens for his

---

197 *b*. Haarlem, Netherlands, 16 February 1848; *d*. Lunteren, Netherlands, 21 May 1935.
198 *b*. Lexington, Kentucky, 25 September 1866; *d*. Pasadena, California, 4 December 1945. A leading 20th century geneticist who established the fact that genes were located in the chromosomes and made the use of fruit flies a common feature in experimental genetics.
199 *b*. Flehingen, Beden, Germany, 9 March 1863; *d*. Chula Vista, California, 24 April 1927. He arrived to the United States at the age of 16.
200 *b*. Gainesville, New York, 19 January 1851; d. Stanford, California, 19 September 1931. Inspired by Louis Agassiz, Jordan devoted his academic research to the study of fishes and accepted Darwinism. He believed that extreme specialization would be followed by 'degeneration' as in the case of cavefishes. He taught Carl H. Eigenmann and indirectly influenced Carl L. Hubbs (Hubbs 1964, Shor 1973).
201 *b*. Monmouth, Illinois, 7 October 1858; d. San Diego, California, 12 January 1947.

comparative studies. He previously had been working on fish reproduction and quickly recognized that the two species of Cuban hypogean fish species known until then were viviparous.

Eigenmann found the localities for the Cuban blind fish to be "monotonous" (Eigenmann 1903) unlike Mammoth Cave, which exhibited great diversity, which is not surprising given the enormous size of the latter. From 1906 to 1907 he conducted laboratory studies in Europe, mostly in Germany, with the Cuban specimens he had collected. From 1898 to 1905 Eigenmann published at least 39 papers and Summaries on cave vertebrates, dealing mostly with developmental and anatomical aspects of vision loss in fishes, salamanders, lizards, and mammals in an attempt to understand the basic process that results in blindness among hypogean vertebrates. He summarized all this research in his *Cave Vertebrates of North America* (Eigenmann 1909).

Although a taxonomist by training, Eigenmann diligently sought explanations for the origin and evolution of the cave fauna. Originally a neo-Lamarckian, Eigenmann thought that the reduction or disappearance of organs among many cave animals was an example of convergent evolution. In other words, the well-defined conditions of the subterranean environment facilitate the evolutionary changes resulting in blindness and depigmentation in a variety of animals. He pointed out that lack of pigmentation had to be understood as the result of a combination of genetically fixed and epigenetically (environmentally influenced) determined characters; in other words, even though a character may be genetically determined, its degree of development can vary when exposed to different amounts of light. For Eigenmann, cave evolution was essentially "degenerative", and all successful cave-invaders had to be somehow "pre-adapted" to that milieu. The origin of caves and that of the blind fauna in them were to him two distinct questions because of his experience with the blind fish found among the rocks of California's seacoast. He insisted on a strong link between ontogeny and phylogeny, and his constant use of terms such as "phyletic degeneration" indicates that he held orthogenetic views. He followed Herbert Spencer"s idea that cave fauna species are not the result of "accidents" but rather the product of an active process of colonization (Eigenmann 1909, Romero 1986b).

At the same time Eigenmann was culminating his blind vertebrates work, his student, Arthur Mangun Banta[202], proposed some variations to the popular explanation of the origin of reduction/loss of phenotypic characters. For Banta "degeneration" of eyes and pigmentation was due to the influence of the environment and such phenomena had to occur before cave colonization could take place and at the embryonic level (Banta 1921). Because they have already suffered "degeneration" they go "voluntarily" into caves and do not return to the surface because they are "unfit" to survive in epigean conditions (Banta 1909, p. 99). Banta, thus, was not a neo-Lamarckian in the sense that he did not believe in disuse as the cause of rudimentation; he even acknowledged that natural selection was the explanatory mechanism for the increased sensory organs of some cave creatures (Banta 1909, p. 104). Although Banta's hypothesis for why cave animals have colonized the hypogean environment (and why they were blind and depigmented) never acquired much credence, his emphasis on the notion of preadaptations, however, became very popular when four years later Cuénot coined the term (see Chapter 2).

In summary, by the beginning of the Twentieth Century, with genetics gaining importance, no new ideas about cave biology were proposed even though the neo-Lamarckian explanations based on use and disuse were discredited. This is not surprising since even the topic of evolution in general languished at this time primarily because it was obvious that no progress was being made, and "Morphology having been

---

202 *b*. near Greenwood, Indiana, U.S., 31 December 1877; *d*. 2 January 1946.

explored in its minutest corners, we turned elsewhere" (Bateson 1922, p. 1412). Lankester, for example, left his professorship at Oxford to become Director of the British Museum of Natural History (Ruse 1996, p. 239) and Eigenmann began a general study of freshwater fish fauna of the western hemisphere.

Now that biospeleology was essentially dead in English-speaking countries, this science would experience a revival in continental Europe, particularly in France, where they had their own brand of neo-Lamarckism and orthogenesis. How was this possible?

Biospeleological Ideas in France and Elsewhere in Continental Europe (1809-1950)

French and French-based researchers from Lamarck to the biospeleologists of the 1950's have had and continue to have a tremendous intellectual influence on biospeleological ideas. Their way of thinking and their terminology have been pervasive in cave biology. To understand why this is so, we must (1) review the political and intellectual environment in France previous to the publication of Darwin's *Origin*; (2) examine how Darwin's book was received and investigate how and (3) why the French developed an evolutionary ideology of their own, particularly when it came to interpreting the nature of cave fauna.

Ideas on evolution (biological and otherwise) in pre-*Origin* France abound, but all have something in common: a strong philosophical rather than an empiric basis. Jean Baptiste Lamarck[203], a physician by training, became first an assistant botanist at the French Royal Botanical Gardens, an active participant of the *Société d'Histoire naturelle*, and then given a position of professor of "insects and worms" at the newly created *Muséum national d'Histoire naturelle*.

Lamarck considered himself a "naturalist-philosopher", and therefore much of his narrative was colored with speculations and metaphysics rather than facts. In addition, his evolutionary views (mostly expressed in his 1809 *Philosophie Zoologique* and the 1815 supplement to the *Histoire naturelle*) were never very well formulated and even sometimes contradictory. To make things worse, Lamarck's writings were translated into numerous languages, but such translations were not always accurate and some of his statements were reproduced out of context which contributed to the general confusion on to what Lamarck really said (Corsi 2005). But one thing is for sure: he was an early organicist and progressionist who viewed nature as being linearly organized and saw today's organisms as the result of increasing complexity (Burkhardt 1977, p. 58 & fol.).

Lamarck was the main (although not the first) advocate of the idea of inheritance of acquired traits concept and of evolution as a goal-oriented process striving towards progressive complexity and perfection. He did not believe in the extinction of species but rather on the constant transformation into new ones

He described a metaphysical "power of life" leading this process of increasing complexity. That, together with the modifying power of the environment was responsible for the life forms we see on earth. Although he never wrote about cave fauna, the case of parasites with simplified organization amused him. Yet, he had a perfect explanation: they appeared primitive because they had been the recent product of spontaneous generation. External circumstances were responsible for deviations from the rule of progression and some contingency (e.g., the disuse of an organ) could alter the path to complexity generating lateral ramifications in his linear view of progression. For him the lack of teeth in whales and eyes in (subterranean) moles were perfect examples. Lamarck had a great influence on many scientists not only at his time but through the Twentieth Century. The progressionists ideas of Lamarck had also a great influence not only in Europe but in America as well where a vigorous neo-Lamarckian school

---

203 *b*. Bazentin-le-Petit, Picardy, France, 1 August 1744; *d*. Paris, France, 28 December 1829.

developed. That school was following Lamarck's tenants with the exception of those that were more mystical in nature (Burkhardt 1977).

Two Lamarck contemporaries would also make their own contributions to the notion of increasing complexity in nature. Cuvier, for example, although a creationist, noticed some "progression" in the succession of the geologic record. Cuvier admitted the existence of anatomical vestiges but did not seek explanations for them. He considered vestigial organs "one of the remarkable peculiarities of natural history" (Coleman 1964, p. 154) and that is as far as he went. Geoffroy Saint-Hillaire[204], a curator of vertebrates at the *Muséum national d'Histoire naturelle*, was a believer in evolution, progressionism, and the Great Chain of Being, always looking for transitional forms (Bourdier 1972b, Appel 1987). He discussed the issue of the origin of vestigial organs from a mystic/religious viewpoint and interpreted them as "disgraces" of natural beauty. Saint-Hillaire, a protégé of Lamarck, was even less materialistic than his mentor and added an aura of mysticism to evolutionary ideas, which in turn were influenced largely by Oken's *Naturphilosophie*.

At this same time French philosophers were thinking along the same lines. For example, Marie-Jean-Antoine-Nicolas de Caritat, Marquis de Condorcet[205], a brilliant mathematician, philosopher, and political activist, infused the idea of progress into virtually all of his historical interpretations. He adopted the concept of inheritance of acquired characters in constructing his vision for the social and organic progressive improvement of humankind, an idea also espoused by other philosophers such as Herbert Spencer[206], Friedrich Engels[207], and Lester Ward[208] (see Condorcet 1793-1794). These ideas strongly influenced the positivist school founded by the French philosopher Auguste Comte[209] and the ideas of another French philosopher, Marcel de Serres[210]. The latter proposed the view that life was a manifestation of progressive perfecting.

---

204 *b*. Etampes, France, 15 April 1772; *d*. Paris, France, 19 June 1844.
205 *b*. Ribemont, Picardy, France, 17 September 1743; *d*. Bourg-la-Reine, France, 28 March 1794. His book *Esquisse d'un tableau historique des progrès de l'esprit humain*, published posthumously in 1795, analyzed human history under the view of progressiveness. For him humanity was destined to achieve an evolutionary apex through the education of the masses. His ideas were mirrored later by Theilhard de Chardin (Granger 1971, Leith 1989, Baker 2004).
206 *b*. Derby, England, 27 April 1820; *d*. Brighton, England, 8 December 1903. He was a Lamarckian who tried to apply evolutionary ideas to support free market ideologies, he believed that humans were on a natural progressionist route and that the state might create obstacles to economic progress by trying to regulate free society. He rejected the notion of special creation and believed that species were the result of modification of preexistent ones. For him evolution (a term he introduced in the biological lexicon) was change from the homogenous to the heterogeneous. He was an agnostic who believed that science and religion were trying to answer different questions and that even if you believe in God that was not necessarily incompatible with the idea of evolution. He later embraced Darwinism but still believed in Lamarckian mechanisms to explain the transformation from simple to complex structures in nature. He believed that individuals also evolved as a consequence of learning from good and bad experiences and that at the end the good learner survived. He introduced the concept of 'survival of the fittest' and the rots of social Darwinism can be traced to him.
207 *b*. Barmen [now part of Wuppertal], Prussian Rhineland, 28 November 1820; *d*. London, England, 5 August 1895. Engels's writings gave the philosophical background to Marxism. His philosophy was based on a materialism that was in accordance with the views of the sciences of the nineteenth century.
208 *b*. Joliet, Illinois, 18 June 1841; *d*. Washington, D.C., 18 April 1913. One of the founders of American sociology he advocated the intervention of the state at humanizing society by eliminating poverty. He also advocated regulation of competition, the establishment of equal opportunities and cooperation. Ward attacked the very notion of social Darwinism, the *laissez-faire* doctrine and determinism. By doing so he turned against Hebert Spencer who he had admired earlier in his career.
209 *b*. Montpellier, France, 17 January 1798; *d*. Paris, France, 5 September 1857. He is considered the father of positivism in philosophy.
210 *b*. Montpellier, France, 3 November 1780; *d*. Montpellier, 22 July 1862. He was a paleontologist and zoologist who believed that the pursuit of truth required violating artificial disciplinary boundaries.

Thus, the intellectual environment in pre-*Origin* France was not anti-evolution as in other parts of Europe and the United States; actually, one can say that no well-educated French person at that time harbored any predisposition against evolution (*transformisme*). In fact, in France, the idea of progression could be traced as far back as the development of Modern Science period (1650-1800) at the time of the Enlightenment and the French Encyclopedism. Lamarck contemporaries, with the exception of Cuvier, embraced some sort of transformism: although they were not sympathetic to (and even ridiculed to certain extent) Lamarck's unfounded speculations, particularly the idea that a new organ could be produced by the "desire" of an organism to create it. However, the French were unprepared to view evolution as a materialistic, random process that excluded any metaphysical explanation. And the way Darwin's *Origin* was translated into French made matters worse.

The Origin was translated into French by Clémence-Augustine Royer[211]. This polymath and feminist writer, was not only a great believer in science, but also thought that women should transform it into "female science." Royer probably first heard of Darwin's new work on evolution through a review of *The Origin* by the Geneva-based Swiss entomologist and paleontologist Françoise Jules Pictet de la Rive[212] while lecturing on Lamarck in Geneva in 1860. Pictet was one of the first to receive a copy of *The Origin of Species* directly from Darwin. As soon as Royer read *The Origin*, she convinced her publisher, Guillaumin, to print the first translation of Darwin's work into French. According to Royer, "It was then [after lecturing in Geneva] that I translated the *Origin of Species* of Ch. Darwin, which had appeared in England, during the same winter in which I had affirmed in my course the doctrine of Lamarck. If I translated Darwin, it was because he had brought new proofs to the support of my thesis." (Harvey 1999). In other words, her interest in translating Darwin was not so much to spread the Britton's gospel, but rather to prove how important Lamarck was as the father of evolution as an idea. And it showed.

With the advice of the French zoologist and early Darwinian enthusiast René-Edouard Claparède[213], who had also enthusiastically reviewed Darwin's book, she translated the third edition of *The Origin* (which was, in terms of explanations on rudimentation, more Lamarckian than the first two editions) adding not only numerous footnotes, but also a lengthy prologue in which she espoused eugenics, being probably the first author to do so by applying Darwin's ideas. Darwin, who had authorized the move to have his book translated into French, was not happy with Royer's preface and footnotes. She not only changed the title of the book, but more significantly, Royer used the word "election" instead of "selection" giving, thus, the impression that nature had a mind of its own directing on purpose evolutionary events.

The title of Darwin's book in French was *De l'origine des espèces, ou Des lois de progrès chez les êtres organizes* ("The origin of species, or the laws of progress among organized beings") giving the impression that Darwin emphasized the idea of progress, a principle for which he was ambiguous at best. Darwin himself, in his correspondence to several of his colleagues such as Jean Louis Armand de Quatrefages[214], Charles Lyell[215],

---

211 *b*. Nantes, Brittany, France, 21 April 1830; *d*. Paris, France; 6 February 1902.

212*b*. Geneva, Switzerland, 27 September 1809; *d*. Geneva, 15 March 1872.

213 *b*. Chancy, Geneva Canton, Switzerland, 24 April 1832 ; *d*. Sienna, Tuscany, Italy, 31 May 1871.

214 *b*. Berthezène, near Valleraugue (Gard), France, 12 February 1810; *d*. Paris, France, 1892. He specialized in invertebrates and was particularly interested in the degeneration (*dégradation*) of structures among organisms, although most of his ideas in this matter were wrong. He opposed Darwin's evolutionary (believed in the fixity of species) ideas but maintained very cordial relationships with him.

215 *b*. Kinnordy, Angus, Scotland, 14 November 1797; *d*. London, England, 22 February 1875. He was the most influential geologist of the 19th. century. His ideas set the stage for Darwin's thinking that life must have been evolving on earth as the geology of our planet had also been changing though long periods of times. He was a

and Asa Gray[216], made it known that he was extremely unhappy with the French translation. Despite this version of *The Origin* being closer to the French state of mind, Darwin sensed that the book had a cold reception in France. In a letter to Quatrefages, a French naturalist who opposed Darwin's ideas on evolution but yet respected him, Darwin wrote, "A week hardly passes without my hearing of some naturalist in Germany who supports my view, & often puts an exaggerated value on my works; whilst in France I have not heard of a single zoologist except M. Gaudry [Albert Jean Gaudry[217]] (and he only partially) who supports my views" (F. Darwin 1896, vol. 2, p. 299). Darwin may have not been happy with this translation; yet, he might not have any other alternatives since he had trouble finding a publisher in France for his book anyway (Herbet 2005).

For years to come, Royer continued publishing and lecturing about Lamarck, her personal hero. She, who was probably the first European woman recognized as a professional anthropologist, had also been an enthusiastic caver.

Royer's translation of *The Origin* was very much celebrated by Étienne Rabaud[218]. Rabaud had been a student of Alfred Girard, the first holder of the Chair of Evolution at the Sorbonne and a rabid Lamarckian. Rabaud became such a fanatical supporter of Lamarck's ideas that by the 1930's he even questioned the value of Darwinism (see, for example, Rabaud 1941). When commenting on Royer's preface, Rabaud was enthusiastic because she had restored Lamarck into public attention.

Were this improper translation and the current intellectual climate the only reasons for the poor reception of Darwin's ideas in France? Not really. Just before the publication of *The Origin*, France had witnessed one of the most public and passionate scientific controversies in history. Between 1858 and 1859 French society was inundated with the tales of the dispute between Félix Archimède Pouchet and Louis Pasteur[219], that is, between the belief in spontaneous generation and the belief that the ability to beget life is an exclusive and continual property of living beings. Although Pasteur won the argument and his was a triumph for science as a method of inquiry, Pouchet's sympathizers also supported agnosticism whereas Pasteur's were more comfortable with religious and metaphysical ideas. Thus, despite the fact that the French were not opposed to evolution as an idea *per se*, the mechanism championed by Darwin, natural selection, reminded them of the agnosticism and materialism attached to spontaneous generation. Thus, the land that had given birth to precursors of evolutionary ideas such as Georges-Louis Buffon[220], Lamarck, and Geoffroy Saint-Hillaire, gave Darwin a cold shoulder, and little public controversy of the book took place.

---

close friend of Darwin and accepted Darwin's evolutionary ideas being one of the few who immediately accepted the notion of natural selection as a major force of evolution (Wilson 1973).

216 *b*. Sauquoit (Paris), Oneida County, New York, 18 November 1810; *d*. Cambridge, Massachusetts, 30 January 1888. A physician who became the prominent American botanist of the 19th. century. He embraced Darwinian evolution corresponding extensively with Darwin but was not enthusiastic about natural selection as its mechanism, to say the least. He tried to reconcile Darwinism with religion through a sort of theistic evolutionism.

217 *b*. St.-Germain-en-Laye, France, 15 September 1827; *d*. Paris, France, 27 November 1908. Affected by the death of his mother when he was very young, he developed a strong mysticism during his entire life. He worked at the *Muséum national d'Histoire naturelle* in Paris and was a great believer of the Great Chain of Being. For him humans were the ultimate example of perfection. He later became a defender of the idea of evolution. Yet his explanation for evolution was mystical: that was designed by God and God rejoiced in his own continuous creation in which God was the only fixed and untransmutable being (Bourdier 1972a).

218 *b*. 1868; *d*. 1956. An anti-Darwinian who taught Pierre-Paul Grassé.

219 *b*. Dole, Jura, France, 27 December 1822, *d*. Chateau Villeneuve-l'Étang, near Paris, France, 28 September 1895. One of the world's most important scientists, he was a chemist by training. Recognized microbes as transmitter of diseases, invented vaccines, and disproved spontaneous generation.

220 *b*. Montbard, Bourgogne [Burgundy], France, 7 September 1707; *d*. Paris, France, 16 April 1788. Well known for his 36-volume *Histoire naturelle (Natural History)* (1749–88). He maintained very advanced evolutionary ideas for his time (Roger 1973, 1997, Farber 1975, Sloan 1976, Eddy 1994).

Other political and social events further cemented the French view of evolution as a mystical idea. One experience that generated a nationwide feeling of disgrace was the political and military humiliation of the French by the Prussians during the 1870-1871 War (Howard 1981). And as in any nation that has been defeated, their people found consolation in mystical nationalistic ideas. The ideas of national destiny and historical progress became strongly rooted in the French psyche and were reinforced through revisions of school curricula. The Spencerian interpretation of "survival of the fittest" became very unpopular: Prussia had developed into an imperialistic and invincible neighbor and looked like "the fittest" to French psyche. Now French intellectuals threw themselves fully into the arms of mysticism to explain their grand views of nature, and evolution was at the center of all this.

It was in this intellectual atmosphere that the seeds for French neo-Lamarckism were planted, and these seeds were sown in abundance by French biospeleologists. The father of these neo-Lamarckian ideas in France was Henri Louis Bergson[221]. Bergson was a philosopher and a mathematician whose ideas on evolution were largely anti-materialistic and sustained that organic evolution was just part of a larger, universal cosmic evolution. A Lamarckian follower regarding the canon of use and disuse and principle that evolution was directed by an internal force which he called *élan vital*. He was fiercely patriotic and opposed Darwinism because he did not accept the notion of an undirected mechanism such as natural selection as the major force of evolution. Part of his popularity was due to the fact that by using the notion of an *élan vital*, he was allowing for a role to be played by religion in the explanation of evolutionary processes (Goudge 1973).

Bergson was familiar with the ideas of Cope and Theodor Gustav Heinrich Eimer[222], a disciple of Rudolf Albert Kölliker[223], who championed the idea of and popularized the term orthogenesis (Eimer 1887-1888). The term orthogenesis was first proposed by the zoologist Johann Wilhelm Haacke[224] (1893). Others used different terms for essentially the same concept: orthoevolution (Plate 1913), nomogenesis (Berg 1926), aristogenesis (H.F. Osborn 1934), and the omega principle (de Chardin 1955). Bergson, an intense French patriot, proposed in 1907 the idea of the *élan vital* or vital impetus (the term is so obscure that it is usually left untranslated, but reminds that of Lamarck's expression of the "power of life"). He used this term to refer to a characteristic of life that, according to him, always pushes life in the direction of complexity. That, for Bergson, was the mechanism of orthogenesis, which moved evolution from the domain of the divine into the natural world. Given that Bergson did not like natural selection as an idea because of its materialistic implications, and at the same time he could not find strong evidence supporting the inheritance of acquired characters, thus, *élan vital* was for him the answer. Of course, and unlike natural selection or the inheritance of acquired characters, since this idea could not be tested, it could not be disproved either.

According to Bergson, both Darwinian evolution and finalism (the idea that evolution has a sense of directedness toward an end and that such a path has already been laid) could coexist. And what is the unifying force behind such a possibility? It cannot be natural selection, of course, since that is based on apparent randomness, but

---

221 *b*. Paris, France, 18 October 1859; *d*. Paris, 4 January 1941.
222 *b*. Stäfa, near Zurich, Switzerland, 22 September 1843; *d*. Tübingen, Germany, 29 May 1898. In 1875, he became a professor of zoology and comparative anatomy at the University of Tübingen. He described orthogenesis as an intrinsic drive in life towards perfection, a form of directed evolution (Churchill 1990). He dismissed natural selection as a major force in evolution while rejecting vitalism at the same time.
223 *b*. Zurich, Switzerland, 6 July 1817; d. Würzburg, Germany, 2 November 1905. He studied under Lorenz Oken, Johannes Müller, and F.G.J. Henle and was greatly influenced by *Naturphilosophie* being a close associate of Nägeli. He embraced evolution but opposed the role of natural selection (Hintzsche 1973).
224 *b*. Clenze, Germany, 23 August 1855; *d*. Lunebourg, Germany, 6 December 1912.

rather it must be a mystical force, *élan vital*. These ideas may have been interpreted as Lamarckian with a religious twist, but that is also unclear: Bergson, a man profoundly concerned about the fate of his fellow Jews, almost became a catholic; it is evident therefore that his religious views were also complex. Bergson's ideas became extremely popular, and other philosophers such as the French Lucien Cuénot[225] expanded them by arguing that species succeed in a particular environment because they were "preadapted." The term he coined was *préadaptation* (Cuénot 1911, vol. IV, p. 306), and it became an extremely popular idea among biospeleologists, many of whom still firmly believe in it today. Needless to say, Cuénot espoused linear evolution, only that in the new era of experimental genetics of early Twentieth Century, he believed that mutation (*sensu stricto*) was the cause of it.

In summary, Bergson was a progressionist but he did not believe that there was a necessarily pre-designed goal; rather that final progression would lead to a less predictable result trying, thus, to taint Darwinism with the very popular idea of progression.

All of these new philosophies of life were developed at the time when speleology in general and biospeleology in particular were becoming sciences in their own right, and all their foundations were being laid by French or France-based naturalists. Such was the case of Édouard-Alfred Martel[226]. Martel was a French lawyer and a geographer by training. He was known for his pioneer work in 1894 on the physiography and accessibility of caves, and he coined the term speleology (in both French and English) in the 1890s. He explored the limestone caves of Cévennes and, with others, made descents into previously unknown caves of Europe, Asia, and America. In 1895 he founded the *Société de Spéléologie* in France. Martel was the judge of the tribunal of commerce in Paris from 1886 until 1899, when he became a professor of subterranean geography at the Sorbonne (the first speleological academic post in the world). He was appointed a member of the staff of the Department of Geological Maps of France in 1901. He is often called "the father of modern speleology" and his publication record includes more than 1,000 articles and books on the subject. In 1904 Armand Viré[227], another Frenchman, coined the term biospeleology (*biospeleologie*). Viré had written his doctoral thesis on cave fauna in 1899 and thereafter established an underground laboratory in the catacombs of Paris.

However, the two figures that would ultimately consolidate biospeleology as a science and give it many of the distinctive features that it has today were Emil G. Racoviță[228] and René Gabriel Jeannel[229]. Racoviță, a Rumanian-born, French-educated naturalist, started exploring caves in the Pyrenees in 1905 together with his protégé Jeannel. Racoviță initiated an extensive international research program under the umbrella of *Biospeologica* (a supplement to the scientific French publication *Archives de Zoologie Experimentale et Generale*), primarily intending to document and collect cave fauna. In 1920 he founded in Cluj, Romania, the world's first speleological institute. He explored 1200 caves in Europe and Africa, collected about 50,000 specimens of cave

---

225 *b*. Paris, France, 21 October 1866; *d*. Nancy, France, 7 January 1951. Cuénot was a brilliant scientist being the first French biologist to accept Darwinism ideas in Lamarckian France although he was not fully convinced of the all-powerful role played by natural selection. He also pioneered genetics studies in France aimed to prove Mendelian inheritance. Yet, he refused to completely accept neo-Darwinism because he maintained a finalistic view of evolution. One of his most lasting influences in biospeleology was the development of his notion of preadaptation according to which new ecological niches were occupied by mutants that already had some characteristics that favor them to colonize such environments (Tétry 1971b).

226 *b*. Pontoise, France, 1 July 1859, *d*. Château de la Garde, near de Montbrison, France, 3 June 1938.

227 *b*. Lorrez-le-Bocage-Préaux, Saine-et-Marne, France, 28 January 1869; d. Moissac, France, 15 July 1951.

228 *b*. Iasi, Romania, 15 November 1868; *d*. Bucarest, Romania, 17 November 1947.

229 *b*. Toulouse, France, 22 March 1879; *d*. Paris, France, 20 February 1965.

animals, and published 66 papers on subterranean fauna totaling almost 6000 pages (Motas 1962). He read and was greatly influenced by Eimer and Cope (on orthogenesis), Packard (on neo-Lamarckism), and Louis Dollo[230] (on general evolutionary ideas). He had a great deal of distaste for the selectionist Weisman (Motas 1962).

Racoviţă's two main publications dealing with biospeleological theory were his 1907 *Essai sur les probleme Biospeologiques* ("Essays on biospeleological problems"), published at the same time that Bergson was proposing his *élan vital* and considered to be the birth certificate of biospeleology as a science) and his little known 1929 *Evolutia si problemele ei* (*Evolution and its problems*) book. In those publications he clearly delineated his evolutionary thought about cave organisms, which can be summarized as follows:

1. All cave organisms were "preadapted" to the cave environment.
2. Function (or lack thereof) creates the organ (or generates its disappearance). He was a strong support of the use vs. disuse concept.
3. Natural selection is of little importance because natural variation is virtually non-existent (he was a staunch typologist).
4. Evolution is directional as evidenced by "phyletic lines."

Similar views were endorsed by his student Jeannel (Jeannel 1950, p. 7) who studied subterranean beetles from Europe and Africa. With Racoviţă he founded in 1907 the journal *Biospeleologica* and in 1926 published *Faune cavernicole de la France*. He considered many of the organisms found in caves as "living fossils", and these ideas continue to have a tremendous impact on biospeleologists all over the world.

Although all this can be presented as a great accomplishment for the French in terms of initiating and developing the systematic study of caves, none of these figures ever embraced any form of Darwinism, but rather different shades of neo-Lamarckism first and different forms of finalism such as orthogenesis and organicism later. Thus, the French biologists who embraced transformism beginning in 1880 did so via neo-Lamarckism while strongly opposing the idea of natural selection (Grimoult 1998, p. 150). This philosophy extended well into the twentieth century with Lucien Cuénot, Maurice Caullery[231] and Jean Rostand[232].

Therefore, the utilization of cave organisms as perfect examples for demonstrating the legitimacy of the French version of neo-Lamarckism seemed to be inevitable, and this is exactly what happened. The main points in common of these French intellectuals were:

1. Acceptance of evolution as a linear phenomenon (orthogenesis) leading to a perfecting complexity in nature
2. Rejection of natural selection as a phenomenon of any relevance
3. Development of finalism, vitalism, organicism, and other expressions of essentialism in biology
4. Utilization of cave organisms as "perfect" examples of these views of life
5. Mutual reinforcement of ideas concerning biospeleological paradigms (blind, depigmented animals) and philosophical notions of progress within the same country: France.

---

230 *b*. Lille, France, 7 December 1857; *d*. Bruxelles, Belgium, 19 April 1931. An engineer turned into biologist, Dollo became famous for the reconstruction of *Iguanodon* fossils in Belgium and for stating the 'Dollo's Law of Irreversibility' according to which organisms never return to their original state, particularly when losing complex structures.

231 *b*. Bergues, France, 5 September 1868; *d*. Paris, France, 13 July 1958. He lectured at the University of Paris (1903) where he taught evolution from a neo-Lamarckian perspective (Tétry 1971a).

232 *b*. Paris, France, 30 October 1894; *d*. Ville d'Avray, France, 4 September 1977. A biologist and philosopher who worked on developmental biology and maintained neo-Lamarckian views of evolution.

The Impact of The Modern Synthesis (1936-1947)

The Modern Synthesis was, without question, the major philosophical and scientific revolution that established evolution as the central idea in biology in the Twentieth Century. It meant that the non-Lamarckian Darwin was rescued; also, that metaphysical ideas in biology were abandoned, and the typological (essentialist) views of life were replaced by populational ones. Of all the major architects of this movement, only one specifically approached the issue of evolution of cave organisms.

Theodosius Dobzhansky[233] (1970, pp. 405-407) put the issue of evolution in caves in its right perspective, and his ideas can be summarized as follows:

1. Evolution is opportunistic
2. Adaptation to a new environment may decrease the importance of some organs/functions which may become vestigial and disappear
3. There are numerous examples of rudimentation and/or loss of organs among both animals and plants
4. Acquisition/enlargement of organs can occur among organisms that otherwise show "regressions" for other organs and/or functions
5. Cave animals provide some of the best examples of the phenomenon of "regression" but it is not unique or exclusive of them: some cave organisms do not display regressions and may be found among non-cave animals.
6. A great deal of variation exists for these characters even within the same species and/or population
7. Both genes and phenotypic plasticity are responsible for troglomorphic characters
8. Neo-Lamarckian explanation aside, two major hypotheses for explaining the genetic mechanisms of rudimentation can be considered: (a) mutation pressure (neutral mutation) if not opposed to natural selection (relaxation of selection); and (b) natural selection directly favoring rudimentation via energy economy or "struggle of the parts." Evidence seems to support the latter, not the former.

The scientific evidence accumulated during the second half of the Twentieth Century supports all these statements (except for 8 or the "struggle of the parts"). The major contributions of Dobzhansky to our understanding of the evolution of cave biota were numerous. The first one was to stress the role played by opportunism in evolution. Opportunism is probably much more important in natural systems than is generally appreciated (Berry 1989). As proven again and again, evolution is a by-product of disrupted communities in which brief opportunities for divergence are created (Dimichele *et al.* 1987). Opportunistic organisms can take advantage of previous conditions (Andersson 1990), to fill empty adaptive zones (Bronson 1979, Benton 1983, Harries 1996), for feeding (Jakšić & Braker 1983), breeding (Tindle 1984), and social behavior (McKenna 1979). Opportunism has also been proven to lead to mutualism (Fiedler 2001), intraspecific parasitism (Tinsley 1990, Field 1992), and reproduction (Kasyanov *et al.* 1997). Opportunism has been described even at the molecular level (Doolittle 1988, Meléndez-Hevia *et al.* 1996, Green 2001) and has also been identified as a major factor for colonizing species (Martín & Braga 1994) particularly in the case of colonization of extreme environments (Tunnicliffe 1991).

As a matter of fact, opportunism is the reason behind life being so ubiquitous on earth: life on earth can be found at naturally extremely low and high temperatures (from

---

233 *b.* Nemirov, Ukraine, Russia, 25 January 1900; *d.* Sacramento, California, 18 December 1975. In 1927 Dobzhansky moved to the United States where he worked with Thomas Hunt Morgan. Although a religious person, he rejected the idea of a god directing the course of nature or a direction in evolution (Hecht and Steere 1970, Ayala 1971).

polar regions to geothermal environments), in both high and low pH and high salinity, including but not limited to hydrothermal vents, freshwater alkaline hotsprings, acidic solfatara fields, anaerobic geothermal mud and soils, acidic sulfur and pyrite areas, carbonate springs and alkaline soils, the cold pressurized depths of the ocean, and soda and highly alkaline lakes (Kristjánsson & Hreggvidsson 1995, Horikoshi & Grant 1998). In fact, today there is an entire branch of biology dealing with what are called extremophiles (term coined by MacElroy in 1974). The discovery of hydrothermal vents in 1977 opened the door to an entirely new set of habitats that did not need light to be self-sustaining. In other words, life has shown an incredible ability to succeed in such a diversity of environments, that some have predicted the presence of life on other planets, including some in our solar system (Nealson & Conrad 1999). Now life forms in caves do not seem so "extreme" nor do we need to use metaphysical explanations to understand their origin and evolution. As George Gaylord Simpson[234], another of the architects of The Modern Synthesis put it, "The course of evolution follows opportunity rather than plan" (Simpson 1949, p. 160).

The second major contribution of Dobzhansky to this issue was to remind biospeleologists that the phenomenon of reduction and/or loss of phenotypic features is not unique to cave organisms and is actually ubiquitous throughout all animal and plant taxa. Other typical animal examples include parasites, deep-sea creatures, and inhabitants of murky waters. Even some parasitic plants have lost chlorophyll. Also, limblessness and flightlessness are common among animals living on small islands and high mountains (Darlington 1943, Byers 1969, Livezy and Humphrey 1986, Roff 1990, Finston & Peck 1995). The loss of limbs among cetaceans and snakes is an example of a major evolutionary novelty by default. Even humans have lost or reduced a number of characters from their ancestors (Diamond & Stermer 1999). Thus, troglomorphisms can be explained using well-known evolutionary mechanisms without the need to resort to neo-Lamarckian explanations or terminology such as "regressive evolution." The problem is that despite Dobzhansky's pointed comments on this issue, the study of this phenomenon has been largely neglected by mainstream evolutionary biologists for at least two reasons: (a) the prevailing idea that evolutionary novelties should result from addition, not subtraction, of characters and (b) the use of this biological phenomenon by neo-Lamarckians to advance their own cause of either inheritance of acquired characters or the notion that evolution has some sort of directionality has made this field little attractive to modern evolutionary biologists (Romero 2001b).

The third major contribution by Dobzhansky was to point out that variability of reduced phenotypic characters is widespread. That is clearly the case, but more importantly, Dobzhansky's statement was a serious blow to typological or essentialist beliefs among biospeleologists: in other words, there is not such a thing as a characteristic "archetype" for cave animals, not all of them are blind and depigmented and when they are, the degree to which such features (and others) are expressed varies greatly.

The final and perhaps more important contribution by Dobzhansky, especially from the mechanistic viewpoint, was his statement that the loss and/or reduction of characters had a genetic basis but was also influenced by phenotypic plasticity. This should not be surprising since there is a correlation between behavioral plasticity and opportunism (Brown 1990, Werdelin 1999, Johnson 2000). Lefebvre and colleagues (1997) found links between opportunism and phenotypic evolution; they also proposed that innovation rate in the field may be a useful measure of behavioral plasticity.

---

234 *b*. Chicago, Illinois, 16 June 1902; *d*. Tucson, Arizona, 6 October 1984. Simpson was one of the most prominent paleontologists of the twentieth century.

Ernst Mayr[235] also stated that "(the) evolutionary phenomena dealing with regression and the loss of structures (...) are entirely consistent with the synthetic theory of evolution" (Mayr 1960, p. 351). One might think that this line of reasoning would have had a major impact on biospeleologists as a whole, but the fact of the matter is that it did not.

For one thing, biospeleology continued to flourish in France and struggled elsewhere. A major speleological journal, *Annales de Speleologie*, was founded in France in 1946, and the first international Speleological Congress took place in France in 1952. But more importantly than that, French evolutionists in general and biospeleologists in particular, rather than softening their neo-Lamarckian and orthogenetic stances, hardened them. We see this rigidity in the writings not only of Lucien Cuénot but also of Jeannel, Maurice Caullery, Jean Rostand, and Pierre-Paul Grassé[236]. They kept espousing neo-Lamarckian explanations on heredity despite all of the evidence to the contrary and more importantly, their firm belief in orthogenetic ideas had now reached an uncompromising finalism, that is, the belief that natural processes, especially evolution, are directed towards some predetermined end or goal by some sort of unexplained or untested force.

This was taken to an extreme by one of the most influential Twentieth Century biospeleologists, Albert Vandel[237]. Vandel championed the idea of organicism and orthogenesis in his writings (duly summarized in his influential 1965, book which was made available in both French and English, see Vandel 1965, pp. 471 and ff. of the English version). According to him, all phyletic lines pass through several successive stages: the stage of the creation, the stage of expansion and diversification, and finally the stage of specialization and senescence. The last stage of this cycle was "regressive or gerontocratic" evolution. He considered cavernicoles good examples of regressive evolution. The title of another influential biospeleological book, *L'Evolution regressive des poissons cavernicoles et abyssaux* ("The regressive evolution of the cave and abyssal fishes") by Georges Thinès[238] (1969) leaves little doubt of the orthogenetic state of mind of this and most other biospeleologists at the time. Probably the most famous orthogenecist of this time was the Jesuit French paleontologist Pierre Teilhard de Chardin[239], who believed that evolution was constantly marching toward some sort of point of perfection (the "Omega point").

---

235 *b*. Kempten, Germany, 5 July 1904; *d*. Bedford, Massachusetts, 3 February 2005. He was a leading evolutionary biologist of the twentieth century and one of the architects of the Modern Synthesis. He was a severe critic of typological thinking and finalistic interpretations of evolution.

236 *b*. Périgueux, France, 27 November 1895; *d*. Paris, France, 9 July 1985. He was mostly known as the editor of the 35-volume *Traité de zoologie*. He did not believe in natural selection and/or mutation as the causes of evolution but rather on an 'internal factor' as the engine of evolutionary change. He claimed that such an 'internal factor' was real, not mystical, and that was different from the mystical vitalism espoused by some of his predecessors.

237 *b*. Besançon, Jura, France, 26 December 1894; *d*. Toulouse, France, 11 October 1980.

238 *b*. Liège, Belgium, 10 February 1923. An experimental psychologist, poet, and essayist.

239 *b*. Sarcenat, France, 1 May 1881; *d*. New York City, New York, 10 April 1955. Chardin was heavily influenced by Henri Louis Bergson and the Bergsonian scholar Eduard Le Roy. His opinions carried some weight since he distinguished himself as a paleoanthropologist. His evolutionary views better articulated in his posthumously published *Le Phénomène humaine* (*The Phenomenon of Man*) contended that cosmic evolution is the process by which God brings into being a 'fullness of Christ' that includes a morally and spiritually mature humanity and a fully developed natural world. For Chardin the evolutionary process is governed by a 'law of complexification' according to which inorganic matter will reach ever more complex forms, resulting in inorganic matter being followed by organic matter and organic matter being followed by conscious life forms. He expected that at some point that this 'complexification' in humans will take men to reach an 'Omega Point' at which Christ's fullness will include as his 'body' a unified humanity that is at peace. Thus, Chardin epitomized the mixture mystic Catholicism and progressionist/positivism views that have dominated many evolutionary conceptions in biospeleology (Olivier 1967, Dobzhansky 1968, Gentner 1968, Potter 1968).

How did all these distinguished French intellectuals and naturalists remain blind to the evidenced being accumulated by biologists elsewhere? Bowler (1983, p. 108) has argued that unlike their British, American, and German counterparts, French biologists of the Darwinian and neo-Darwinian eras were rather isolated from their colleagues elsewhere and also seemed to be content with Cuvier's legacy, and since Cuvier had beaten Lamarck in the argument about evolution, why bother to discuss the ideas of a Britton in this regard? Also, French biologists had remained closely tied to the morphological-systematic tradition of Cuvier and Geoffroy Saint-Hillaire and were totally uninterested in other areas such as ecology or developmental biology and maintained a fixed, descriptive view of life. However, having said that, I want to mention two other factors, which I have pointed out previously, that also contributed to the French view of life: their feelings of nationalism and catholic mysticism. After all, no one prevented them from reading Dobzhansky, Mayr, Simpson, or any other major contributor to The Modern Synthesis.

It is interesting to note that the only French scientists that embraced the new populational view of evolution were not biologists but mathematicians: these rare exceptions were the population geneticists Georges Téissier[240] and Philippe L'Heritier[241] who, because they worked entirely outside the realms of biology and in a field (mathematics) that did not need metaphysics to achieve its goals, were free to pursue the mathematical population ideas of Ronald Fisher[242] and Sewall Wright[243] who so greatly contributed to our current ideas in evolution.

How did the non-French thinkers and biospeleologists respond when faced with the clear contrast of ideas between The Modern Synthesis on one side and neo-Lamarckism and orthogenesis on the other? Not particularly well. First of all, many philosophers of the 1920s and '30s, such as Samuel Alexander[244], a British realist metaphysician, and Jan Smuts[245], the South African statesman, continued to support orthogenetic theories. The same can be said of later philosophers such as Alfred North Whitehead[246] with his theory of organisms and Mihály Polanyi[247] with his theory of personal knowledge.

Biospeleology in other countries was delayed and anemic and therefore leaned heavily on French ideas and concepts in its development. In the United States, for example, very little had been done since the early Twentieth Century when Eigenmann published his 1909 book on the cave vertebrates of America. In fact, what was accomplished in the remainder of the first half of the Twentieth Century was by foreign

---

240 *b*. Paris, France, 19 February 1900; *d*. Roscoff, France, 7 January 1972. He profoundly influenced many French scientists including Jacques Monod.
241 *b*. Ambert (Puy de Dôme), France, 1906; *d*. 1990.
242 *b*. East Finchley, London, United Kingdom, 17 February 1890; d. Adelaide, Australia, 29 July 1962.
243 *b*. Melrose, Massachusetts, 21 December 1889; *d*. Madison, Wisconsin, 3 March 1988.
244 *b*. Sydney, New South Wales, Australia, 6 January 1859; *d*. Manchester, England, 13 September 1938. He was a philosopher that proposed the idea of 'emergent evolution' according to which evolution allows for the appearance of certain features such as consciousness due to some reorganization of pre-existing features. These ideas were an extension of Henri Bergson's *Créative Evolution* (1907).
245*b*. Bovenplaats, near Malmesbury, Cape Colony, South Africa, 24 May 1870; *d*. Doornkloof, Irene, near Pretoria, South Africa, 11 September 1950. He was a soldier, statement and scholar who in 1926 published a book *Holism and Evolution* in which he proposed that nature had the tendency of creating wholes that were greater than the parts through the process of 'creative evolution' by which he was espousing an orthogenetic view of evolutions.
246 *b*. Ramsgate, Kent, England, 15 February 1861; *d*. Cambridge, Massachusetts, 30 December 1947. A mathematician by training became very interested in speculative metaphysics dealing with the issue of the role played by constructions of mathematics, science, and philosophy in the nature or things.
247 *b*. Budapest, Hungary, 12 March 1891; *d*. Northampton, England, 22 February 1976. He was trained as a physician worked on philosophy and social sciences. His philosophical work was full of examples from natural sciences.

researchers such as the Spaniard Ignacio Bolívar[248] and the French Jeannel who extensively explored U.S. caves in 1928 (the results were published in 1931). After that, a few taxonomists showed sparse interest in some cave groups (see Barr 1966) but without contributing anything to biospeleological theory. In fact, the National Speleological Society (NSS) was not founded until 1941; that is, 47 years after the founding of its French counterpart, and as Barr (1966, p. 16) himself put it, "for the first 15 years of its existence, the society had little effect on cave biology."

The first American scientist who started to look at cave organisms from a non-orthogenetic stance was Charles Marcus Breder[249] whose behavioral, physiological, and ecological studies on cave fishes are still cited in the literature. However, since he was not a cave explorer, his contributions have been largely ignored by "hard core" speleologists (Barr, 1966, does not even mention him in his history of cave biology of the U.S.); this is an interesting phenomenon that permeates biospeleology to this date: on one side of the fence are the cave explorers/scientists still strongly influenced by orthogenetic ideas on cave fauna; on the other side we find the "outside" scientists who just happen to study cave organisms because they find them interesting, not because these scientists happen to be spelunkers.

In fact, it was not until the 1960s that the first modern generation of American biologists began making contributions to biospeleology beyond a purely taxonomic level. Names such as Thomas Poulson, David Culver, Thomas Barr, John Holsinger, and Kenneth Christiansen are the first to come to mind, but they are not the only ones. Yet, although they did not subscribe to Vandel's extreme orthogenetic interpretation of cave fauna, they were certainly ambiguous about the importance of natural selection, heavily utilized orthogenetic concepts and jargon such a preadaptation and "regressive evolution," and rarely (if ever) mentioned opportunism or phenotypic plasticity as mechanisms directly involved in the evolution of cave fauna. Unfortunately, it seems that even today, biospeleology has not recovered from the distractions of its slow, stumbling beginning and still fails to fully embrace modern evolutionary and ecological theory.

The Roots of Current Intellectual Inertia

The effects of this intellectual inertia continue to be pervasive. In France natural selection has yet to become central to evolutionary discussions; French evolutionary biologists seem to have jumped from neo-Lamarckism right into molecular evolution. Fortunately, the French molecular biologist and Nobel Prize winner Jacques Monod[250], one of the best spokespeople for the latter, wrote a very strong argument against finalism and other forms of teleology in his 1970 *Le hasard et la necessité* ("Chance and Necessity"). But somehow these and other strong arguments against metaphysical biology have yet to fully impact biospeleology, even in Anglo-Saxon countries. As Mayr (1982, p. 516) put it: "To convince someone who is not familiar with the evolutionary mechanisms that the world is not predetermined and –so to speak- programmed seems hopelessly difficult."

---

248 *b*. Madrid, Spain, 9 November 1850; *d*. Mexico City, Mexico 19 November 1944. He was an entomologist who described more than 1,000 species of insects, some of them from caves.
249 *b*. Jersey City, New Jersey, 25 June 1897; *d*. Englewood, Florida, 28 October 1983. Breder led the Renaissance of the study of cavefishes by using *Astyanax fasciatus* as prime research subject. He was the dominant figure in hypogean fish research in the 1940s and 1950s (Romero 1984, Romero 1986a, Romero 2001a).
250 *b*. Paris, France, 9 February 1910; *d*. Paris, 31 May 1976. He read Darwin at an early age which motivated him to become a biologist. He was influenced by George Teissier, among others. He was extraordinary in comparison to his fellow French biologists in that he viewed evolution as the result of chance, not as a predetermined phenomenon in the best of the neo-Darwinian tradition.

Even current American biospeleologists have not escaped the shadow of neo-Lamarckism and orthogenesis as evidenced by their uncritical use of concepts and terms such as "preadaptation" and "regressive evolution", not only have such intellectual schools of thought been created in the United States, but also, they used cave organisms to epitomize these ideas. What the French did, as true developers of biospeleology as a science, was to color their explanations with additional metaphysical auras.

Unfortunately, the belief that evolution has a direction, such as toward complexity, is deeply rooted, although no one has proven that such is the case (see, for example, Maynard Smith 1970 for a discussion on this). Biospeleology is a science characterized by confusion on both terminology and concepts. I believe this confusion is the result of biospeleology being a science whose pioneers and major practitioners having been very resistant to any neo-Darwinian ideas espoused, rather, neo-Lamarckian views in conjunction with the related concepts of orthogenesis, organicism, and other forms of finalisms.

Although cave organisms are extremely interesting and deserve much more attention, biospeleological phenomena can be explained using current biological ideas without the need to invoke metaphysical explanations of any kind. Does that mean that biospeleology requires a new paradigm? Not really. All that we need to do is find explanations for the phenomena that occur in caves via the scientific body of information available in modern biology. The loss or simplification of phenotypic characters is neither unique nor exclusive to hypogean organisms. Many cave organisms represent excellent examples of natural selection by means of phenotypic plasticity, and those hypogean organisms and their habitats also represent excellent subjects in natural laboratories for the test and expansion of current and new ideas in modern biology.

Literature Cited

Agassiz, E.C. 1890. *Louis Agassiz. His life and correspondence.* Boston: Houghton, Mifflin and Company.

Agassiz, L. 1847 [1848]. [Plan for an Investigation of the Embryology, Anatomy and Effect of Light on the Blind-fish of the Mammoth Cave, *Amblyopsis spelaeus*]. *Proceedings of the American Academy of Arts and Sciences* **1**:1-180.

Agassiz, L. 1851. Observations on the blind fish of the Mammoth cave. *American Journal of Science* **11**:127-128.

Alberti, L. 1550. *Descrittione di tvtta Italia.* Bologna: Anselmo Giaccarell.

Anderson, L. 1976. Charles Bonnet's taxonomy and Chain of Being. *Journal of the History of Ideas* **37**:45-58.

Andersson, M. 1990. Evolution: a case of male opportunism. *Nature* **343**:20.

Anonymous. 1833. Cabinet of nature. Cavern of the guacharo. *Monthly Repository* **4**:24-28.

Anonymous. 1981. Stephen L. Bishop. 1821 – 1857. Explorer and Guide. Mammoth Cave. *Journal of Spelean History* **15**:11.

Anonymous. 1992. Bishop, Stephen, pp. 82-83, *In: Kentucky encyclopedia* (John E. Kleber, Ed.). Lexington, Ky.: University Press of Kentucky.

Appel, T.A. 1987. *The Cuvier-Geoffroy Debate: French Biology in the Decades Before Darwin.* New York: Oxford University Press.

Argyll, Duke of. 1867. *The Reign of Law.* London: Alexander Strahan.

Artigas, M., T.F. Glick & R.A. Matínez. 2006. *Negotiating Darwin. The Vatican Confronts Evolution, 1877-1902.* Baltimore: The Johns Hopkins University Press.

Ayala, F. 1971. Dobzhansky, Theodosius, pp. 233-242, *In:* C.C. Gilliespie (Ed.). *Dictionary of Scientific Biography*, Vol. 4. New York: Scribner.

Baker, K.M. 2004. On Condorcet's "sketch". *Daedalus* **133**:56-64.

Banta, A.M. 1909. *The Fauna of Mayfield's Cave*. Washington, D.C.: Carnegie Institution of Washington.

Banta, A.M. 1921. An eyeless daphnid, with remarks on the possible origin of eyeless cave animals. *Science* **53**:462-463.

Barr, T.C. 1966. Evolution of Cave Biology in the United States, 1822-1965. *National Speleological Society Bulletin* **28**:15-21.

Barr, T.C. 1986. Mammoth Cave in the years 1836-1855. *Journal of Spelean History*. **20**:39-40.

Bellés, X. 1991. Survival, opportunism and convenience in the processes of cave colonization by terrestrial faunas. *Oecologia Aquatica* **10**:325-335.

Bellon, R. 2001. Joseph Dalton Hooker's ideals for a professional man of science. *Journal of the History of Biology* **34**:51-82.

Benton, M.J. 1983. Dinosaur success in the Triassic: a noncompetitive ecological model. *Quarterly Review of Biology* **58**:29-55.

Berg, L.S. 1926. *Nomogenesis; or, evolution determined by law*. London: Constable.

Berry, R.J. 1989. Ecology: where genes and geography meet. Presidential address to the British Ecological Society, December 1988. *Journal of Animal Ecology* **58**:733-759.

Besson, J. 1569 [1969]. *L'art et science de trouver les eaux et fontaines cachees soubs terre : autrement que par les moyens vulgaires des agriculteurs et Architectes*. Orléans: E. Gibier. [facsimile reproduction by Editions Coral, Columbus, Ohio].

Bourdier, F. 1971. Cuvier, Georges, pp. 521-528, *In*: C.C. Gilliespie (Ed.). *Dictionary of Scientific Biography*, Vol. 3. New York: Scribner.

Bourdier, F. 1972a. Gaudry, Albert Jean, pp. 295-297, *In*: C.C. Gilliespie (Ed.). *Dictionary of Scientific Biography*, Vol. 5. New York: Scribner.

Bourdier, F. 1972b. Geoffroy Saint-Hilaire, Étienne, pp. 355-358, *In*: C.C. Gilliespie (Ed.). *Dictionary of Scientific Biography*, Vol. 5. New York: Scribner.

Bowler, P.J. 1983. *The Eclipse of Darwinism. Anti-Darwinian Evolution Theories in the Decades Around 1900*. Baltimore: The John Hopkins University Press.

Bowler, P.J. 2005. Revisiting the Eclipse of Darwinism. *Journal of the History of Biology* **38**:19-32.

Burkhardt, F. & S. Smith (Eds.). 1987. *The correspondence of Charles Darwin. Volume 3. 1844-1846*. Cambridge: Cambridge University Press.

Burkhardt, F. & S. Smith (Eds.). 1987. *The correspondence of Charles Darwin. Volume 5. 1851-1855*. Cambridge: Cambridge University Press.

Burkhardt, F. & S. Smith (Eds.). 1990. *The correspondence of Charles Darwin. Volume 6. 1856-1857*. Cambridge: Cambridge University Press.

Burkhardt, R.W. 1977. *The Spirit of the System. Lamarck and Evolutionary Biology*. Cambridge: Harvard University Press.

Byers, G.W. 1969. Evolution of wing reduction in crane flies (Diptera: Tipulidae). *Evolution* **23**:346-354.

Chen, Y.-R., J.-X. Yang & Z.G. Zhu, 1994. A new fish of the genus *Sinocyclocheilus* from Yunnan with comments on its characteristic adaptation (Cypriniformes: Cyprinidae). *Acta Zootaxonomica Sinica* **19**:246-253.

Coleman, W. 1964. *Georges Cuvier, Zoologist: A Study in the History of Evolution Theory*. Cambridge: Harvard University Press.

Colp, R. 1986. "Confessing a murder": Darwin's first revelations about transmutation. *Isis* **77**:9-32.

Condorcet, J.-A.N. de 1793-4. *Esquisse d'un Tableau Historique des Progrès de l'Espirit Human*. Paris: Agasse

Cope, E.D. 1864. On a blind silurid from Pennsylvania. *Proceedings of the Academy of Natural Sciences of Philadelphia* **1864**:231-233.

Cope, E.D. 1896. *The Primary Factors of Organic Evolution*. Chicago: Open Court.

Corsi, P. 2005. Before Darwin: transformist concepts in European natural history. *Journal of the History of Biology* **38**:67-83.

Cuénot, L. 1911. *La genesis de las especes animals*. Paris: Librairie Félix Alcan.

Darwin, C. 1859. *On the Origin of the Species by Means of Natural Selection*. London: J. Murray.

Darwin, C. 1861. *On the Origin of the Species by Means of Natural Selection*. (3rd Ed.). London: J. Murray.

Darwin, F. 1896. *The Life and Letters of Charles Darwin*. New York: D. Appleton and Company, 2 vols.

Davidson, J.P. 1997. *The Bone Sharp. The Life of Edward Drinker Cope*. Philadelphia: The Academy of Natural Sciences of Philadelphia.

De Beer, G. 1973. Lankester, Edwin Ray, pp. 26-27, *In*: C.C. Gillispie (Ed.). *Dictionary of Scientific Biography*, Vol. 8. New York: Scribner.

De Chardin, T. 1955. *Le phénomè humain*. Paris: Du Seuill.

DeKay, J. E. 1842. *Zoology of New York or the New-York Fauna*, Part IV, Fishes. Albany: W. & A. White & J. Visscher.

Desmond, R. 1972. Hooker, Joseph Dalton, pp. 488-492, *In*: C.C. Gillispie (Ed.). *Dictionary of Scientific Biography*, Vol. 6. New York: Scribner.

Dexter, R.W. 1965. The "Salem secession" of Agassiz zoologists. *Essex Institute Historical Collections* **101**:27-39.

Dexter, R.W. 1979. The impact of evolutionary theories on the Salem Group of Agassiz Zoologists (Morse, Hyatt, Packard, Putnam). *Essex Institute Historical Collections* **115**:144-171.

Diamond, J. & D. Stermer. 1999. Evolving backward. *Discover* **19**:64-68.

Dimichele, W.A., T.L. Phillips, & R.G. Olmstead, R.G. 1987. Opportunistic evolution: abiotic environmental stress and the fossil record of plants. *Review of Palaeobotany & Palynology* **50**:151-178.

Dobzhansky, T. 1968. Teilhard de Chardin and the orientation of evolution. *Zygon* **3**: 242-258.

Dobzhansky, T. 1970. *Genetics of the Evolutionary Process*. New York: Columbia University Press.

Doolittle, R.F. 1988. Lens proteins. More molecular opportunism. *Nature*. **336**:18.

Eddy, J.H. 1994. Buffon's *Histoire naturelle*. History? A critique of recent interpretations. *Isis* **85**:644-661.

Eigenmann, C.H. 1898. A new blind fish. *Proceedings of the Indiana Academy of Sciences* (897):231.

Eigenmann, C.H. 1903. In search of blind fishes in Cuba. *World Today* **5**:1131-1136.

Eigenmann, C.H. 1909. *Cave Vertebrates of America. A study in degenerative evolution*. Washington, D.C.: Carnegie Institution of Washington.

Eigenmann, C.H. 1919. *Trogloglanis pettersoni* a new blind fish from San Antonio, Texas. *Proceedings of the American Philosophical Society* **58**:397-400.

Ewan, J. 1975. Rafinesque, Constantine Samuel, pp. 262-264, *In*: C.C. Gillispie (Ed.). *Dictionary of Scientific Biography*, Vol. 11. New York: Scribner.

Farber, P.L. 1975. Buffon and Daubenton: divergent traditions within the *Histoire naturelle*. *Isis* **66**:63-74.

Gentner, D.R. 1968. The scientific basis of some concepts of Pierre Teilhard de Chardin. *Zygon* **3**:432-441.

Goode, C.E. 1986. *World wonder saved. How Mammoth Cave became a national park*. Mammoth Cave, Kentucky: The Mammoth Cave National Park Association.

Goudge, T.A. 1973. Bergson, Henri Louis, pp. 8-12, *In*: C.C. Gilliespie (Ed.). *Dictionary of Scientific Biography*, Vol. 2. New York: Scribner.

Granger, G. 1971. Condorcet, Marie-Jean-Antoine-Nicolas, Marquis de, pp. 383-399, *In*: C.C. Gillispie (Ed.). *Dictionary of Scientific Biography*, Vol. 3. New York: Scribner.

Greene, J.C. 1959. *The Death of Adam: Evolution and its Impact on Western Thought*. Ames: Iowa State University Press.

Grimoult, C. 1998. *Évolutionnisme et fixisme en France. Histoire d'un combat 1800-1882*. Paris: CNRS Éditions.

Hecht, M.K. & W.C. Steere. 1970. *Essays in Evolution and Genetics in Honor of Theodosius Dobzhansky*. New York: Appleton-Century-Crofts.

Heuss, T. 1991. *Anton Dohrn: A Life for Science*. New York: Springer-Verlag.

Hill, C.R. 1974. *The Sources and Influence of the Descrittione di tutta Italia of Fra Leandro Alberti*. Ph.D. Dissertation. University of Edinburgh.

Hintzsche, E. 1972. Henle, Friedrich Gustav Jacob, pp. 268-270, *In*: C.C. Gillispie (Ed.). *Dictionary of Scientific Biography*, Vol. 6. New York: Scribner.

Hintzsche, E. 1973. Koellikeer, Rudolf Albert von, pp. 437-440, *In*: C.C. Gillispie (Ed.). *Dictionary of Scientific Biography*, Vol. 7. New York: Scribner.

Horder, T. J. 1998. Why do scientists need to be historians? *Quarterly Review of Biology* **73**:175-187.

Howard, M. 1981. *The Franco-Prussian War: The German Invasion of France, 1870-1871*. London: Routledge.

Hubbs, C.L. 1964. David Starr Jordan. *Systematic Zoology* **13**:195-200.

Humboldt, A. von 1793. *Florae fribergensis specimen, plantas cryptogamicas praesertim subterraneas exhibens*. Berolini: H.A. Rottmann.

Humboldt, A. von 1805, Mémoire sur une nouvelle espèce de pimelode, jetée par les volcans du Royaume de Quito, *In*: Voyage de Humboldt et Bonpland, Deuxième partie. *Observations de Zoologie et d'Anatomie comparée*. Paris. Vol 1, edited by A von Humboldt & A. Bompland.

Humboldt. A. von. 1817. Mémoire sur le Guacharo de la caverne de Caripe. *Recueil d´Observations de Zoologie et d´Anatomie* no. 2.

Jackson, J.R. & W.C. Kimler. 1999. Taxonomy and the personal equation: the historical fates of Charles Girard and Louis Agassiz. *Journal of the History of Biology* **32**:509-555.

Jeannel, R.G. 1950. *La marche de l'évolution*. Paris: Presses universitaires de France.

Južnič, S. 2006. Karst research in the 19th century – Karl Dežman's (1821-1889) work. *Acta Carsologica* **35/1**:139-148.

Kircher, A. 1665. *Mundus subterraneus, in XII libros digestus; quo divinum subterrestris mundi opificium, mira ergasteriorum naturæ in eo distributio, verbo pantámorphou Protei regnum, universæ denique naturæ majestas & divitiæ summa rerum varietate exponuntur*. Amsterdam: J. Janssonium and E. Weyerstraten.

Kuhn, T.S. 1970. *The Structure of Scientific Revolutions*. Chicago: The University of Chicago Press.

Lankester, E.R. 1880. *Degeneration: a chapter in Darwinism*. London: Macmillan.

Lankester, E.R. 1893. Blind animals in caves. *Nature* **47**:389.

Laurenti, J.N. 1768. *Specimen medicum, exhibens synopsin reptilium emendatam cum experiment is circa venena et antidote reptilium Austria Corum*. Wien: Joan Thomae.

Leith, J.A. 1989. L'Évolution de l'idée de progrès a travers l'histoire. *Transactions of the Royal Society of Canada* **4**:3-8.

Lister, M. 1674. An account of two uncommon mineral substances, found in some coal and iron-mines of England; as it was given by the intelligent and learned Mr. Jessop of Bromhal in York-Shire to the ingenious Mr. Lister, and by him communicated to the publisher in a letter of January 7. 1663/74. *Philosophical Transactions of the Royal Society of London* **8**:6179-6181.

Longrás Otín, L. 2002. Francisco de Tauste (1626-1685), pp. 9-38, *In: Arte y Bocabvlario de la lengva de los Indion Chaymas, Cvmanagotos, Cores, Parias, y otros diversos de la Provincia de Cvmana, o Nueva Andalvcia* (M.A. Pallarés Jiménez, Ed.). Zaragoza, Spain: Instituto Aragonés de Antropología.

Lurie, E. 1960. *Louis Agassiz. A Life in Science.* Chicago: University of Chicago Press.

Lurie, E. 1970. Agassiz, Jean Louis Rodolphe, pp. 72-74, *In*: C.C. Gilliespie (Ed.). *Dictionary of Scientific Biography*, Vol. 1. New York: Scribner.

Maynard Smith, J. 1970. Time in evolutionary process. *Studium Generale* **23**:266-272.

Mayr, E. 1960. The emergence of evolutionary novelties, pp. 349-380, *In*: S. Tax (Ed.). *The Evolution of Life. Its Origin, History, and Future.* Chicago: The University of Chicago Press.

Mayr, E. 1982. *The Growth of Biological Thought: Diversity, Evolution, and Inheritance.* Cambridge: Belknap Press.

Montalembert, M.-R. 1748. Observations de physique générale. *Histoite de la Academie Royale des Sciencies* **1748**:27-28.

Morgan, R. 1943. Caves in world history. *Bulletin of the National Speleological Society* **5**:1-16.

Morris, P.J. 1997. Louis Agassiz's arguments against Darwinism and his additions to the French translation of the *Essay on Classification. Journal of the History of Biology* **30**:121-134.

Motas, C. 1962. Emil G. Racovitza: Founder of biospeleology. *National Speleological Society Bulletin* **24**: 3-8.

Mystakidou, K., E. Tsilika, E. Parpa, E. Katsouda & L.Vlahos. 2004. Death and grief in the Greek culture. *Omega, Journal of Death and Dying* **50**:23-34.

Nealson, K.H. & P.G. Conrad, P.G. 1999. Life: past, present and future. *Philosophical Transactions of the Royal Society of London Series B-Biological Sciences* **354**: 1923-1939.

Olivier, G. 1967. Teilhard de Chardin et le transformisme. *Annales de l'Université de Paris* **37**:358-365.

Oppenheimer, J.M. 1982. Ernest Heinrich Haeckel as an intermediary in the transmutation of an idea. *Proceedings of the American Philosophical Society* **126**:347-355.

Packard, A.S. 1871. The Mammoth Cave and its inhabitants. *American Naturalist* **5**:739-761.

Packard, A.S. 1888. On certain factors of evolution. *American Naturalist* **22**:808-821.

Packard, A.S. & F.W. Putnam. 1872. *The Mammoth Cave and its inhabitants.* Salem: Naturalist Agency.

Pilet, P.E. 1973. Bennet, Charles, pp. 286-287, *In*: C.C. Gilliespie (Ed.). *Dictionary of Scientific Biography*, Vol. 2. New York: Scribner.

Poey, F. 1858. *Memorias sobre la historia natural de la isla de Cuba, 2 volúmenes.* La Habana: Barcina.

Potter, V.R. 1968 Teilhard de Chardin and the concept of purpose. *Zygon* **3**:367-376.

Putnam, F.W. 1872. The blind fishes of the Mammoth Cave and their allies. *American Naturalist* **6**:6-30.

Putnam, F.W. 1875. 1875. On some of the habits of the blind crawfish, *Cambarus pellucidus*, and the reproduction of lost parts. *Proceedings of the Boston Society of Natural History* **18**:16-19.

Rabaud, E. 1941. *Introduction aux sciences biologiques.* Paris: Armand Colin.

Rafinesque, C.S. 1822. On two salamaders of Kentucky. *Kentucky Gazette, Lexington* (n.s.) **1**:3.

Rafinesque, C.S. 1832. The caves of Kentucky. *Atlantic Journal and Friend of Knowledge* **1**:27-30.

Richards, R. 1992. *The Meaning of Evolution. The Morphological Construction and Ideological Reconstruction of Darwin's Theory*. Chicago: The University of Chicago Press.

Richards, R. 2002. *The Romantic Conception of Life. Science and Philosophy in the age of Goethe*. Chicago: The University of Chicago Press.

Risse, G.B. 1971. Döllinger, Ignaz, pp. 146-147, *In*: C.C. Gilliespie (Ed.). *Dictionary of Scientific Biography*, Vol. 4. New York: Scribner.

Roger, J. 1973. Buffon, Georges-Louis Leclerc, Compte de, pp. 576-582, *In*: C.C. Gilliespie (Ed.). Dictionary of Scientific Biography, Vol. 2. New York: Scribner.

Roger, J. 1997. *Buffon. A Life in Natural History*. Ithaca: Cornell University Press.

Romero, A. 1984. Charles Marcus Breder, Jr. 1897-1983. *National Speleological Society News* **42**:8.

Romero, A. 1986a. Charles Breder and the Mexican blind characid. *National Speleological Society News* **44**:16-18.

Romero, A. 1986b. He wanted to know them all: Eigenmann and his blind vertebrates. *National Speleological Society News* **44**:379-381.

Romero, A. 1999a. The blind cave fish that never was. *National Speleological Society News* **57**:180-181.

Romero, A. 1999b. Myth and reality of the alleged blind cave fish from Pennsylvania. *Journal of Spelean History* **33**:67-75.

Romero, A. 2000. The speleologist who wrote too much. *National Speleological Society News* **58**:4-5.

Romero, A. 2001a. Evolution is opportunistic, not directional. *BioScience* **51**:2-3.

Romero, A. 2001b. Scientists prefer them blind: the history of hypogean fish research. *Environmental Biology of Fishes* **62**:43-71.

Romero, A. 2002c. The life and work of a little known biospeleologist: Theodor Tellkampf. *Journal of Spelean History* **36**:68-76.

Romero, A. 2002*d*. Between the first blind cave fish and the last of the Mohicans: the scientific romanticism of James E. DeKay. *Journal of Spelean History* **36**:19-29.

Romero, A. 2009. *Cave Biology: Life in Darkness*. 291 pp. Cambridge, UK: Cambridge University Press.

Romero, A. & Z. Lomax. 2000. Jacques Besson, cave eels and other alleged European fishes. *Journal of Spelean History* **34**:72-77.

Romero A. & K.M. Paulson. 2001. Humboldt's alleged subterranean fish from Ecuador. *Journal of Spelean History* **36**:56-59.

Romero, A. & A. Romero. 1999. On Cope, caves, and skeletons in the closet. *National Speleological Society News* **57**:341-343.

Romero, A. & J. Woodward. 2005. On white fish and black men: did Stephen Bishop really discover the blind cave fish of mammoth cave? *Journal of Spelean History* **39**:23-32.

Romero, A.; Y. Zhao & X. Chen. 2009. The hypogean fishes of China. *Environmental Biology of Fishes* **86**(1):211-278.

Rossmässler, E.A. 1838-1844. Iconographie der Land- und Süßwassermollusken, mit vorzüglicher Berücksichtigung der europäischen noch nicht abgebildeten Arten. (1) 2. - pp. (4+44 pp.), (4+46 pp.), (4+15 pp.), (4+37 pp.), Taf. 31-60. Dresden, Leipzig: Arnold.

Rudwick, M.J.S. 1997. *Georges Cuvier, fossil bones, and the geological catastrophes: new translations and interpretations of the primary texts*. Chicago: University of Chicago Press.

Sanford, W.F. 1965. Dana and Darwinism. *Journal of the History of Ideas* **26**:531-546.

Schiner, J.R. 1854. Fauna der Adelsberg, Lueger und Magdalener-grotte. In *Die Grotten und Hölen von Adelsberg, Lueg, Planina und Lass,* edited by A. Schmidl. Wien: Braunmüller.

Schiödte, J.C. 1849. *Specimen faunæ subterraneae. Bidrag til den underjordiske fauna.* Denmark: Kjöbenhavn.

Schmidt, F. 1832. Beitrag zu Krain's fauna. *Illÿrfches Blatt.* **21**:9-10

Serres, M. de. 1851. *Du perfectionnement graduel des etres organises.* Bordeaux: Lafargue.

Shaw, T.R. 1992. *History of cave science. The exploration and study of limestone caves, to 1900.* Broadway: Sydney Speleological Society.

Shor, E.N. 1973. Jordan, David Starr, pp. 169-170, *In*: C.C. Gillispie (Ed.). *Dictionary of Scientific Biography*, Vol. 7. New York: Scribner.

Simpson, G.G. 1949. *The Meaning of Evolution.* New Haven: Yale University Press.

Sloan, P.R. 1976. The Buffon-Linnaeus controversy. *Isis* **67**:356-375.

Smith, D.C. & H.W. Brown, Jr. 2000. Louis Agassiz, the Great Deluge, and early Maine geology. *Northeastern Naturalist* **7**:157-177.

Stanton, W. 1971. Dana, James Dwight, pp. 549-554, *In*: C.C. Gillispie (Ed.). *Dictionary of Scientific Biography*, Vol. 3. New York: Scribner.

Stone, A.J. 1995. *Images from the underworld: Naj Tunich and the tradition of Maya cave painting.* Austin: University of Texas Press.

Tauste, F. 1678. *Relación de las Misiones de los Religiosos Capuchinos en la provincia de Cumaná.* (unpublished manuscript).

Taylor, G. 1970. Bentham, George, pp. 614-615, *In*: C.C. Gilliespie (Ed.). *Dictionary of Scientific Biography*, Vol. 1. New York: Scribner.

Tellkampf, T. 1844a. Beschreibung einiger neuer in der Mammuth-Höhle in Kentucky aufgefundener Gattungen von Gliederthieren. *Archives Vereins Freund Naturliche Mecklenburg* **10**:318-322.

Tellkampf, T. 1844b. Uber den blinden Fisch der Mammuthhöhle in Kentucky. *(Muller's) Archives fur Anatomie und Physiologie* **1844**:381-395.

Tétry, A. 1971a. Caullery, Maurice, pp. 148-149, *In*: C.C. Gillispie (Ed.). *Dictionary of Scientific Biography*, Vol. 3. New York: Scribner.

Tétry, A. 1971b. Cuénot, Lucien, pp. 492-494149, *In*: C.C. Gillispie (Ed.). *Dictionary of Scientific Biography*, Vol. 3. New York: Scribner.

Unwin, R.W. 1995. A provincial man of science at work: Martin Lister, F.R.S., and his illustrators 1670-1683. *Notes and Records of the Royal Society of London* **49**:209-230.

Uschman, G. 1972. Haeckel, Ernst Heinrich Philipp August, pp. 6-11, *In*: C.C. Gillispie (Ed.). *Dictionary of Scientific Biography*, Vol. 6. New York: Scribner.

Valvasor, J.W. 1687. An extract of a letter written to the Royal Society out of Craniola, being a full and accurate description of the wonderful Lake of Kirknitz in that country. *Philosophical Transactions of the Royal Society of London* **16**:411-427.

Valvasor, J.W. 1689. *Die Ehre des Herzogthums Krain.* Ljubljana, Slovenia: Endter.

Vandel, A. 1965. *Biospéologie: la biologie des animaux cavernicoles.* Paris: Gauthier-Villars.

Viré, A. 1904. La biospéologie. *Comptes rendus de la Académie des Sciences du Paris* **139**:

Wallace, D.R. 1999. *The bonehunters' revenge. Dinosaurs, greed, and the greatest scientific feud of the gilded age.* Boston: Houghton Mifflin Company.

Warren, L. 2004. *Constantine Samuel Rafinesque. A voice in the American Wilderness.* Lexington: The University Press of Kentucky.

Wilson, L.G. 1973. Lyell, Charles, pp. 563-577, *In*: C.C. Gillispie (Ed.). *Dictionary of Scientific Biography*, Vol. 8. New York: Scribner.

Wright, S. 1931. Evolution in Mendelian populations. *Genetics* **16**:97-159.

Wyman, J. 1843. Description of a 'blind fish' from a cave in Kentucky. *American Journal of Science* **45**:94-96.

Wyman, J. 1851. [Account of dissections of the blind fishes (*Amblyopsis spelaeus*) from the Mammoth Cave, Kentucky]. *Proceedings of the Boston Society of Natural History* **3**:349, 375.

Wyman, J. 1854a. the eyes and organs of hearing in *Ambyopsis spelaeus*. *Proceedings of the Boston Society of Natural History* **4**:149-151.

Wyman, J. 1854b. On the eye and the organ of hearing in the blind fishes (*Amblyopsis spelaeus* DeKay) of the Mammoth Cave. *Proceedings of the Boston Society of Natural History* **4**: 395-396.

Xie, Y.J. 1540. *Report on the Alu Cave*. Manuscript. (in Chinese).

Chapter 4. Scientists Prefer them Blind: The History of Hypogean Fish Research [251]

Summary

The history of hypogean fish research has been strongly influenced by neo-Lamarckism, typological thinking, and orthogenesis. Only in the last few decades have neo-Darwinism made any in-roads in the research approach to this subject. The majority of the most distinguished and productive hypogean fish researchers have used their research subjects to confirm their own views of evolution rather than to use those subjects as a spring of knowledge to enrich mainstream biological thought. Of these views, I found that the most perversive of all is the notion of evolutionary "progress" that has led many researchers to envision hypogean fishes as prime examples of "regressive evolution." I propose that the utilization of hypogean fish for the study of convergent evolution should catapult these subjects of research into prime objects of evolutionary studies.

Introduction

The development of a particular area of science is deeply influenced by its history. Through historical studies, one can better perceive the impacts of fashion, error, and the effects of intellectual inertia (Horder 1998). By doing that the researcher can also project the best venues to develop future lines of research.

In this paper I analyze the scientific views on the nature of hypogean fish. I will contend that both neo-Lamarckism (including orthogenesis) and typological thinking have prevented the use of hypogean fishes as one of the prime examples to explain some general biological phenomena in evolutionary biology as it is the case for convergent evolution. Another question is how hypogean fish research fits within the broader context of the history of three sub-disciplines: ichthyology, speleology, and evolutionary biology. This is a rather difficult task due to lack of up-to-date contextual material.

For example, a comprehensive history of ichthyology has yet to be written. Although some attempts have been made, none of these give us the entire picture. They are either too old (Cuvier 1828 [1995], Günther 1880, Jordan 1905), limited in scope geographically (Myers 1964, Hubbs 1964) or conceptually (Pietsch & Anderson 1997). Thus, there is not an appropriate background to refer to in this regard. The history of speleology has faced similar problems. The only attempt to summarize the history of this field is that of Shaw (1992), yet his emphasis is on geomorphology as he recognizes (Shaw 1992:iv). Barr's (1966) short article on the history of cave research is confined to the United States. The role played by study of cave organisms in general on the neo-Lamarckian movement in the United States as well as the idea of progress is better known (e.g., Ruse 1966). More recently, Romero (2009, Chapter 1) came up with a comprehensive analysis of the history of biospeleology that has been later updated in Chapter 4 of the present book.

Methods

---

[251] Based on Romero, A. 2001. Scientists prefer them blind: The history of hypogean fish research. *Environmental Biology of Fishes* **62**(1-3):43-71.

I compiled 1,617 published references on hypogean fishes. They were obtained by using standard bibliographical sources, including the *Zoological Record* up to 1998, the *Biological Abstracts* up to June 2000, Dean's A Bibliography of Fishes (Dean 1916-1923), Current References in Fish Research (Cvancara[252]), as well as other bibliographical data accumulated by the author in the last 40 years, all of them deposited in the Aldemaro Romero Jr. Collection at the Lovejoy Library of Southern Illinois University Edwardsville[253]. These references were analyzed by author, topic, year of publication, and by country/institution by using Procite 5 [254]. I emphasized studies on hypogean fishes that show some degree of troglomorphy (blindness, depigmentation). The study of non-troglomorphic hypogean fishes, despite their tremendous potential for understanding the processes of cave colonization and evolution, has been largely overlooked by scientists and, thus, have remained mostly anecdotal (Poly 2001). Although this type of compilations can never be fully complete, I am confident that I studied and analyzed all relevant sources.

In order to provide adequate historical and geographical context, I have added the dates and places of birth and death in parenthesis after mentioning, for the first time, all major players in the history of hypogean fish research. Secondary players are only given the years of birth and death. All scientific names of fishes are given following Romero (2001a). After that, a great number of new species were described, particularly from China (Romero *et al.* 2009). Sometimes those names are followed by a parenthesis to mention the name with which the species was originally described. For referential purposes, I present the chronology of the discovery of valid species/populations of troglomorphic fishes up to 1999 in Table 1.

Results

I identified six periods of hypogean fish research: (1) pre-Linnean (1541-1752), (2) first discoveries and research (1805-1864), (3) American neo-Lamarckism (1868-1919), (4) dominance of typological thinking (1921-1940), (5) American renaissance (1936-1960), and (6) times of philosophical conflict (1960-1990). Notice that between (1), (2), (3), and (4) there are gaps. This is because during those in-between years nothing significant happened in the field. On the other hand, there is an overlap between (5) and (6); this is because the American and European schools that dominated each of those periods respectively were very much independent of each other, not only geographically but also conceptually (see description below).

Pre-Linnean Times (1541-1752)

The first known printed reference of a hypogean fish is that of the hyaline fish (*Sinocyclocheilus hyalinus*), reported by Jie (1541) for the Alu Limestone caves (103° 45'E, 24° 33'N) in what is today Luxi County, Yunnan, China. However, specimens of the hyaline fish were not collected for scientific research purposes until 1991 and the description of the species was not published in the scientific literature until three years later (Chen *et al.* 1994).

There are at least three more pre-Linnean printed references to hypogean fishes, all of them from Europe. Given that they have been cited as precursors in the history of biospeleology, it is important to analyze their validity as true scientific developments.

---

[252]Cvancara, V.A. 1976-1998. Current references in fish research. 23 vols.
[253] https://www.siue.edu/lovejoy-library/romero/
[254]2000. ProCite Version 5.0. ISI ResearchSoft, Berkeley.

The first is by Jacques Besson[255]. He was an engineer and mathematician with no formal training in the natural sciences (Romero & Lomax 2000). Besson (1569: 41) reported little eels ("*petites anguilles*") in a cave stream. Although Shaw (1992:227) claims that such observation took place "in a cave stream in France," Besson did not give a locality of where he made that observation. Besson does not describe the fish as being blind and/or depigmented (extraordinary characteristics even to the casual observer). He may have seen common eels, *Anguilla anguilla*, or a species of some of the European freshwater fishes with eel-like bodies that are sympatric with the areas he used to travel (France and Switzerland). Those fish families include Petromizonidae, Cobitidae, Siluridae, and Clariidae (Blanc *et al.* 1971).

The second was Athanasius Kircher[256]. This prolific Jesuit priest polymath wrote, in what is probably the first printed work on speleology (Fig. 4.1.), that

"There is also in the landscape of Krain [Carniola?] close to the town Haubach a huge field from which each year during Spring time a large body of water containing fish bursts forth with the result that in a few days it transforms the field into a lake teeming with fish (...) in Switzerland rivers rise from the caves of the mountains, that flow from May until September, but stop the rest of the time (...) as they come out of the mountains, are full of fish, which is clear proof that they [the fish] emerge from subterranean waters along the rivers (...) it is not implausible that, as under the earth all kind of fishes occur and live" (Kircher 1665, 2:85).

These references to subterranean fishes, however, are vague, unsubstantiated, and given Kircher's reputation as an uncritical repeater of other people's tales, highly suspect (Romero 2000a). Furthermore, he made no reference to features typically associated to troglomorphic fishes: blindness and depigmentation.

The third pre-Linnean reference to subterranean fishes in Europe was by Marc-René Marquis de Montalembert[257] (Fig. 4.2.). He was an aristocrat, military man, and engineer known for his design of fortifications. He reported a blind, subterranean fish in a spring at Gabard, Angoumois, near one his estates in southwestern France. He noted: "it is common to fish either blind or one-eyed pike; one-eyed ones always miss the right eye and among the blind ones, the right eye seems further reduced than the left eyes" (Montalembert 1748). He left no drawings, much less preserved specimens. He said that what he saw was a pike. That, by itself, is not surprising. The pike, *Esox lucius* is, by far the most common freshwater fish of the Northern Hemisphere. The fact that this fish can be identified as a pike despite being blind is also not surprising. Many hypogean fishes are almost identical to their surface, epigean forms except for the reduction of eyes and pigmentation. But Montalembert never mentioned depigmentation in his description. Furthermore, he says that some of the fish lacked one eye and, when that was the case, it was always the right eye. Troglomorphic fish generally show the same degree of reduction in both eyes. Finally, the location mentioned by Montalembert cannot be found today nor has any true blind cavefish ever been described from Europe (Romero 1999a).

Thus, all pre-Linnean reports of European troglomorphic hypogean fishes are unsupported by scientific evidence. Two of them (Besson 1569, Kircher 1665) do not even describe them with the typical troglomorphic features while the third (Montalembert 1748) is suspect. The most compelling evidence that the reports by these three authors are inaccurate is the fact that there is no confirmed report of troglomorphic fish in Europe to this date. All the European citations display the characteristic of their times: vague

---

[255] *b*. Colombières, France, 1530?; *d*. Orléans, France, 1573
[256] *b*. Geisa, Germany, 2 May 1602; *d*. Rome, 28 November 1680
[257] *b*. Angoulême, France, 16 July 1714; *d*. Paris, 29 March 1800

(Besson), uncritical and scholastic (Kircher), or fascinated with monstrosities (Montalembert). Among all pre-Linnean references, only Jie's (1541) has been substantiated by facts.

Figure 4.1. Athanasius Kircher (public domain image).

Figure 4.2. The only known color portrait of Marc René de Montalembert (public domain picture).

First Discoveries and Research (1805-1864)

This period is characterized by two factors: The first is the discovery and publication in the scientific literature of the first species of troglomorphic fishes. The second is the use of morphological studies aimed to describe in detail their different features. During this period all troglomorphic fishes were found either in the United States or Cuba. For the United States, the years preceding the American Civil War (1861-1865) were characterized by explorations by both individuals acting independently and state-sponsored surveys.

An early published post-Linnean record of a possible hypogean fish is that of Alexander von Humboldt (1769-1859) who described *Astroblepus cyclopus* (= *Pimelodus cyclopum*) which, according to him, could be found in the subterranean waters in the Andes of Quito at the basin of Río Esmeraldas, Ecuador (Humboldt 1805). However, he never saw that fish in any hypogean environment. Furthermore, this is actually an eyed, pigmented species with ample distribution in northwestern South America (Burgess 1989:448).

The first published, true record of a troglomorphic fish in the Western Hemisphere is probably that of James Flint who visited Indiana in 1820 and recorded that "a Colonel C – [sic] of Indiana told me that a settler in his neighbourhood [sic] digging a well, penetrated into a stream of water, and found blind fishes in it." He added as a footnote that "Since the above was written, a notice of blind fishes has appeared (if I mistake not) [sic] in the memoirs of the Wernerian Society of Edinburgh" (Flint 1822:256). I reviewed the entire collection of this journal, which consists of eight volumes published between 1808 and 1838, and did not find such reference to that blind fish.

The first published report of a troglomorphic fish sighted in its natural environment in the Western Hemisphere was, probably, by Robert Davidson (1808-1876). In October 1836 he visited Mammoth cave in Kentucky accompanied by Stephen Bishop, a self-educated black slave who guided visitors through the cave. He reported that "*white fish* [258] were found here without eyes" whose existence was already known by some of the locals (Davidson 1840:54-56). Others have pointed out that it was on 20 September 1838 that the Echo River in the Mammoth cave was discovered and in it, a blind fish (Soule 1982, Romero & Woodward 2005).

The first time that a troglomorphic fish was mentioned in the scientific literature was in a short note in the Proceedings of the Academy of Natural Sciences of Philadelphia (Anonymous 1842). There it was reported that a W. T. Craigie donated to the Academy at the 24 May 1842 meeting, a specimen of "a small white fish, also eyeless (presumed to belong to a subgenus of *Silurus*), both taken from a small stream called the 'River Styx' in the Mammoth Cave, Kentucky, about two and one-half miles from the entrance." At the collection of the Academy there are three specimens of *Amblyopsis spelaea* in alcohol, that appear linked to this donation. Two are catalogued as ANSP 7964 collected by W.T. Craige, and the other, ANSP 7964, collected by "Mrs. C.H.Graff, Messrs. Craige & Lambert." All three specimens were captured in Mammoth Cave, but no date is given.

The first scientific description of a troglomorphic fish was carried out by James Ellsworth De Kay[259] (Fig. 4.3). Son of an American captain stationed in Lisbon, De Kay studied medicine at Yale and Edinburgh, Scotland. He was one of the founders of the Academy of Medicine and an active member of both the New York Lyceum of Natural History and the American Association for the Advancement of Science. He was the first of the first-generation, American-born naturalists making important contributions in ichthyology (Hubbs 1964, Romero 2002a).

---

[258] Italics in the original.
[259] *b*. Lisbon, Portugal, 12 October 1792; *d*. Oyster Bay, Long Island, New York, 21 November 1851.

Figure 4.3. James Ellsworth DeKay (public domain drawing).

His work can be framed within the movement that started in the 1940's when several states of the United States inaugurated natural history surveys and published catalogues of the local faunas (Coe 1918). DeKay was selected to contribute a book on the zoology of New York state. In it, De Kay (1842:187-188) gave the first recognized scientific description of a troglomorphic fish, the northern cavefish, *Amblyopsis spelaea* (= *Amblyopsis spelaeus*) (Fig. 4.4). The description was not very detailed nor of a great quality. This could have been due to the fact that it was based on a poor specimen in the Cabinet of the Lyceum of Natural History of New York (Putnam 1872) or to the fact that DeKay was not a trained ichthyologist (Smallwood 1941:163-164).

Although the northern cavefish was captured in the River Styx, Mammoth Cave, Kentucky, DeKay included that and many other non-New York species because "It cannot therefore fail to be perceived that the Ichthyology of New-York will embrace a very large proportion of the Fishes of the United States" (DeKay 1842:iv). He actually placed this new species under a list of fishes under the subheading "(EXTRA-LIMITAL)" [sic]. The specimen, which originally belonged to the Cabinet of the Lyceum of Natural History of New York, cannot be located today and is presumed lost.

DeKay would never write on *Amblyopsis* again; however, this fish caught the attention of a number of anatomists who immediately began to study it. The first of those was Jeffries Wyman[260] (Fig. 4.5). He graduated in medicine from Harvard in 1837 and studied under George Cuvier and Richard Owen (Gifford 1967). Wyman helped to lay the foundations of comparative anatomy in the U.S. He was responsible -together with

---

[260] *b*. Chelmsford, Middlesex, Massachusetts, 11 August 1814; *d*. Bethlehem, New Hampshire, 4 September 1874.

Louis Agassiz (Fig. 4.6) and Asa Gray- for making Harvard the most important center for the study of natural history in the United States. He is little remembered today mostly because he was a very modest man who avoided generalizations.

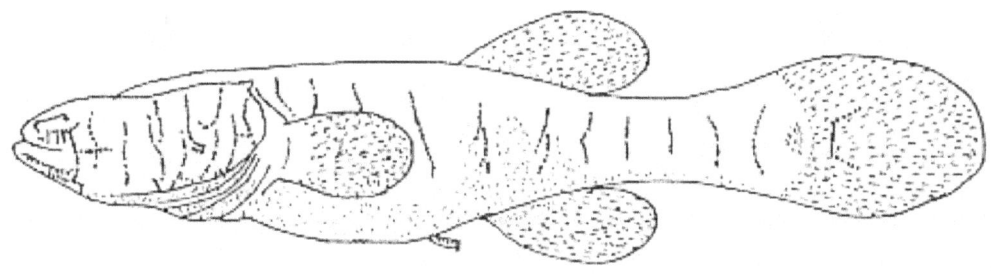

Figure 4.4. Amblyopsis spelaea from Romero & Bennis 1998.

Wyman, together with Agassiz and Gray, embraced the philosophical, or transcendental anatomy, i.e., the search for ideal patterns of structure in nature (Appel 1988). Thus, the discovery of a blind cavefish attracted Wyman's attention and he described *A. spelaea* in much more detail (Wyman 1843) to the point that some have mistakenly referred to him as the first person who described a blind cavefish (e.g., Gurnee 1992).

Figure 4.5. Jeffries Wyman (public domain drawing)

Figure 4.6. Louis Agassiz (public domain picture)

In his first paper on *A. spelaea*, Wyman reported that "On the most careful dissection no traces of eyes were found" (Wyman 1843). Later he wrote that "The optic lobes existed; according to the general rules of physiology these should not exist; as they bare strict relation to the sense of sight, which receives its nerve from them (...) Here the optic lobes were not so large as the allies fishes, but yet they were of good size, and nearly as large as the cerebral lobes" (Wyman 1851). He later re-examined three specimens and found imperfect eyes covered by areolar tissue and skin, and hence unable to see. He proposed that this imperfection of the eyes "might be owing to a want of stimulus through a series of generations" and that the organ of vision, however imperfect, "is constructed after the type of the eyes of other vertebrates" (Wyman 1854a). He also pointed out numerous structures without evident functions, organs that were of morphological rather than physiological value (Wyman 1954b). He produced the most detailed drawings of the internal anatomy of *A. spelaea* to that date (Wyman 1872).

For Wyman, *A. spelaea* was an excellent subject of study in his quest for evidence of a common plan underlying the differences caused by adaptive modifications. Although he quickly converted to evolutionism, he did not accept natural selection as its mechanism. He even regarded Agassiz as backwards for his refusal to accept evolution (Appel 1988). Yet, of all his papers on *A. spelaea* were devoid of evolutionary speculations, something that characterized most of his writings.

The next naturalist to study *A. spelaea* was August Otto Theodor Tellkampf[261]. He "Americanized" his name by adding either "A." or "G." as his middle initial. In 1838 he received a doctorate in Medicine from the University of Würzburg, Bavaria, and immigrated into the Unites States in 1839, where he practiced medicine in Cleveland and New York. He had some interests in cave fauna, having visited Mammoth cave in October 1842 (Tellkampf 1844), and described several species of invertebrates. He was a member of the Lyceum of Natural History of New York.

Tellkampf contributed detailed descriptions of *A. spelaea* and concluded that its eyes and those of blind cave crayfishes had become rudimentary as a result of disuse: "While it is true, in general, that all animals retain their essential form, and that no species passes over into another by transformation, we know that less material changes or form are produced by external influences such as changes in climate or food, lasting though many generations of the same species." But if the lack of light could only produce change within a species, then where was the original unmodified species? Tellkampf remained cautious on this point: the relationship of the blind fauna to unmodified species could not be settled until "such species, corresponding with them in all essential points, are found" (Tellkampf 1844, Romero 2002d).

European-based researchers also started to show interest in *A. splaea*. By January 1844 a specimen arrived to Edinburgh, Scotland, that was collected by townsman Gordon A. Thompson. William Thompson (1805-1852), then president of the Belfast Natural History and Philosophical Society, reported it as "perhaps the first examples of their respective species brought thence to Europe" (Thompson 1844).

Thus, during the first years of research into *A. spelaea*, this fish was considered to be largely a curiosity with varying significance. It was not until 1847 that the first published insight on the potential importance of troglomorphic fishes to biological research was published. Interestingly enough, it came from an unrepentant anti-evolutionist: Jean Louis Rodolphe Agassiz[262] (Fig. 4.6). During the 5 October 1847 meeting of the American Academy of Arts and Sciences, Agassiz proposed a "Plan for an investigation of the embryology, anatomy and effect of light on the blind-fish of the Mammoth Cave, *Amblyopsis spelaeus*" (Agassiz 1847). There he suggested that by studying this fish "there was an opportunity to settle, by actual experiment, the extent of physical influences in causing organized beings to assume their peculiar and distinctive characteristics in relation to the media in which they live." Agassiz, the creationist, was not thinking in terms of the environment influencing evolution, but rather the effects of the environment on development. He proposed to raise individuals of *A. spelaea* under different light conditions (dark, moderate, and intense light) and see if "there is an eye

---

[261] *b*. Heinde, Germany, 27 April 1812; *d*. Hannover, Germany, 7 September 1883.
[262] *b*. Motieren-Vuly, Switzerland, 28 May 1807; *d*. Cambridge, Massachusetts, 14 December 1873.

formed in the dark to ascertain when and how (the pigmentation) disappears, as it is entirely wanting in the full-grown individuals, and again notice the differences in this respect between specimens growing under the influence of light" (*op. cit.*).

He never carried out those experiments, yet he kept insisting on the importance of *A. spelaea* in biological research: "You asked me to give my opinion, respecting the primitive state of the eyeless animals of the Mammoth Cave. This is one of the most important questions to settle in natural history, and I have several years ago, proposed a plan for its investigation which, if well conducted would lead to as important results, for it might settle, once forever, the question, in what condition and where the animals now living on the earth, were first called into existence. But the investigation would involve such long and laborious researches, that I doubt it will ever be undertaken. (...) If physical circumstances ever modified organized beings, it should be easily ascertained here." Despite these difficulties he remained optimist: "Whoever would settle the question by direct experiment might be sure to earn the everlasting gratitude of men of science, and here is a great aim for the young American naturalist who would not shrink from the idea of devoting his life to the solution of one great question" (Agassiz 1851). These words may not have fallen into a vacuum since, as we will see below, several of his students showed a great deal of interest in cavefishes.

Despite his insight, there is no question that Agassiz maintained an anti-evolutionist view until the very end of his life. On the rudimentary organs, he wrote "The organ remains, not for the performance of a function, but with reference to a plan" (Agassiz 1859: 11). He considered *A. spelaea* to be an "aberrant cyprinodont (...) created under the circumstances in which they now live" (Agassiz 1851). And he never changed his mind: "Have fishes descended from a primitive type? So far am I from thinking this possible, that I do not believe there is a single specimen of fossil or living fish, whether marine or fresh-water, that has not been created with reference to a special intention and a definite aim" (Agassiz 1885, 1:392-393).

The next scientist who would make significant contributions to the study of hypogean fishes was Felipe Poey[263] (Romero 2007). He was a lawyer in Cuba, where tales of troglomorphic fishes were around since at least 1831. Poey secured the specimens and described two new species: *Lucifuga subterraneus* and *Lucifuga dentatus* (= *Stygicola dentatus*) (Poey 1858, 2:100) (Fig. 4.7.).

---

[263] *b*. La Habana, Cuba, 26 May 1799; *d*. La Habana, 28 January 1891.

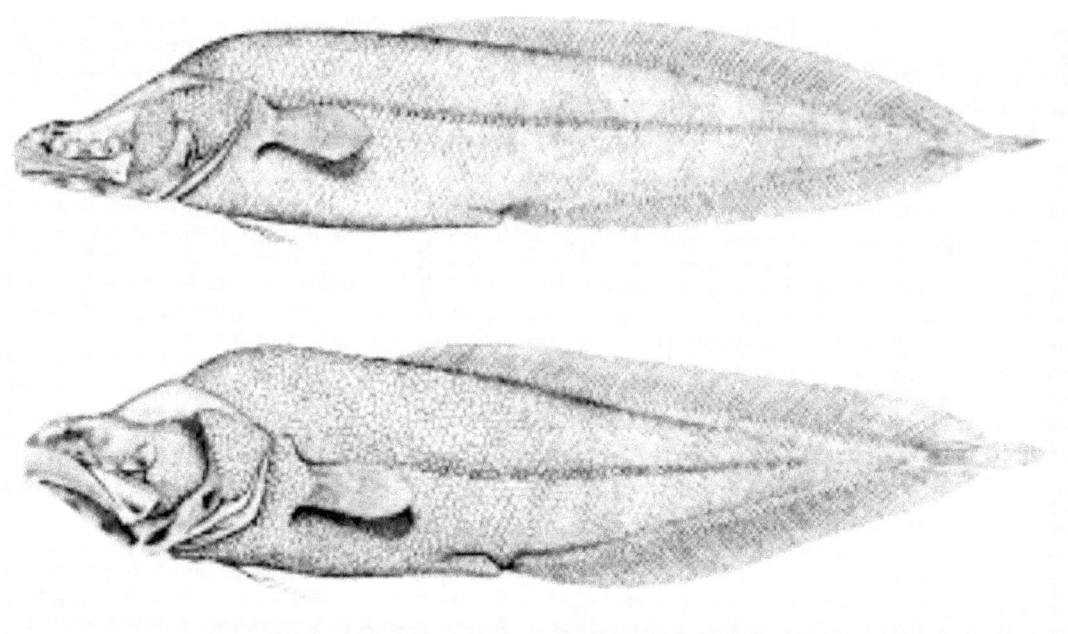

Figure 4.7. A 1902 illustration of *Lucifuga subterraneus* (top) and *Lucifuga dentata* by C.H. Kennedy. Courtesy of the Smithsonian Institution, NMNH, Division of Fishes.

The descriptions of these two species were extraordinary on several accounts. Not only they were very detailed and precise from the external and internal anatomical viewpoints, but there was also a wealth of information on their behavior, habitat, and history. Further, based on specimens he received from the United States, he described *A. spelaea*, as a point for meaningful comparisons. The descriptions of the Cuban species were far superior to those published on *A. spelaea* by his American and European counterparts. Poey also demonstrated full familiarity with the published literature on cavefishes up to that time.

It would be easy to label Poey as an "isolated genius" (*sensu* Beddall 1983) from social, scientific, and geographic viewpoints. He lived in a country which, at that time, was still a colony of Spain that lacked strong academic institutions. Poey himself was the founder of the first natural history museum in Havana and the first who taught zoology at the University of Havana. Yet he was not isolated from the scientific community. Poey was in contact with the most prominent contemporary researchers of cavefishes including Girard, Gill, Packard, and Putnam (see below). The latter visited Poey in Cuba in 1886 (Cockerell 1920). He received specimens of *A. spelaea* which he examined for his comparative studies adding information and making corrections to previous observations (Poey 1858, 2:104-106). He also provided American institutions such as the Museum of Comparative Zoology with specimens of the Cuban hypogean fish (Putnam 1872, Romero 2014).

The next species was described by Charles Frédéric Girard[264]. He had been brought to the United States by Louis Agassiz in 1847 and later worked at the Smithsonian Institution until 1860. While there, he was given some specimens collected by a J. E. Younglove "from a well near Bowling Green, Ky." He gave those specimens a new species status, *Typhlichthys subterraneus*, which he included in the family Amblyopsidae (Girard 1859). This new species seemed to have "characters apparently transitory" between *A. spelaea* and the other species of the amblyopsid family known at

---

[264] *b*. Mulhouse, France, 8 March 1822; *d*. Neulilly-sur-Seine, France, 29 March 1895.

that time: *Chologaster cornutus*, an epigean species. *A. spelaea* lacked eyes but had ventral fins; *T. subterraneus* lacked both eyes and ventral fins, while *C. cornutus* had eyes and lacked ventral fins.

The Civil War broke out while he was in Paris where he stayed and supported the confederate cause by sending drugs, medical supplies, and arms. After the war he stayed in France where he practiced medicine (Jackson & Kimler 1999). Apparently, he maintained some interest on cavefishes, because later in life he published a number of popular articles on the topic (e.g., Girard 1888).

With these species at hand, blindness and depigmentation among fishes had become synonymous with cave life, which, in turn, led to inaccurate assumptions and misleading information. That was the case with a new "species" of cavefish described by Edward Drinker Cope[265]. He published a paper on what he thought to be a new species and genus of troglomorphic fish, "*Gronias nigrilabris*" (Fig. 4.8), from Pennsylvania (Cope 1864). Although Cope did not present any evidence that such fish had been captured in the hypogean environment, he was quick to suggest that such fish "is supposed to issue from a subterranean stream, said to traverse the Silurian limestone in that part of the Lancaster county, and discharge into the Conestoga." Cope was known for his hasty conclusions and the superficiality of some of his work (Romero & Romero 1999). Further studies have shown that the specimens on which he based this description were specimens of *Ictalurus nebulosus* that had eyes present which were asymmetrically developed --probably as a result of a teratological condition. Unfortunately, his assertion on this fish continued to be repeated in the literature until recently (see Romero 1999b for full history of this misconception).

Thus, by the end of this period what we have are the descriptions of four species of troglomorphic fishes, two from the United States and two from Cuba as well as number of morphological descriptions about them. Absent from this period were evolutionary discussions. That would soon change.

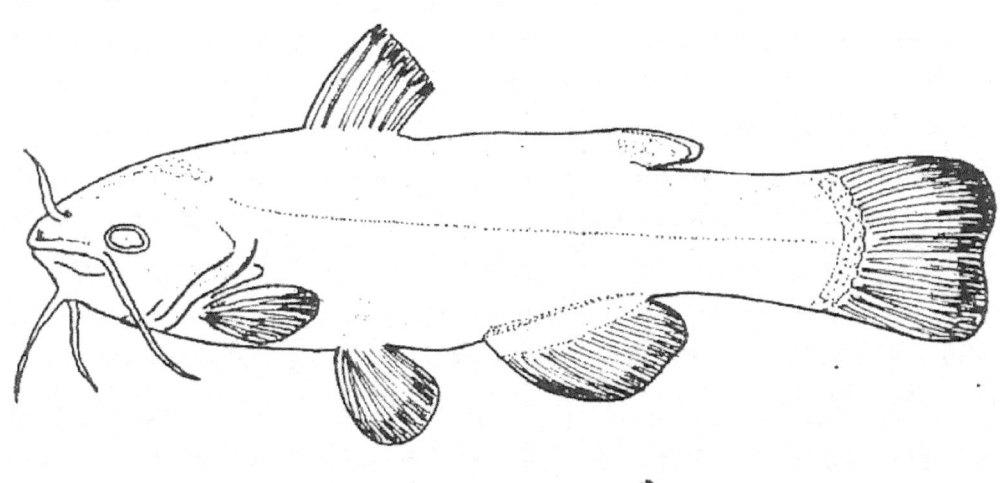

Figure 4.8. "Gronias nigrilabris". Drawing by Olga Mayayo.

American Neo-Lamarckism (1868-1919)

---

[265] *b*. Philadelphia, Pennsylvania, 28 July 1840; *d*. Philadelphia 12 April 1897.

The divide between this and the previous era was marked by two major historical events. The first was the American Civil War (1861-1865), which paralyzed scientific field activity in those areas where hypogean fish had been found. The second was the publication of *On the Origin of Species* (Darwin 1859). This period was also marked by the dominance, for the first time, of a generation of American-born ichthyologists. Together with a few European expatriates, they began cataloguing the fish fauna of North America at the same time that the United States began the systematic exploration of the West through efforts such as the Mexico Boundary and Pacific Railroad Surveys (Hubbs 1964). Although some new species were described during this time, the intellectual era of American neo-Lamarckism was characterized by the usage of cave fauna in general, and troglomorphic fishes in particular, to advance this view of evolution.

Charles Darwin[266] wrote on cave fauna based on the observations made of European cave animals by the Danish naturalist Jorgen C. Schiodte (1815-1884), and the observations of American cave fauna of James Dwight Dana (1813-1895) as explicit support for evolution. He noted that cave fauna was more closely related to the fauna of the surrounding regions than elsewhere, as is the case for fauna of other isolated habitats. Thus, he argued that the cave fauna descended from the fauna of their surrounding region "the colonists having been subsequently modified and better fitted to their new homes" (Darwin 1859: 403). At first Darwin considered the mechanisms of both, natural selection and disuse to explain troglomorphic features, i.e., enlargement of some sensory systems and appendages for the former, and blindness and depigmentation for the latter. To Darwin this suggested a "contest ... between selection enlarging and disuse alone reducing these organs" (*op. cit.*: 296). However, in the third edition of the *Origin* (Darwin 1861) he de-emphasized the importance of natural selection, eliminating the speculation of a "contest" between selection and disuse. This is important to keep in mind because, as we will see below, the members of the American neo-Lamarckian school closely followed Darwin in their interpretation of the mechanisms producing troglomorphic characters.

Darwin went back to much earlier explanations of evolutionary mechanisms and shielded the understanding of the evolution of cave fauna from more modern interpretations. This gave researchers of cave organisms arguments to support the use and disuse explanation. In many ways, Darwin maintained a modified version of the Great Being of Chain (Bowler 1984:55-59), which was championed by the Swiss naturalist Charles Bonnet (1720-1793) and the French Philosopher Jean-Baptiste Robinet (1735-1820). They were followed by Jean-Baptiste, Chevalier de Lamarck[267], again firmly planted into the Chain of Being tradition. Lamarck argued that organisms experience "needs" (*besoins*) which were brought about by the environment and triggered fluids (including electricity) which, when circulated in the body, enlarged or developed the appropriate organ. In "higher" animals, a crucial causal factor was the "inner consciousness" (*sentiment interieur*), which makes parts respond and develop. This resulted in the inheritance of derived characters. As we will see, even long after of the development of "regressive evolution") would continue to be strongly imbedded in the research of hypogean fishes and in cave fauna in general.

The neo-Lamarckian period of hypogean fish research was dominated by three figures: Packard, Putnam, and Eigenmann. Alpheus Spring Packard, Jr.[268] (Fig. 4.9) studied under Agassiz and was his assistant from 1862 to 1864 at the Museum of Comparative Zoology. He received his M.D. from the Maine Medical School at Bowdoin College in 1864. After breaking with Agassiz in the so-called "Salem secession" he went

---

[266] *b*. Shrewsbury, England,12 February 1809; *d*. Downe, Kent, England, 19 April, 1882.
[267] *b*. Bazentin-le-Petit, Picardy, France, 1 August 1744; *d*. Paris, France, 28 December 1829.
[268] *b*. Brunswick, Maine, 19 February 1839; *d*. Providence, Rhode Island, 14 February 1905.

on to become a leading figure of American neo-Lamarckism which he championed from his positions at the Boston Society of Natural History, the Peabody Academy of Sciences at Salem, Massachusetts, and at Brown University (Dexter 1965, Bocking 1988). It was Packard who coined the term "neo-Lamarckism" and called Lamarck "the real founder of organic evolution" (Packard 1901:v).

In 1867, together with Edward Sylvester Morse and Alpheus Hyatt --also a former Agassiz students-- Packard founded *The American Naturalist,* the journal that published the most articles on cave fauna during the nineteenth century. He first examined Mammoth Cave specimens after the Indianapolis meeting of the American Association for the Advancement of Science when many of the participants visited that cave. He published an account of its fauna the same year. The Mammoth cave fauna, including its fish, convinced him of their usefulness as a demonstration of evolution. "Our western naturalists will thoroughly explore the sinks and holes in the cave country of the western and middle states. The subject is one of the highest interests in a zoological point of view, and from the light it throws in the doctrine of evolution" (Packard 1871). In 1874 he was associated with the Kentucky Geological Survey, which intensified his interest on fauna of Mammoth Cave and other caverns in the Midwest on which he produced several publications (e.g., Packard 1888).

Packard, like Cope, thought that cave fauna was of very recent origin and that the loss of certain organs was compensated by the hypertrophy of others. Packard's views were not in opposition to those of Darwin himself, but to the neo-Darwinians like August Weismann (1834-1914) and Edward Ray Lankester[269]. Lankester, influenced by German biologist Felix Anton Dohrn[270], wrote that blindness among cave animals was due to a special kind of natural selection. He began with the assumption that some animals are, by chance, born with defective eyes. Occasionally a few animals, some of which have normal eyes and some defective eyes, fall or are swept into caves. In each generation, those that have good eyes were able to see the light and escape, and eventually only those that are blind remained in the cave (Lankester 1893). Lankester also believed that you can find organisms degenerating ontogenetically and phylogenetically. He defined "degeneration" as "a loss of organization making the descendent far simpler or lower in structure than its ancestor," a phenomenon that he found widespread. "Any new set of conditions occurring to an animal which render its food and safety very easily attained, seem to lead as a Rule of Degeneration" (Lankester 1880:33).

The other leading figure of this time was Frederic Ward Putnam[271] (Fig. 4.10). Like Packard, Putnam studied under Agassiz and was his assistant until a split from him during the "Salem Secession." He was much more of an ichthyologist than Packard was, occupying such positions as curator of ichthyology and/or vertebrates in many institutions including the Boston Society of Natural History, the Essex Institute, the Peabody Academy of Science, and Harvard's Museum of Comparative Zoology. Of all of these experiences, it was his position in 1874 as Assistant of the Kentucky Geological Survey that brought him in direct contact with hypogean fishes.

---

[269] *b*. London, England, 15 May 1847; *d*. London, 15 August 1929.
[270] *b*. Stettin (Szczecin), Pomerania, Prussia, 29 September 1840; *d*. Munich, Germany, 26 September 1909.
[271] *b*. Salem, Massachusetts, 16 April 1839; *d*. Cambridge, Massachusetts, 14 August 1915.

Figure 4.9. Alpheus Drinker Packard. Courtesy of the Ernst Mayr Library of the Museum of Comparative Zoology, Harvard University © President and Fellows of Harvard College.

Like Packard, Putnam first visited Mammoth cave to collect fishes in 1871 after the adjournment of the meeting of the American Association for the Advancement of Science. He returned in 1874 following an invitation from Nathaniel S. Schaler, another of Agassiz's students who, as director of the Kentucky State Geological Survey, appointed Putnam as special assistant to the Survey that year. He made the first large collection of Mammoth cavefishes from Kentucky (22 in total) in October 1874, some of which were displayed alive at the Essex Institute. Interestingly, the first public exhibition of a live amblyopsid took place not in the United States but in the Dublin Zoological Gardens, Ireland, sometime in 1870 (Baird 1872).

Although less known than Packard because he was less inclined to provide sweeping generalizations or engage in theoretical discussions, Putnam expressed a high degree of critical thinking. For example, he criticized Cope's interpretation that *A. spelaea* was able to survive in those waters because it had "the projecting under jaw and upward direction of the mouth renders it easy for the fish to feed at the surface of the water (...) This structure also probably explains the fact of its being the sole representative of the fishes of subterranean waters. No doubt many other forms were carried into the caverns since the waters first found their way there, but most of them were like those of our present rivers, deep waters or bottom feeders. Such fishes would starve in a cave river, where much of the food in carried to them on the surface of the stream..." He asked, then: where are the surface forms of the "surface feeders"? How come other surface feeders are not found in caves? He did not understand that Cope made this statement when he himself had described an alleged "cavefish" ("*Gronias nigrilabris*") from Pennsylvania which was a bottom feeder and that the blind cavefishes from Cuba were bottom feeders as well. He also said that studies of stomach contents in *A. spelaea* had shown that they ate essentially crayfish and other fishes. He asked that, if blindness was the direct result of darkness, as some contended, how come *Chologaster agassizzi* lives in caves but has eyes? (Putnam 1872).

Putnam (1872) also stated that "The blind fish of the Mammoth Cave has from its discovery been regarded with curiosity by all who have heard of its existence, while anatomists and physiologists have considered it as one of those singular animals whose special anatomy must be studied in order to understand correctly facts that have been demonstrated from other sources; and, in these days of the Darwinian and development theories, the little blind fish is called forth to give its testimony, pro or con." He viewed the amblyopsids as former marine and saltwater estuary fishes that were slowly trapped in that geographical area. He substantiated this hypothesis by pointing out that the eyed amblyopsid *C. cornutus*, for him an intermediate form between *A. spelaea* and *T. subterraneus*, and their eyed ancestors, was "now living in the ditches of the rice fields of South Carolina, under very similar conditions to those under which others of the family may have lived in long preceding geological time; and to prove that the development of the family was not brought about by the subterranean conditions under which some of the species now live, we have the ones with eyes living with the one without, and the South Carolina species to show that a subterranean life is not essential to the development of the singular characters which the family possess" He further supported this hypothesis by mentioning the that Cuban blind cavefishes belonged to the genera "with their nearest representative in the family a marine form, and with the whole family of cods and their allies, to which group they belong, essentially marine." Putnam (*op. cit.*) also described a new species of amblyopsid, *Chologaster agassizi*, as the fifth troglomorphic fish. By 1874, the year of his last publication on hypogean fishes, Putnam switched careers from ichthyology to archaeology after he was appointed as the Curator of the Peabody Museum of American Archaeology and Ethnology at Harvard University.

All discussions on the evolution of cave fauna in general and hypogean fish in particular, at the time of Packard and Putnam, took place under the influence of the so-called "Hyatt-Cope" position or school which was based on drawing parallels between embryology and phylogeny. Cope had little time for natural selection (e.g., Cope 1864) and seeing adaptation as part of the picture, fell for Lamarckism. Central to Cope's vision of the biological world was the way in which organisms fall in trends. "The method of evolution has apparently been one of successional increment or decrement of parts along definite lines" (Cope 1896: 24). This is what was late called orthogenesis, the view that evolution has a momentum of its own that carries organisms along certain tracts. Cope also proposed the "Law of the Unspecialized": when an organism has gone too far down a particular phylogenetic path (as cave species do, according to him), especially a particular adaptively specialized path, it can never pull itself out and evolve into an altogether new and fruitful form (Cope 1896: 172-174).

Figure 4.10. Frederick Ward Putnam. Courtesy of the Ernst Mayr Library of the Museum of Comparative Zoology, Harvard University © President and Fellows of Harvard College.

Aplheus Hyatt[272] (Fig. 4.11), whose more prominent position was that of director of the Boston Museum of Natural History, visited Mammoth Cave in September 1859, much earlier that his contemporary colleagues, and collected seven specimens of *T. subterraneus* that Putnam later studied (Bocking 1988). Hyatt's ideas can be summarized as follows:(1) just as the individual eventually declines into old age and senescence, so also the group declines into old age and senescence; (2) before extinction, there is degeneration of species; (3) change is a function of the speeding ("acceleration") or slowing ("retardation") of development which, in turn, is propelled by use and disuse (Hyatt's "Law of Acceleration," is a direct descendant of Agassiz's "Principle of Recapitulation"); (4) degeneration is therefore a virtually inevitable outcome of the evolutionary process, since any organism tends to collapse into old age, and it is a matter of time before this decay is delayed in development until adulthood (for a summary of Hyatt's ideas see Brooks 1909). Thus, Hyatt worked within a progressionist framework, a kind of Americanized *Naturphilosophie*. After all, the kind of evolutionism expressed by Agassiz's students and those under their influence (which was quite vast) had more to do with the transcendentalism of the German *Naturphilosophie* than to the natural theology of essential Darwinism. This is an idea that, as we will see, would be largely embraced by the European neo-Lamarckians of the Twentieth Century.

Figure 4.11. Alpheus Hyatt. Courtesy of the Ernst Mayr Library of the Museum of Comparative Zoology, Harvard University © President and Fellows of Harvard College.

---

[272] *b.* Washington, D.C., 5 April 1838; *d.* Cambridge, Massachusetts, 15 January 1902.

By the late Nineteenth Century, discussions on evolution in general languished primarily because it was obvious that no progress was being made, and "Morphology having been explored in its minutest corners, we turned elsewhere" (Bateson 1922). Thus, evolutionary morphology stalled and fell to the status of a second-rate science. Many morphologists went into the kind of science practiced in museums. Lankester, for example, moved from being a professor at Oxford to Director of the British Museum of Natural History (Ruse 1996:239).

This era of cave fauna research in the United States closed in the 1870's, within a conceptual framework of Blauplane, homologies, and parallelisms between embryonic and phylogenetic development while natural selection was at best ignored, at worst scorned.

Thus, cave fauna, with their bizarre adaptations to an extreme environment, were interpreted by neo-Lamarckians as providing excellent evidence for the effect of the environment upon the evolution of organism. Cave fauna could demonstrate the power of the physical environment as an evolutionary factor, capable of maintaining an organism's adaptation to its changing surroundings. Packard attributed more evolutionary importance to the direct effect of the environment than to the effect of changes in habits. Cave research therefore helped to establish that the changing environment could be responsible for the underlying progressive trends in evolution that Packard saw in his work in embryology.

Figure 4.12. Carl Eigenmann. Photo courtesy of the National Museum of Natural History, Smithsonian Institution.

By the 1870's Wyman, Packard, Putnam, and the rest stopped publishing on hypogean fishes. The topic seemed to have been conceptually exhausted. Nearly 20 years were to pass before anyone one else would show an interest in the topic.

The next relevant figure in the history of hypogean fish research was Carl H. Eigenmann[273] (Fig. 4.12). Influenced by David Starr Jordan (a student of Louis Agassiz's son, Alexander), he became a biologist particularly interested in fishes. His first experience with troglomorphic fishes took place in 1886 while at Indiana University he received a living blind fish taken from a well in Corydon, Indiana. The next year he married Rosa Smith, the first woman president of the scientific society Sigma-Xi. In 1889 they established residence in California where he was named Curator of the San Diego Natural History Society. There, his wife introduced him to the blind goby found among the rocks of the California rocks, *Typhlogobius californiensis* (see Eigenmann 1890 for a historical account of this encounter and how much impressed him). In 1891 he was appointed Professor of Zoology at Indiana University, a perfect location to study the blind vertebrates of the caves in the nearby areas. This motivated him to devote a substantial part of his scientific career to the study of blind vertebrates, most of them in caves (Romero 1986b).

Between 1887 and 1909, much of his work was devoted to comprehending the process of the loss of visual structures in cave vertebrates. In May 1896 he visited Dalton's Spring (actually a cave-stream) where he secured 20 specimens of *A. spelaea*. This became his favorite collecting locality. In 1898, Eigenmann published the description of a new species of cavefish with rudimentary eyes from south-western Missouri, *Amblyopsis rosae* (=*Typhlichthys rosae*), which he named after his wife.

Eigenmann extensively visited the caves of Indiana, Kentucky, Texas, and Missouri in search of specimens for his work and in March 1902, he visited Cuba for the first time to secure cave specimens for his comparative studies. He had been working on fish reproduction in the past and quickly recognized that these two Cuban hypogean fish species were viviparous.

Contrary to Mammoth cave, Eigenmann found the localities for the Cuban blind fish "monotonous" (Eigenmann 1903). From 1906 to 1907 he did many laboratory studies in Europe, mostly in Germany, with the Cuban specimens collected. During this period, he made plans to visit the Yucatán peninsula in Mexico because of persistent reports of a varied cave fauna in that part of the world. From 1898 to 1905 Eigenmann published at least 39 papers on cave vertebrates, dealing mostly with developmental and anatomical aspects of vision loss in fishes, salamanders, lizards, and mammals as an attempt to understand the underlying process of blindness among hypogean animals. All this research was summarized in his "Cave Vertebrates of North America" (Eigenmann 1909), a 341-page, 30-plate monograph.

Although a taxonomist by training, Eigenmann quickly sought answers to the issues of the origin and evolution of the cave fauna. Originally a neo-Lamarckian, Eigenmann thought that the reduction or disappearance of organs among cave animals was a case of convergent evolution, i.e., well-defined conditions of the subterranean environment facilitates the evolutionary changes leading to blindness and depigmentation in a variety of vertebrate and invertebrate organisms which come to inhabit them. He was quick to point out that lack of pigmentation had to be understood as the combination of genetically fixed and epigenetically (environmentally-influenced) characters; in other words, that although a character was genetically determined, its degree of development may vary under certain light conditions. For Eigenmann cave evolution was essentially "degenerative" and all successful cave-invaders were somehow "pre-adapted" to that medium. The origin of caves and of blind fauna were two distinct

---

[273] *b.* Flehingen, Beden, Germany, 9 March 1863; *d.* Chula Vista, California, 24 April 1927.

questions because of his experience with the blind fish found among the rocks of California's seacoast. He insisted on a strong link between ontogeny and phylogeny and his constant use of terms such as "phyletic degeneration" indicates that he held orthogenetic views. He followed Herbert Spencer's (1820-1903) idea that cave faunas were not the result of "accidents" but rather the product of an active process of colonization (Eigenmann 1909).

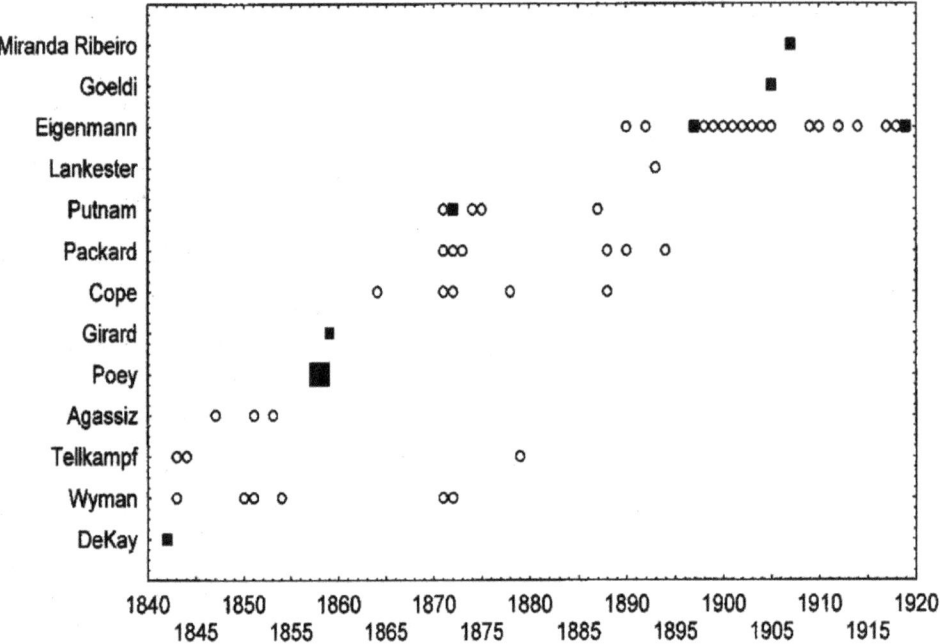

Figure 4.13. Time-line graph of major papers on hypogean fishes published between the first scientific description of a troglomorphic one, *Amblyopsis spelaea* in 1842, to right before the beginning of the typological era. Squares represent papers describing new and valid species. The larger square for Poey represents a single paper describing two species. Open circles represent papers dealing with anatomy and evolution. Notice the divide created by the American Civil War (1861-1865) which interrupted virtually all research efforts.

Eigenmann's last contributions in the field were the descriptions of a new species of blind fish, *Trogloglanis pattersoni*, from the artesian waters of San Antonio, Texas (Eigenmann 1919). This description was based on a specimen collected by G.W. Brackenridge of San Antonio, who gave the fish to J.T. Patterson, a professor at the University of Texas, who in turn sent it to Eigenmann at Indiana University.

Two more troglomorphic species were discovered during this period: *Phreatobius cisternarum* by Goeldi (1905) and *Pimelodella kronei* (*Typhlobagrus kronei*) by Miranda-Ribeiro (1907), both from Brazil. Probably due to geography, the discovery of these species never made any important impact on general discussions of biological concepts.

Thus, from 1866 on, the neo-Lamarckian school flourished and found among American naturalists interested in cave fauna its strongest supporters (Figs. 4.12 and 4.13). Although Barr (1966) called the beginning of this era the "Neolamarckian revolt against Darwin's theory of evolution by means of natural selection," the fact of the matter is that those neo-Lamarckians were very much in line with what Darwin himself said, particularly in later editions of his *Origin of Species*.

Dominance of Typological Thinking (1921-1940)

This period is characterized by three trends: (1) The incremental discovery of new species, all of them outside the of United States, which gave a broader perspective not only geographically but also taxonomically to the diversity of hypogean fishes; most of these new species belonged to fish families with no previous hypogean representative. (2) The absence of researchers who devoted considerable part of their research to this topic as Eigenmann, Wyman, Packard, and Putnam had done before. During this period all authors of hypogean fish papers spent very little time on the subject, mostly because they were studying the fish fauna of a particular area where troglomorphic fish were found. The only possible exceptions to this were Jacques Pellegrin[274] and Luisa Gianferrari[275] who published a handful of papers on the subject. Yet, those papers were the product of their interest on the fish fauna of a particular geographical region more than the study of hypogean fishes *per se*. (3) Most of the new species/populations discovered were given a new generic status based almost exclusively on the fact that they lacked eyes and pigmentation.

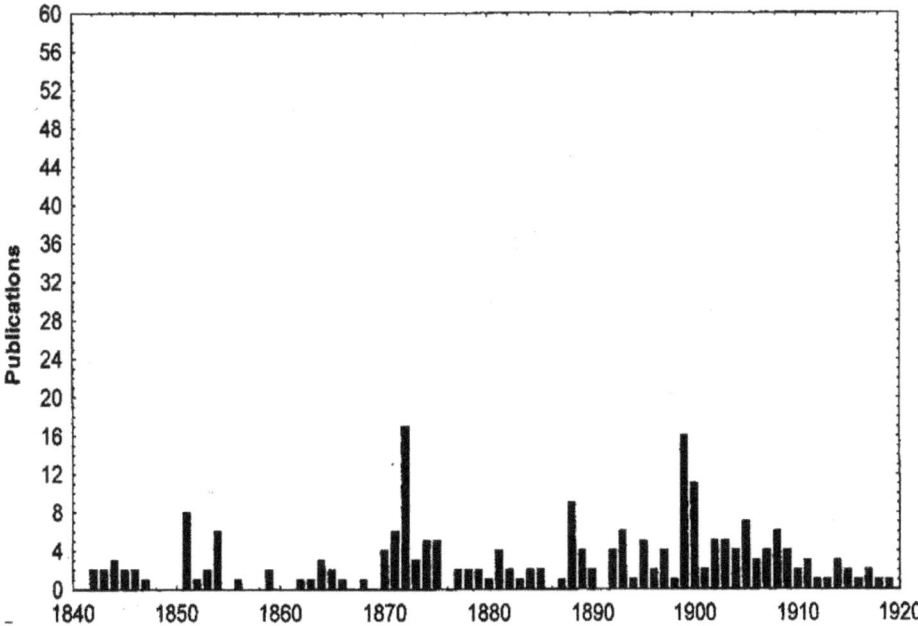

Figure 4.14. Number of publications on hypogean fishes between 1842 and 1919. The "pulse" for the 1870's represents the high point of the neo-Lamarckian school's interest in these animals. The one for the early 1900's is almost entirely due to the intense publishing activity by Carl H. Eigenmann.

Of the 12 new species/populations described during this period (Appendix 1, Fig. 4.15), six were from Africa, three from Mexico, two from Japan and one from the Trinidad, W.I. Most of the new discoveries in Africa were propelled by the presence of European colonial powers in that part of the world. The ones in Mexico were carried out by American-born researchers venturing for the first time outside the United States and the ones from Japan and Trinidad were described by British researchers residing in the

---

[274] *b*. Paris, France, 12 June 1873; *d*. Paris, 12 August 1944.
[275] *b*. 1890; *d*. 1977

United Kingdom. The specimens from Africa were collected by colonial authorities in those countries. The ones in Mexico were described by U.S. ichthyologists doing field work in that country and the two from Japan were re-descriptions based on specimens collected by a Japanese researcher. The one in Trinidad was collected by a local natural historian.

The first description of a new species of troglomorphic fish for this period epitomizes the trend set by European researchers when dealing with African hypogean species. When King Leopold II of Belgium foresaw the importance of scientific research Africa, he instructed voyagers and explorers to collect natural history information about Congo. Not surprisingly, many Belgian officials collected and brought to Europe unknown zoological specimens that had never been studied by science. One of those specimens fell in the hands of George Albert Boulenger[276] (Fig. 4.16). He worked from 1880 until his retirement in 1920 at the British Museum. By then, Boulenger had published 877 papers totaling more than 5,000 pages, as well as 19 monographs on fishes, amphibians, and reptiles in which he described 1,096 species of fish, 556 species of amphibians, and 872 species of reptiles. He was considered the world's expert on African freshwater fishes despite the fact he had never been to Africa. Boulenger's only previous stint as researcher of cave animals had been a short note on the eye development of the cave salamander *Proteus* (Boulenger 1893).

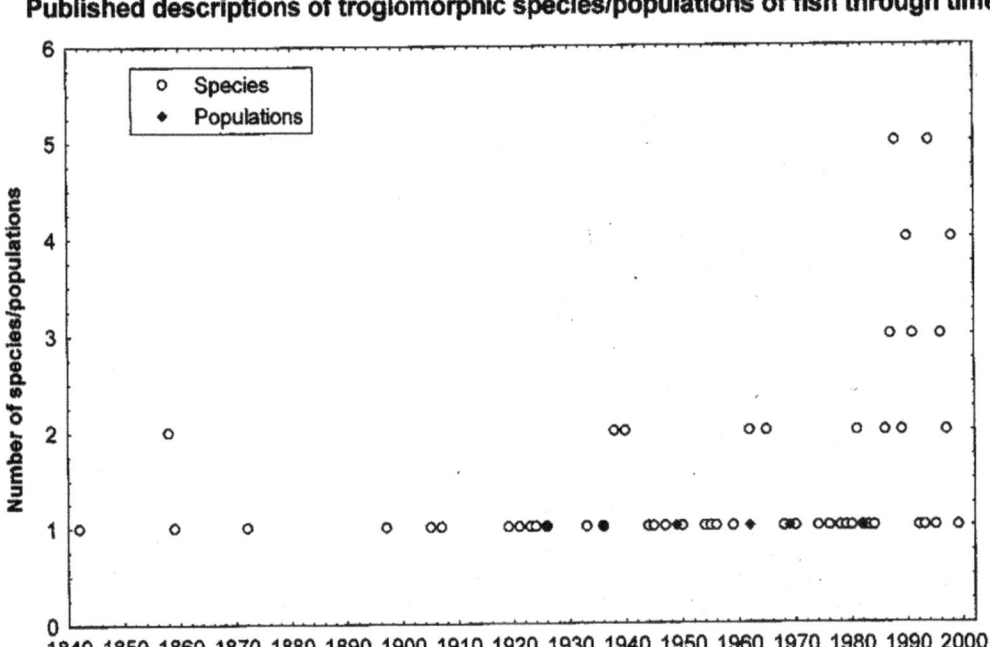

Figure 4.15. Published scientific descriptions of troglomorphic species/populations of troglomorphic fishes. There has been a steady number of new descriptions since the 1920's when the typological started with description of new species, mostly from Africa. The increase number in the late 1980's and Throughout the 1990's and beyond is due to new discoveries in Asia, particularly in China (Romero *et al.* 2009).

---

[276] *b.* Brussels, Belgium, 19 October 1858; *d.* Saint Malo, France, 23 November 1937.

In 1921, Boulenger received a strange fish specimen from Congo. It was eyeless and lacked pigmentation. He was quick to recognize its uniqueness since it did not seem closely related to any extant African epigean species. He named it *Caecobarbus geertsii*, from *caeco* = blind, *barbus* = barb, and *geertsii*, honoring a mysterious person, M. Geerts, who provided him with the specimen (Boulenger 1922). This species of fish may have been originally collected in 1915 by M. Delporte, probably the first European to see this fish. Soon this species became a celebrity in the world of cavefish research. Being easy to transport alive, many European scientists started to work on it. It became so famous that in May 1951, it was exhibited at the New York Aquarium (Romero & Benz 2000).

The same can be said of other African hypogean fishes. Three were discovered in Somalia, collected by local colonial Italian officials, but then described by ichthyologists in museums of natural history of Italy (Gianferrari 1923, Vinciguerra 1924, Di Caporiacco 1926). Another was collected by a French explorer in Madagascar and sent to France for identification (Petit 1933) and the last from Africa was collected in South Africa by a local physician and sent to the British Museum of Natural History (Trewavas 1936).

Figure 4.16. George Albert Boulenger. Photo courtesy of the National Museum of Natural History, Smithsonian Institution

The case of the hypogean fish from Trinidad had a different twist. It was collected by a local and well-known naturalist, Frederick William Urich[277], who sent the specimens to London for identification. When received by the British Museum of Natural History in July 1924, they were given to John Richardson Norman[278], who had published a few papers on the fishes of the nearby island of Tobago. Besides certain reduction in the eye size, this fish was extremely similar to *Rhamdia quelen*, a well-known epigean fish common to northeastern South America and Trinidad. Fearing that the specimen could represent an accident of nature rather than a true new fish species, Norman requested two more specimens and Urich complied. He later published its report and named the fish *Caecorhamdia urichi*, (*caeco* = blind; *rhamdia* = the genus of a catfish to which this cavefish seemed most related to; *urichi* = honoring Urich, the collector) (Norman 1926). Since then, this fish species has consistently appeared in the lists of blind cavefishes of the world although today they are considered individuals of the common catfish *R. quelen*, a species of nocturnal habits showing different degrees of (but never complete) blindness and depigmentation (Romero & Creswell 2000, Romero *et al.* 2002). This would be the first report of a cavefish population of an epigean species with certain degrees of troglomorphic features (Fig. 4.17).

Figure 4.17. Probably the first picture taken at the entrance of the Cumaca Cave in Trinidad, W.I., in 1911. Standing to the right is former U.S. President Theodore Roosevelt. It is believed that one of the people accompanying him was local and well-known naturalist, Frederick William Urich, who sent the specimens of the cavefish to London which was late described as *Caecorhamdia urichi*, today considered a cave population of *Rhambdia quelen* (see Romero *et al.* 2002).

---

[277] *b.* 1870; *d.* Port-of-Spain, Trinidad, 22 July 1937.
[278] *b.* 1899; *d.* 1944.

As previous examples demonstrate, during this period almost every cavefish population that was discovered, regardless of how close it looked to its epigean ancestor, had a new genus and species erected for it based on its taxonomic characters of blindness and depigmentation. This was the typological approach prevalent at that time, a derivative of the essentialist philosophy that has dominated Western thinking that was not replaced until after The Modern (Evolutionary) Synthesis of the late 1930's and early 1940's. As we will see later, such thinking has been retained to certain extent in some quarters of hypogean fish research.

Besides the re-descriptions by Charles Tate Regan[279] of two Japanese hypogean fishes (Regan 1940), all the other discoveries during this period took place in Mexico by American-born researchers. Since they operated intellectually and geographically independently from their European counterparts, and because their discoveries opened new doors to hypogean fish research, they deserve to be treated separately despite an overlap of a few years with the efforts of their colleagues elsewhere.

American Renaissance (1936-1960)

The resurgence of hypogean fish research took place by the time the typological period started to wind down. Although the discovery of new species continued in Asia, Africa, and Australia, it was the work of American-born scientists doing field studies outside the United States and experimental work in American institutions that introduced a renewed interest in the subject, particularly because aspects other than taxonomy and morphology were investigated. This period is characterized by more comprehensive studies that included ecology, physiology, and behavior. This was, in part, due to the fact that the species more suitable for this kind of studies was discovered at the very beginning of this era.

In 1936, Salvador Coronado, a young Mexican in charge of the Fish Culture Station at Almoloya del Rio near Mexico City, sent 75 fish collected at La Cueva Chica (the little cave) to C. Basil Jordan, a fish dealer from Dallas, Texas. Coronado thought those animals were particularly interesting: pinkish and blind. Jordan was impressed because all of the fish arrived to Texas alive, something particularly interesting for someone whose business largely depended upon the ability for a fish to survive transportation.

Not being able to determine the species, Jordan sent (at least some of) the fish to William Thornton Innes (1874-1969), a well-known aquarist and aquarium writer. Strongly suspecting that he had a new species in his hands, Innes remitted on November 1936 the specimens together with Jordan's notes to Carl Leavitt Hubbs[280] (Fig. 4.18). Hubbs was, by that time, one of the most respected American ichthyologists who had been doing field work in Mesoamerica and whose knowledge on freshwater fishes taxonomy was considered to be virtually encyclopedic.

It took Hubbs only a few weeks upon receiving the material to publish a joint article with Innes describing this fish from the characid family as a new genus and species: *Anoptichthys* (eyeless fish) *jordani* (honoring Jordan). As Hubbs himself put it, this was the "most surprising, by far subterranean fish belonging to the family Characidae, of which no blind representative has ever been seen before" (Hubbs & Innes 1936).

Hubbs' surprise was manyfold. First, it was most unusual to capture so many individuals of a cavefish species in a single locality; the amblyopsids, by that time the best-known cavefish family, were not so abundant. Second, the fact that all 75 individuals had arrived in the U.S. alive and were easily kept in captivity said something about the

---

[279] *b.* Sherborne, Dorset, England, 1 February 1887; *d.* 12 January 1943.
[280] *b.* Williams, Arizona, 18 October 1894; *d.* La Jolla, California, 30 June 1979.

potential of this species as a laboratory research subject (Innes 1937). Third, this cave characin did not grossly display the hyperdeveloped sensory organs quite common among other cave animals. As a matter of fact, it only differed from its likely ancestor, *Astyanax fasciatus*, in lacking eyes and pigmentation. Yet, in line with the typological thinking of the times, it was given a new generic and specific status.

Figure 4.18. Carl Leavitt Hubbs at his laboratory at the Scripps Institution of Oceanography. This picture was taken in 1945, shortly after he had made his major contributions to hypogean fish research (photo courtesy of the Scripps Institution of Oceanography Library).

This Mexican cavefish was so amusing that in both Mexico and the United States a great deal of interest arose. A group of the Mexican *Escuela Nacional de Ciencias Biológicas* that included, among others by José Álvarez[281] and Bibiano Fernández Osorio y Tafall[282], began the exploration of the whole cave system for the area which yielded about 30 localities containing this fish. As a matter of fact, two new populations, in addition to the La Cueva Chica one, were given new specific names: *Anoptichthys antrobius* for the La Cueva El Pachón population (Alvarez 1946) and *A. hubbsi* for the La Cueva de los Sabinos population (Alvarez 1947). As more cave populations were discovered, it became evident that this flow of specific names was leading nowhere and that all the troglomorphic fish should be considered as a single species.

Hubbs' hypogean cave research also comprised the Yucatán peninsula. Although there had been rumors of blind fishes from the cenotes (sink holes) of Yucatán (Girard 1888, Eigenmann 1909), it was not until the 1930's that organized scientific expeditions were carried out to explore this unique subterranean habitat. There, Hubbs found a number of epigean fish living in the cenotes, some of which showed certain degree of eye reduction and depigmentation and erected new subspecies status for them. But, more importantly, he also described two new troglomorphic species: *Ogilbia pearsei* (=*Typhlias*

---

[281] *b*. Spain, 1903; *d*. Mexico, 1986.
[282] *b*. Pontevedra, Spain, 3 December 1902; *d*. Ciudad de México, Mexico, 13 August 1990.

pearsei) and *Ophisternon infernale* (=*Pluto infernalis*) in a monographic publication in which he also reviewed all known information on cavefishes to that date (Hubbs 1938).

But it was his discovery of the cave characid the one that revitalized hypogean fish research. Hubbs and Innes (1936) of the troglomorphic population of *Astyanax fasciatus* (=*Anoptichthys jordani*) paper immediately attracted the attention of non-taxonomists. One of them was Myron Gordon (1899-1959) an internationally known fish geneticist who had been associated with the New York Aquarium since 1938. Since the late 1920's he had a particular interest on the inheritance of normal and abnormal pigment cell growth, especially melanoma tissues, in fishes. Thus, in 1939 he visited the cave in which the fish had been discovered in Mexico. There he collected more individuals which he took to New York. This fish elicited the interest of his New York colleagues, among them Edward Bellamy Gresser (1898-1951), a practicing physician and a professor of ophthalmology at New York University. He used the laboratories of the New York Aquarium to study the histology of the cave *Astyanax* and published an article about it (Gresser & Breder 1940). However, it was the co-author of that article who actually showed an intense interest on that species.

Charles Marcus Breder Jr.[283] (Fig. 4.19), was the director of the New York Aquarium in 1939. After learning about the newly described fish and hearing Gordon"s account of his recent trip to Mexico, Breder took the initiative of organizing and leading an expedition to Mexico to perform field studies, obtain enough ecological information for a cave habitat display for the Aquarium, to shoot a documentary to be presented at the 1941 annual meeting of the New York Zoological Society, and, most importantly, to bring back enough fish to conduct extensive laboratory research (Breder 1942, Romero 1984, 1986a).

In January 1940, Breder held a meeting with other scientists where the whole expedition, known as "The Aquarium Cave Expedition to Mexico," was organized. By 11 March 1940, the group was already in Ciudad Vallés, nearby La Cueva Chica, the fish type locality. Despite the fact that most members of the expedition suffered from "tropical fevers" (possibly histoplasmosis) after the trip, the expedition was a complete success (Bridges 1954).

Unlike many hypogean fish researchers before him, Breder was not particularly interested in systematics. Yet, in 1944 he was offered the chair of the Department of Fishes of the American Museum of Natural History in 1944 and retired in 1965 (Atz 1986). Between 1940 and 1954 Breder alone (sometimes coauthoring with Gresser or Priscilla Rasquin, an Ichthyologist at the American Museum of Natural History) published 17 papers (168 printed pages of information) on which *Astyanax* was the major subject of research. Most of the research concentrated on behavior, particularly responses to light and chemicals. But he also published on the fish sensory organs, metabolism, genetics, ecology and evolution. Because the cave and the surface forms interbreed to produce fertile hybrids, he was the first who strongly suspected that the blind depigmented form was nothing but a remarkably locally-adapted (ecotone) population of the surface species *Astyanax fasciatus* well before advanced genetic techniques proved it. Many of Breder's contributions are still cited and several of his associates and students embarked in researching this cavefish. There is no question that Breder was the dominant figure in hypogean fish research in the 1940's and 50's, and his papers are still frequently cited.

In summary, this period was characterized by the non-taxonomical research of Breder and his associates first at the New York Aquarium, and later at the American Museum of Natural History. Yet, not until the late 1950's did a number of European researchers also become interested in the genetics and evolution of these fishes which

---

[283] *b.* Jersey City, New Jersey, 25 June 1897; *d.* Englewood, Florida, 28 October 1983.

would, ultimately, signal the rebirth of neo-Lamarckian ideas about the evolution of cave fauna in general and hypogean fishes in particular.

Figure 4.19. Charles Breder (right) and Salvador Coronado during their first expedition to "La Cueca Chica" in Mexico in 1940. Photograph by Mary Edwards, courtesy of the Wildlife Conservation Society.

Times of Philosophical Conflict (1960-1990)

From 1960 on, not only the number of papers on hypogean fishes increased (Fig. 4.20), but also two major schools of thought started to developed very strongly. One was the school of orthogenesis which, essentially, utilized cave organisms as prime examples for "regressive evolution," and the neo-Darwinian school, which tried to use the new concepts and approaches provided by The Modern Synthesis in order to explain the evolution of those very same organisms.

Let us begin with the orthogentic school. The notion of progress has been deeply embedded in evolutionary thought since pre-Darwinian times, and it was usually referred to as a matter of levels of complexity (Nisbet 1979, Ruse 1996). For Karl Ernst von Baer (1792-1876), for example, "The more homogeneous the whole mass of the body is, so much the lower is the grade of its development. Then grade is higher when nerves and muscles, blood and cell-substance, are sharply distinguished" (von Baer 1828: 207). We can also see the concept of "progress" also being used in the social sciences, particularly by those who held Marxist ideas (e.g., Demoor *et al.* 1897, Guillaume 1971).

How was this idea embraced by students of cave faunas? To answer, we need to look back to the American neo-Lamarckian school.

As we mentioned earlier, Cope introduced the idea that evolution was directed by trends and that "The method of evolution has apparently been one of successional increment or decrement of parts along definite lines" (Cope 1896:143). He, like other neo-Lamarckians, had little regard for natural selection; thus, these "natural" trends explained how evolution worked overall. That was the foundation of "orthogenesis," i.e., the idea that there is a directionality in evolution. When an organism went too far in a particular direction or adaptively specialized path, like cave organisms, it could not reverse into a new form (Cope 1896: 172-174).

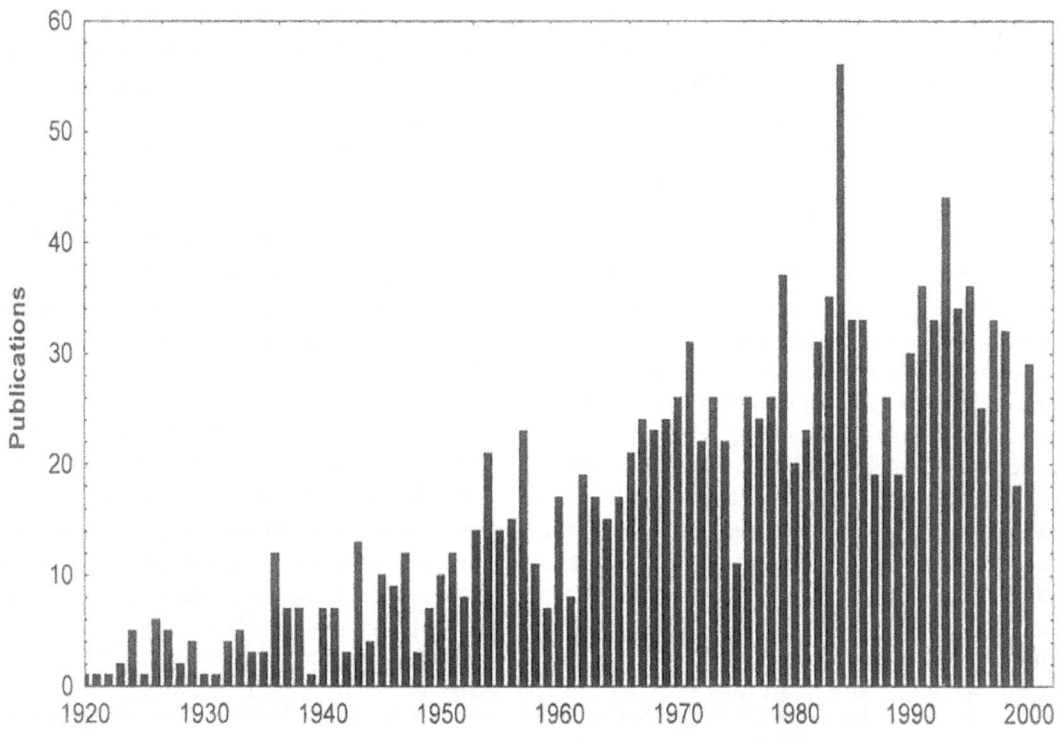

Figure 4.20. Number of publications on hypogean fishes from 1920 to 2000. The robustness of the publication numbers from 1960's on, is due in good part to the research activity by the Hamburg School, whose interest was prompted by the pioneering research by Curt Kosswig in genetics.

With the re-discovery of Mendel's laws and under the influence of The Modern Synthesis, orthogenetic ideas advocated by neo-Lamarckians were dismissed for a long time. That is one of the reasons we see very little, if any, evolutionary discussions from the end of the neo-Lamarckian school of thought until the 1960's when the books by Vandel (1964) and Thinès (1969), made their mark. For Albert Vandel (1894-1980), the explanation of the evolution of cavernicoles was neither neo-Lamarckism (for which no empirical evidence could be found) nor neo-Darwinism (for which French researchers had little sympathy at that time). The explanation was "organiscism" which basically meant that "all phyletic lines pass through several successive stages: the stage of the creation (...) The stage of expansion and diversification, and finally the stage of specialization and senescence (...) The first stage of this cycle is a manifestation of progressive evolution and the last of regressive or gerontocratic evolution. The evolution of cavernicoles provides a particularly good example of regressive evolution which can occur only in the final stages of the evolution of animal lines" (Vandel 1964:463). Curiously, these authors never acknowledged that these ideas were neo-Lamarckian in nature and were born in the United States, but rather preferred to trace themselves back to French authors such as Henri Bergson (1859-1941), René Gabriel Jeannel (1879-1965), or Pierre Teilhard de Chardin (1881-1955).

It was a German researcher who popularized the whole notion of regressive evolution among cave animals in general and cavefishes in particular: Curt Koswigg[284]. Since the 1930's he had been working on cavefishes and showed an interest in evolutionary issues, but it was a paper he wrote while in exile in Turkey that set the basis for his studies on "regressive" evolution (Kosswig 1949; for a more update and accesible version see Kosswig 1960). Basically, he used the results of his own Mendelian studies to explain the rudimentation or loss of structures among cave animals. Without being a neo-Lamarckian, the fact that he used modern (at that time) genetic techniques to make his point gave respectability to his work. On the other hand, there is an underlying notion in his work about directionality in evolution which also sets him away from contemporary neo-Darwinian thinking. As a matter of fact, the notion of "regressive evolution" is still the central theme of the most productive group of hypogean fish researchers at the Zoological Museum and Zoological Institute at the University of Hamburg. This idea of "regressive evolution" has been the source of monographic texts and, as Thinès (1969) book's title, the central idea for describing (although not necessarily explaining) the common features among troglomorphic organisms.

The major problem with "regressive evolution" is that it gives the impression that the evolution of cave organisms is the result of non-Darwinian mechanisms and, thus, set apart from mainstream evolutionary thought as evidenced by the attempts to explain phenotypic simplification by means of neutral mutations (e.g., Wilkens 1988). However, this view of cave life had three difficulties: (1) shared characteristics among troglomorphic organisms that are phyologenetically unrelated, seem to be an excellent example of convergent evolution, and thus, easily explained by using neo-Darwinian mechanisms; (2) the fact that the gain and loss of characters in evolution is a common phenomenom. For example, nobody has had the anti-anthropocentric temerity to label *Homo sapiens* as the result of "regressive evolution" because that species lost the tail, most of the pelage, and the ability to graciously jump from one tree to another from its ancestors; (3) if this evolution is regressive, then the question of "regressive to where?" remains unanswered (for further discussion see Romero 1985).

From the neo-Darwinian side, this period begins with the work of Thomas Layman Poulson (*b*. 1934), who, in the early 1960's started to apply new ecological concepts to the study of cave adaptation for amblyopsids, a fish family whose study had been largely neglected since the times of Eigenmann. Poulson took a closer look at some of the adaptations to this environment such as metabolic rate, life history, sensory morphology and foraging behavior, and tried to draw a step-by-step description of that process from a holistic viewpoint by comparing different species within the same family. This approach was in tune with the neo-Darwinian view of evolution in which adaptationism and gradualism play a major descriptive role (Poulson 1963, 1964). However, he would later take a view that combined both neutralism and selectionism when trying to explain the evolution of troglomorphic features (Poulson 1985).
Another influential American biospeleologist of this period is David Clair Culver (*b*. 1944). Although he worked on cave crustaceans and not on cavefishes, his ideas were, and continue to be, influential in the field. He also started to take a look of the phenomenon of troglomorphy from an ecological viewpoint and apply island

---

[284] *b*. Berlin, Germany, 30 October 1903; *d*. Hamburg, Germany, 29 March 1982.

biogeography theory to a system that, ecologically speaking, was largely isolated (Culver 1976). Although he initially took a neutralist stance (Culver & Fong 1986), his more recent studies convinced him that selection does play a major in the convergence of troglomorphic features (e.g., Culver *et al*. 1995).

One of the most influential papers for that this time of conflict did not come from hypogean fish specialists but rather from people trying to apply molecular techniques to understanding genetic variation. Such was the case in the study of evolutionary genetics of the cave *Astyanax fasciatus* (Avise & Selander 1972). Largely based on the Master's thesis by John Charles Avise (*b*. 1948) directed by his adviser Robert Keith Selander[285], this work demonstrated that highly divergent morphologies (epigean vs troglomorphic) can be achieved with very little genetic differentiation. This, in turn, questioned the validity of the typological thinking of assigning new species, and especially new generic status, to any troglomorphic population based only on blindness and depigmentation.

This new genetic information signaled the possibility that major phenotypic modifications via phenotypic plasticity could be achieved by relatively minor genetic changes, as long as those changes affected developmental genes (Romero & Green 2005). This notion has been supported by recent studies and is invigorating the study of cave fishes as models in developmental evolutionary biology (e.g., Yamamoto & Jeffery 2000).

Conclusions

The history of hypogean fish research has been marked by rather precise intellectual contexts. Each of the periods described above has been heavily influenced by particular lines of thoughts that were very popular at that time. It seems that most researchers have used hypogean fishes as examples to justify their own philosophical stances rather than to use hypogean fishes for the sake of scientific discovery. The use of these fishes, first to support a neo-Lamarckian view of evolution, then to practice a typological view of classification; and, third, to demonstrate that they must represent some sort of stage in alleged evolutionary trends, have marred what may have been an excellent topic for more mainstream evolutionary studies. This latter view has, probably, been the most perverse of all given the jargon used by most researchers ("regressive", "degenerative", etc.) to describe their evolutionary history.

But we should not fault them for that. After all, the idea of progress has been the backdrop to evolutionary thinking and not until recently have we started to realize that the notion of evolutionary progress is a fallacy (Maynard Smith & Szathmary 1995:4). In my view, it is a cultural construct that fits perfectly with anthropocentric ideas that have dominated humans' notions of themselves since the dawn of history. This is an idea that has been central in the battles science has fought to enlighten humankind, from geocentrism to creationism. The idea that, somehow, humans represent the zenith of evolution and that anything else is inferior, anything not aiming there is "regressive" or "degenerative" (for a full discussion on this issue see Romero 2006).

However, if we see troglomorphic fish, and troglomorphic fauna in general for that matter, from a less philosophically charged perspective, we will find that they are part of nature and their characters that amaze us from them, mostly their blindness and depigmentation, are part of common biological phenomena found everywhere in nature. Further, that the loss of certain phenotypic characters can take place thanks to relatively minor genetic changes.

Also, the increasing interest on non-troglomorphic hypogean fishes should shed much light on our understanding of evolution as a dynamic, non-stopping process. The finding of genetically very undistinguishable epigean and hypogean populations as well

---

[285] *b*. 21 July 1927, *d*. 14 June 2015.

as hypogean species that seem to be in *status nascendi* is also important (e.g., Borowsky & Mertz 2001, Burr *et al.* 2001). It tells us that the separation between both epigean and troglomorphic populations is in many cases very tenuous. To apply typological thinking by creating artificial divides to create species just for the sake of it, will mask the reality of the fluidity in nature, that nothing stands stills as originally understood by Heracleitus[286].

When we see troglomorphic fishes we should see them as normal parts of nature, as life trying to occupy every possible space; after all, evolution is opportunistic (Dobzhansky 1970:405). This is a view more in tune with the Red Queen Hypothesis (Van Valen 1973), according to which organisms will keep changing attempting to survive both as individuals and/or through their genes. Troglomorphic fishes, by evolving from their epigean ancestors, are just trying to make the best of an environment lacking visual information.

And that brings us to the last point. One of the reasons why the notion of evolution by means of natural selection has taken so long to be accepted, and is still debated today, is because of misunderstandings at different levels of how natural selection works. Students of the hypogean fauna have historically missed the point that it is precisely this fauna that represents one of the prime examples of convergent evolution. It also does not help the fact that the study of convergence and the constraints of form have never been the subject of a single synthesis (Conway Morris 1998:13).

The study of hypogean fishes has the potential to help to understand better how convergence works.

Acknowledgments

Andrew Miller compiled most of the bibliography in the data base. Graham Proudlove provided copy of the Chen *et al.* article. Jerry Reedy of Macalester College revised A.R. translation of Kircher's passages on cavefishes. Herman W. de Swart secured the Dutch translation of Kircher's *Mundus Subterraneus* at the library of the National Museum for the History of Science in Boerhaave, Leiden, Holland, and sent me translations of some its portions. Janis Langins of the University of Toronto provided a copy of Montalembert's portrait. The following people provided me with information or hints where to find information about Theodore A. Tellkampf: Toby A. Appel of Yale University; Jocelyn K. Wilk of the Columbia University Archives & Columbiana Library; Robert Young of the Museum of Comparative Zoology at Harvard University; Lynn K. Nyhart of the University of Wisconsin-Madison, Edward T. Morman of the New York Academy of Medicine, Uwe Kunert from Germany; Frederick B. Churchill of Indiana University; Mary P. Winsor of the University of Toronto. William G. Saul of the Academy of Natural Sciences provided useful information on the specimens originally classified as "*Gronias nigrilabris*" that are deposited at the Academy's collection. Mark Sabaj of Department of Ichthyology of The Academy of Natural Sciences of Philadelphia provided information about the *A. spelaea* at that institution. Olga Mayayo provided the drawing of "*Gronias nigrilabris*." G. Teugels, of the Musée Royal de l'Afrique Centrale, in Tervuren, Belgium, provided useful information on Georges Boulenger. Steven Johnson Librarian and archivist of the Wildlife Conservation Society provided useful information on Mayron Gordon. Dr. James W. Atz read the section on C. M. Breder and made valuable suggestions. Jessica Romero secured a copy of the entire collection of the Memoirs of the Wernerian Society in search of an alleged first printed reference to a Kentucky "eyeless" fish. Karen Lucatelli read the manuscript and made valuable suggestions.

---

[286] *b.* Ephesus, Ionia, Persian Empire, ca. 435 BCE; *d.* Ephesus, Ionia, Delian League, ca. 475 BCE.

Table 1. Chronological order of discoveries of troglomorphic species/populations of hypogean fishes (*sensu* Romero 2001) from 1842 to 1999. Since then, a great number of new cave fish species have been described for China (Romero *et al.* 2009). Following an (*) is the author that reported the first hypogean/troglomorphic population of an already described epigean species.

Amblyopsis spelaea DeKay 1842
Lucifuga subterraneus Poey 1858
Lucifuga dentatus Poey 1858
Typhlichthys subterraneus Girard 1859
Chologaster agassizii Putnam 1872
Amblyopsis rosae (Eigenmann 1897)
Phreatobius cisternarum Goeldi 1905
Pimelodella kronei (Miranda-Ribeiro 1907)
Trogloglanis pattersoni Eigenmann 1919
Caecobarbus geertsii Boulenger 1921
Uegitglanis zammaranoi Gianferrari 1923
Phreatichthys andruzzii Vinciguerra 1924
Rhamdia quelen (Quoy & Gaimard 1824) (* Norman 1926)
Barbopsis devecchii Di Caporiacco 1926
Typhleotris madagascariensis Petit 1933
Clarias cavernicola Trewavas 1936
Astyanax fasciatus (Cuvier 1819) (* Hubbs & Innes 1936)
Ogilbia pearsei (Hubbs 1938)
Ophisternon infernale (Hubbs 1938)
Luciogobius pallidus Regan 1940
Luciogobius albus Regan 1940
Iranocypris typhlops Bruun & Kaiser 1944
Milyeringa veritas Whitley 1945
Satan eurystomus Hubbs & Bailey 1947
Horaglanis krishnai Menon 1950
Prietella phreatophila Carranza 1954
Typhlogarra widdowsoni Trewavas 1955
Garra barreimiae Fowler & Steinitz 1956
Typhleotris pauliani Arnoult 1959
Ophisternon candidum (Mees 1962)
Astroblepus pholeter Collette 1962
Monopterus eapeni (Eapen 1963)
Ogilbia galapagosensis (Poll & LeLeup 1965)
Stygichthys typhlops Brittan & Böhlke 1965
Trichomycterus chaberti Durand 1968
Nemacheilus evezardi Day 1872. (* Thines 1969?)
Lucifuga spelaeotes Cohen & Robins 1970
Speoplatyrhinus poulsoni Cooper & Kuehne 1974
Paracobitis smithi (Greenwood 1976)
Ophisternon aenigmaticum Rosen & Greenwood 1976
Pterocryptis cucphuongensis (Mai 1978)
Triplophysa gejiuensis (Chu & Chen 1979)
Caecocypris basimi Banister & Bunni 1980
Oreonectes anophthalmus Zheng 1981
Lucifuga simile Nalbant 1981
Rhamdia laticauda (Kner 1858) (*Greenfield *et al.* 1982)

Typhlobarbus nudiventris Chu & Chen 1982
Nemacheilus starostini Parin 1983
Triplophysa xiangxiensis (Yang, Yuan & Liao 1986)
Sinocyclocheilus anatirostris Lin & Luo 1986
Garra dunsirei Banister 1987
Nemacheilus sijuensis Menon 1987
Ancistrus cryptophthalmus Reis 1987
Sinocyclocheilus anophthalmus Chen, Chu, Luo & Wu 1988
Sinocyclocheilus cyphotergous (Dai 1988)
Schistura oedipus (Kottelat 1988)
Lucifuga teresinarum Diaz Perez 1988
Sinocyclocheilus microphthalmus Li 1989
Nemacheilus troglocataractus Kottelat & Géry 1989
Sinocyclocheilus angularis Zheng & Wang 1990
Triplophysa yunnanensis Yang in Wu 1990
Schistura jarutanini Kottelat 1990
Sundoreonectes tiomanensis Kottelat 1990
Poropuntius speleops (Roberts 1991)
Caecogobius cryptophthalmus Berti & Ercolini 1991
Glossogobius ankaranensis Banister 1994
Sinocyclocheilus hyalinus Chen & Yang in Chen, Yang & Zhu 1994
Protocobitis typhlops Yang, Chen & Lan 1994
Ancistrus galani Perez & Viloria 1994
Astroblepus riberae Cardona & Guerao 1994
Prietella lundbergi Walsh & Gilbert 1995
Oxyeleotris caeca Allen 1996
Eigenmannia vicentespelaea Triques 1996
Trichomycterus itacarambiensis Trajano & de Pinna 1996
Ancistrus formoso Sabino & Trajano 1997
Cryptotora thamicola (Kottelat 1998)
Paracobitis longibarbatus Chen, Yang, Sket & Aljancic 1998
Monopterus roseni Bailey & Gans 1998
Pterocryptis buccata Ng & Kottelat 1998
Troglocychlocheilus khammouanensis Kottelat and Brehier 1999

Literature Cited

Agassiz, E. (Ed.). 1885. *Louis Agassiz: His life and correspondence.* 2 Vols. London: Macmillan and Company, Vol. 1, 794 pp.

Agassiz, L. 1847 [1848]. [Plan for an investigation of the embryology, anatomy and effect of light on the blind-fish of the Mammoth Cave, *Amblyopsis spelaeus*]. *Proceedings of the American Academy of Arts and Sciences* **1**:180.

Agassiz, L. 1851. Observations on the blind fish of the Mammoth cave. *American Journal of Science* **11**:127-128.

Agassiz, L. 1859. *An Essay on Classification.* London: Longman, Brown, Green, Longmans, Roberts. 381 pp.

Alvarez, J. 1946. Revisión del género *Anoptichthys* con descripción de una especie nueva (Pisc., Characidae). *Anales de la Escuela Nacional de Ciencias Biológicas de México* **4**: 263-282.

Alvarez, J. 1947. Descripción de *Anoptichthys hubbsi* caracínido ciego de la Cueva de los Sabinos, S.L.P. *Revista de la Sociedad Mexicana de Historia* Natural **8**: 215-219.

Anonymous. 1842. [Mammoth Cave Blind Crayfish and Fish]. *Proceedings of the Academy of Natural Sciences of Philadelphia* **1**:175.

Appel, T.A. 1988. Jeffries Wyman, Philosophical Anatomy, and the scientific reception of Darwin in America. *Journal of the History of Biology* **21**:69-94.

Atz, J.W. 1968. C.M. Breder, Jr. 1897-1983. *Copeia* **1968**:853-856.

Baird, S.F. 1872. Living eyeless fish. *Annals and Records of Science and Industry* **1871**:266.

Barr, T.C. 1966. Evolution of cave biology In the United States, 1822-1965. *National Spelelogical Society Bulletin* **28**:15-21.

Bateson, W. 1922. Evolutionary faith and modern doubts. *Science* **55**:1412.

Beddall, B.G. 1983. The isolated Spanish genius -myth or reality? Félix de Azara and the Birds of Paraguay. *Journal of the History of Biology* **16**:225-258.

Besson, J. 1569 [1969]. *L'art et science de trouver les eaux et fontaines cachees soubs terre: autrement que par les moyens vulgaires des agriculteurs et architectes.* Orléans: E. Gibier, 83 pp. Reprinted by Editions Coral, Columbus Ohio.

Blanc, M., J.-L. Gaudet, P. Banarescu & J.-C. Hureau. 1971. *European inland fish: a multilingual catalogue.* Fishing News (Books) Ltd., London, 149 pp.

Bocking, S. 1988. Alpheus Spring Packard and cave fauna in the evolution debate. *Journal of the History of Biology* **21**:425-456.

Borowsky, R. & L. Mertz. 2001. *Schistura oedipus* (Cypriniformes; Balitoridae): One cave fish species or a cluster in status nascendi? *Environmental Biology of Fishes* **62**:225-231.

Boulenger, G. 1893. Blind animals in caves. *Nature* **47**:608.

Boulenger, G. 1922. Description d'un poisson aveugle decouvert par M.G. Geerts dans la grotte de Thysville (Bas-Congo). *Revue Zoologique Africaine* **9**:252-253,

Bowler, P.J. 1984. *Evolution. The history of an idea.* Berkeley: University of California Press, 412 pp.

Burr, B.M., G.L. Adams, J. Ktejca, R.J. Paul & M.L. Warren. 2001. Cavernicolous sculpins of the *Cottus carolinae* species group in Perry County, Missouri: distribution, external morphology, and conservation status review. *Environmental Biology of Fishes* **62**:279-296.

Breder, C. M. 1942. Descriptive ecology of La Cueva Chica, with especial reference to the blind fish, *Anoptichthys. Zoologica* **27**: 7-15.

Bridges, W. 1954. *Zoo expeditions.* New York: Curator Publications, 191 pp.

Burgess, W.E. 1989. *An atlas of freshwater and marine catfishes: a preliminary survey of the Siluriformes.* T.F.H. Publications, Neptune City, New Jersey, 784 pp.

Chen, Y.-R.; J.-X. Yang and Z.-G. Zhu. 1994. A new fish of the genus *Sinocyclocheilus* from Yunnan with comments on its characteristic adaptation (Cypriniformes: Cyprinidae). *Acta Zootaxonomica Sinica* **19**:246-253 (In Chinese).

Cockerell, T.D.A. 1920. Biographical memoir of Alpheus Spring Packard 1839-1905. *Nat. Acad. Sci. Biog. Mem.* **9**:181-236.

Coe, W. R. 1918. A century of zoology in America. pp.:391-438. In: E. S. Dana (ed.) *A century of science in America.* Yale University Press, New Haven.

Conway Morris, S. 1998. *The crucible of creation. The Burgess Shale and the rise of animals.* Oxford University Press, Oxford. 242 pp.

Cope, E.D. 1864. On a blind silurid from Pennsylvania. *Proc. Acad. Nat. Sci. Phil.* **1864**:231-233.

Cope, E.D. 1896. *The primary factors of organic evolution.* Open Court, Chicago. 547 pp.

Culver, D.C. 1976. The evolution of aquatic cave communities. *Amer. Nat.* **110**:955-957.

Culver, D.C. & D. W. Fong. 1986. Why all cave animals look alike. *Stygicola* **2**:208-216.

Culver, D.C., T.C. Kane & D.W. Fong. 1995. *Adaptation and natural selection in caves.* Harvard University Press, Cambridge. 223 pp.

Cuvier, G. 1828 [1995]. *Historical portrait of the progress of ichthyology: from its origins to*

our own time. Johns Hopkins University Press, Baltimore. 366 pp.
Darwin, C. 1859. *On the origin of the species by means of natural selection.* J. Murray, London. 502 pp.
Darwin, C. 1861. *On the origin of the species by means of natural selection.* J. Murray, London. 538 pp.
Davidson, R. 1840. *An excursion to the Mammoth Cave, and the barrens of Kentucky. With some notices of the early settlement of the state.* A. T. Skillman & son. Lexington, 148 pp.
Dean, B. 1916-1923. *A bibliography of fishes.* American Museum of Natural History, New York, Vol. 1, 718 pp., Vol 2. 702 pp., Vol 3, 707 pp.
DeKay, J. E. 1842. *Zoology of New York or the New-York Fauna. Part IV Fishes.* Albany: W. & A. White & J. Visscher, 566 pp.
Demoor, J., J. Massart & É. Vandervelde. 1897. *L'évolution régressive en biologie et en sociologie.* Félix Alacn, Ed., Paris. 324 pp.
Dexter, R.W. 1965. The "Salem secession" of Agassiz zoologists. *Essex Inst. Hist. Coll.* **101**:27-39.
Di Caporiacco, L. 1926. Un nuovo genere di Ciprinide somalo delle acque di pozzo. *Monit. Zool. Ital.* **37**:23-25.
Dobzhansky, T. 1970. *Genetics and the evolutionary process.* Columbia University Press, New York. 505 pp.
Eigenmann, C.H. 1890. The Point Loma blind fish and its relatives. *Zoe* **1**:65-96.
Eigenmann, C.H. 1903. In search of blind fishes in Cuba. *World Today* **5**:1131-1136.
Eigenmann, C.H. 1909. *Cave vertebrates or America. A study in degenerative evolution.* Carnegie Institution of Washington, Washington, D.C. 241 pp.
Eigenmann, C.H. 1919. *Trogloglanis pettersoni* a new blind fish from San Antonio, Texas. *Proc. Amer. Phil. Soc.* **58**:397-400.
Flint, J. 1822. *Letters from America, containing observations on the climate and agriculture of the western states, the manners of the people, the prospects of emigrants, &c., &c.* [sic] W. & C. Tait, Edinburgh, 330 pp.
Gianferrari, L. 1923. *Uegitglanis zammaranoi* un nuovo siluride cieco africano. *Soc. Ital. Sci. Nat. Milan.* **62**:1-3.
Gifford, G.E. 1967. An American in Paris, 1841-1842: four letters from Jeffries Wyman. *J. Hist. Med. Allied Sciences* **22**:274-285.
Girard, C.F. 1859. Ichthyological Notes. *Proc. Acad. Nat. Sci. Phila.* **1859**:63-64.
Girard, C. F. 1888. Les poissons souterrains du nouveau monde. *Le Naturaliste* **10**:222.
Gresser, E. B. & C.M. Breder. 1940. The histology of the eye of the cave characin, *Anoptichthys jordani. Zoologica* **25**:113-116.
Goeldi, E.A. 1905. Nova zoologica aus der Amazonas-Region. *Compt. Rend. 6th. Congr. Inter. Zool.* **1905**: 542-549.
Guillaume, J.-L. 1971. *La marche au socialism. Évolution progressive ou régressive?* Centre d'Études Politiques et Civiques, Paris. 52 pp.
Günther, A.C. L. G.1880. *An introduction to the study of fishes.* A. and C. Black, Edinburgh. 720 pp.
Gurnee, R. 1992. A brief history of cave studies in the United States before 1887 (16th to the 19th Century). *J. Spelean Hist.* **26**:11-20.
Horder, T. J. 1998. Why do scientists need to be historians? *Q. Rev. Biol.* **73**:175-187.
Hubbs, C. L. 1938. Fishes from the caves of Yucatan. *Carnegie Inst. Wash. Publ.* (491): 261-295.
Hubbs, C.L. 1964. History of ichthyology in the United States after 1850. *Copeia* **1964**:42-60.
Hubbs, C.L. & W. T. Innes. 1936. The first known blind fish of the family Characidae: a new genus from Mexico. *Occ. Pap. Mus. Zool.* (342):1-7.

Humboldt, A. von. 1805. Quatrième mémoire, Sur une nouvelle espèce de pimelode, jetée par les volcans du royaume de Quito. Pp. 40-48. *In*: *Voyage de Humboldt et Bonpland, Deuxième partie. Observations de Zoologie et d'Anatomie comparée.* Paris, F. Schoell.

Innes, W. T. 1937. A cavern characin *Anoptichthys jordani*, Hubbs and Innes. *Aquarium* **5**:200-202.

Jackson, J.R. & W. C. Kimler. 1999. Taxonomy and the personal equation: The historical fates of Charles Girard and Louis Agassiz. *J. Hist. Biol.* **32**:509-555.

Jie, Yi-Jing. 1541. [Report on the Alu cave]. Publisher's name and place of publication unknown. (In Chinese).

Jordan, D.S. 1905. The history of ichthyology, an address by David Starr Jordan. *Proc. Amer. Assoc. Avd. Sci.* **51**:427-456.

Kircher, A. 1665. *Mundus subterraneus, in XII libros digestus; quo divinum subterrestris mundi opificium, mira ergasteriorum naturæ in eo distributio, verbo pantámorphou Protei regnum, universæ denique naturæ majestas & divitiæ summa rerum varietate exponuntur.* J. Janssonium & E. Weyerstraten, Amsterdam, 2 vols.

Kosswig, C. 1949. Phänomene der regressiven Evoltion im Lichte des Genetik. Communs. *Fac. Sci. Univ. Ankara* **2**:110-150.

Kosswig, C. 1960. Darwin und die degenerative evolution. *Abhandl. Verh. Naturw. Ver. Hamburg* **4**:21-42.

Lankester, E.R. 1880. *Degeneration: A chapter in Darwinism.* Macmillan, London. 75 pp.

Lankester, E.R. 1893. Blind animals in caves. *Nature* **47**: 389.

Maynard Smith, J. & E. Szathmáry. 1995. *The major transitions in evolution.* Oxford University Press, New York. 346 pp.

Miranda Ribeiro, A. 1907. Una novedade ichthyologica. *Kosmos* **4**:21-22.

Montalembert, M.-R. 1748. Observations de Physique Générale. *Hist. Acad. Roy. Sci.* **1748**:27-28.

Myers, G.S. 1964. A brief sketch of the history of ichthyology in America to the year 1850. *Copeia* **1964**:33-41.

Nisbet, R. 1979. *History of the idea of progress.* Basic Books, Inc., New York. 370 pp.

Norman, J.R. 1926. A new blind catfish from Trinidad, with a list of the blind cave-fishes. *Ann. Mag. Nat. Hist.* **18**:324-331.

Packard, A. 1871. The Mammoth cave and its inhabitants. *Amer. Nat.* **5**:739-761.

Packard, A. 1888. The cave fauna of North America with remarks on the anatomy of the brain and origin of the blind species. *Mem. Nat. Acad. Sci.* **4**:1-156.

Packard, A. 1901. *Lamarck, the founder of evolution, his life and work.* Longmans, Green, and Co., New York. 451 pp.

Petit, G. 1933. Un poisson cavernicole aveugle des eaux douces de Madagascar. *Compt. Rend. Hebd. Séanc. Acad. Scienc.* **4**:347-348.

Pietsch, T.W. & W. D. Anderson, Jr. (ed.) 1997. *Collection building in ichthyology and herpetology.* American Society of Ichthyologists and Herpetologists, Lawrence, 593 pp.

Poey, F. 1858. *Memorias sobre la historia natural de la isla de Cuba.* Barcina, Habana. 2 vols.

Poly, W. J. 2001. Nontroglobitic fishes in Bruffey-Hills Creek Cave, West Virginia, and other caves worldwide. *Env. Biol. Fish.* **62**:73-83.

Poulson, T.L. 1963. Cave adaptation in amblyopsid fishes. *Amer. Mdl. Nat.* **70**:257-290.

Poulson, T.L. 1964. Animals in aquatic environments: animals in caves. pp. 749-771. *In*: D.B. Dill, E.F. Adolph & C.G. Wilber (ed.) *Handbook of Physiology*. Section 4: adaptation to the environment. American Physiological Society, Washington, D.C.

Poulson, T.L. 1985. Evolutionary reduction by neutral mutations: plausibility arguments and data from amblyopsid fishes and linyphiid spiders. *Nat. Speleol. Soc. Bull.* **47**:109-117.

Putnam, F.W. 1872. The blind fishes of the Mammoth cave and their allies. *Amer. Nat.* **6**:6-30.

Putnam, F.W. 1874. [The blind fish and some of the associated species of the Mammoth cave, Kentucky, probably of marine origin]. *Bull Essex Inst.* **6**:191-200.

Regan, C.T. 1940. The fishes of the Gobiid genus *Luciogobius* Gill. *Ann. Mag. Nat. Hist.* **5**:462-465.

Romero, A. 1984. Charles Marcus Breder, Jr. 1897-1983. *Nat. Speleol. Soc. News* **42**:8.

Romero, A. 1985. Can evolution regress? *Natl. Speleol. Soc. Bull.* **47**:86-88.

Romero, A. 1986a. Charles Breder and the Mexican Blind Characid. *Nat. Speleol. Soc. News* **44**: 16-18

Romero, A. 1986b. He wanted to know them all: Eigenmann and his blind vertebrates. *National Speleological Society News* **44**:379-381.

Romero, A. 1999a. The blind cave fish that never was. *National Speleological Societ News* **57**:180-181.

Romero, A. 1999b. Myth and reality of the alleged blind cave fish from Pennsylvania. *J. Spelean Hist.* **33**(4):67-75.

Romero, A. 2000. The speleologist who wrote too much. *Nat. Speleol. Soc. News* **58**:4-5.

Romero, A. 2001. It is a wonderful hypogean life. an introduction to the biology of hypogean fishes. *Env. Biol. Fish.* **62**:13-41.

Romero, A. 2002. The life and work of a little known biospeleologist: Theodor Tellkampf. *Journal of Spelean History* **36**(2):68-76.

Romero, A. & K. Benz. 2000. The unsung heroes of speleology. *Nat. Speleol. Soc. News* **58**:106, 126.

Romero, A. & J. E. Creswell. 2000. In search of the elusive "eyeless" cave fish Trinidad, W.I. *National Speleological Society News* **58**(10):282-283.

Romero, A.; Y. Zhao & X. Chen. 2009. The hypogean fishes of China. *Environmental Biology of Fishes* **86**(1):211-278.

Romero, A. & Z. Lomax. 2000. Jacques Besson, cave eels, and other alleged European cave fishes. *Journal of Spelean History* **34**(2):72-77.

Romero, A. & A. Romero. 1999. On Cope, caves, and skeletons in the closet. *National Speleological Society News* **57**:341-343.

Romero, A., A. Singh, A. McKie, M. Manna, R. Baker, K. M. Paulson, & J. E. Creswell. 2002. Replacement of the Troglomorphic Population of *Rhamdia quelen* (Pisces: Pimelodidae) by an Epigean Population of the Same Species in the Cumaca Cave, Trinidad, West Indies. *Copeia* **2002**(4):938-942.

Ruse, M. 1996. *Monad to man. The concept of progress in evolutionary biology.* Harvard University Press, Cambridge. 629 pp.

Shaw, T.R. 1992. *History of cave science. The exploration and study of limestone caves, to 1900.* Sydney Speleological Society, Broadway, N.S.W., Australia, 338 pp.

Smallwood, W.M. 1941. *Natural history and the American mind.* Columbia University Press, New York. 445 pp.

Soulé, G. K. 1982. A Mammoth cave chronology. *J. Spelean Hist.* **16**:3-9.

Tellkampf, T. 1843. Beschreibung einiger neuer in der Mammuth-Hole in Kentucky aufgefundener Gattungen von Gliederthieren. *Arch. Vereins Freund Natur. Mecklenburg* **10**:318-322.

Tellkampf, T. 1844. Uber den blinden Fisch der Mammuthhole in Kentucky. Muller's *Arch. Anat. Phys.* **1844**:381-395.

Tellkampf, T. 1870. Note respecting the eyes of *Amblyopsis spelaeus*. *Ann. Lyceum Nat. Hist. N.Y.* **9**:150-152.

Trewavas, E. 1936. Dr. Karl Jordan's expedition to South-West Africa and Angola. The fresh water fishes. *Novit. Zool.* **40**:63-76.

Thompson, W. 1844. Notice of the blind fish, cray-fish, and insects from the mammoth cave, Kentucky. *Ann. Mag. Nat. Hist.* **13**:111-113.

Van Valen, L. 1973. A new evolutionary law. *Evol. Thor.* **1**:1-30.

Vinciguerra, D. 1924. Descrizione di un ciprinide cieco proveniente dalla Somalia Italiana. *Ann. Mus. Civ. Stor. Nat. Genoa.* **51**:239-243.

von Baer, K.E. 1828.Uber *Entwickelungsgeschichte der thiere. Beobachtung und reflexion.* Bei Den Gebtdern Borntrager, Konigsburg, 315 pp.

Wilkens, H. 1988. Evolution and genetics of epigean and cave *Astyanax fasciatus* (Characidae, Pisces). Support for the neutral mutation theory. *Evol. Biol.* **23**: 271-367.

Wyman, J. 1843. Description of a 'Blind Fish,' from a cave in Kentucky. *Amer. J. Sci.* **45**:94-96.

Wyman, J. 1851. [Account of dissections of the blind fishes (*Amblyopsis spelaeus*) from the Mammoth cave, Kentucky]. *Proc. Boston Soc. Nat. Hist.* **3**:349, 375.

Wyman, J. 1854a. The eyes and organs of hearing in *Amblyopsis spelaeus*. *Proc. Boston Soc. Nat. Hist.* **4**:149-151.

Wyman, J. 1854b. On the eye and the organ of hearing in the blind fishes (*Amblyopsis spelaeus* DeKay) of the Mammoth Cave. *Proc. Boston Soc. Nat. Hist.* **4**:395-396.

Wyman, J. 1872. Notes and drawings of the rudimentary eye, brain, and tactile organs of *Amblyopsis spelaeus*. *Amer. Natur.* **6**:16-20.

Yamamoto, Y.& Jeffery, W. R. 2000. Central role for the lens in cave fish eye degeneration. *Science* **289**:631-633.

Chapter 5. When Whales Became Mammals: The Journey of Cetaceans from Fish to Mammals and Intellectual Inertia [287]

Summary

The recognition of cetaceans (whales and dolphins) as mammals by the scientific community took a long time. It was not until the 10th edition of Linnaeus's *Systema Naturae* that they were recognized as a natural group and totally separate from fishes. This is puzzling given that for about 2,000 years before Linnaeus's work many naturalists had identified a number of characteristics of these animals that clearly placed them closer to land mammals (or "viviparous quadrupeds") than to fish. In this chapter I survey pre-Linnean descriptions and classifications of cetaceans and explore several explanations for this case of intellectual inertia. Since Linnaeus was not an evolutionist, we cannot support the idea that lack of evolutionary thinking prevented the understanding of the proper place of cetaceans in animal classification. I believe that a combination of environmental classification and scholasticism led to their misclassification for centuries. Linnaeus great contribution (although heavily influenced in this case by others) was to clearly differentiate between analogy and homology.

Introduction

Cetacea (whales and dolphins) is a natural group that has for centuries generated a great deal of misunderstanding and controversy regarding its proper place in natural classification. As late as 1945 Simpson wrote that "Because of their perfected adaptation to a completely aquatic life, with all its attendant conditions of respiration, circulation, dentition, locomotion, etc., the cetaceans are on the whole the most peculiar and aberrant of mammals."

Although both molecular and paleontological data have provided a much better understanding of the placement of this group among mammals, there is no question that despite being studied and dissected by dozens of naturalists since Aristotle, these animals were always misclassified for about 2,000 years. This group provides an interesting case study for intellectual inertia in the history of science. In other words, why did so many scientists misplace this group in the natural classification despite the fact that they themselves were gathering critical information that showed the close relationship these animals had to what we know today as mammals?

The aim of this chapter is to explore this question. To that end I will (1) survey the naturalists who studied cetaceans providing clues of their true nature, (2) describe the intellectual environment in which their conclusions were made, and (3) discuss the factors behind this intellectual inertia.

For the purpose of this chapter, I have only taken into consideration works that had some scientific basis and/or that in some ways influenced the process of placing cetaceans as mammals. Authors are enumerated based on the date of the major publication they produced on cetaceans. For synonyms in names of marine mammals through time see Artedi (1738) and Linnaeus (1758).

---

[287] Based on Romero, A. 2012. When whales became mammals: the scientific journey of cetaceans from fish to mammals in the history of science. Pp. 4-30. *In*: Romero, A. & E.O. Keith (Eds.). 2012. *New Approaches to the Study of Marine Mammals*. Rijeka, Croatia: InTech.

Ancient Times

Aristotle

Aristotle[288] was the son of Nicomachus, the personal physician of King Amyntas of Macedon and Phaestis, a wealthy woman[289]. Nicomachus may have been involved in dissections (Ellwood 1938, p. 36), a key tool in Aristotle's biological studies, particularly on marine mammals. Aristotle lost both his parents when he was about 10 and from then on, he was raised of his uncle and/or guardian Proxenus, also a physician (Moseley 2010, p. 6). Early Greek physicians known as asclepiads usually taught their children reading, writing, and anatomy (Moseley 2010, p. 10).

In 367 BCE Aristotle moved to Athens to study at Plato's Academy, and later travelled throughout Asia Minor and studied living organisms while at the island of Lesbos (344-342 BCE) where he collected a lot of information about marine mammals. He later created his own philosophical school, the *Lyceum*, in Athens where most of his written work was produced between 335 and 323 BCE.

Aristotle is the first natural historian from whom we have any extensive work. One of his surviving opuses is *Historia Animalium* (inquiry about animals)[290]. There he classified animals as follows (beginning from the top): "blooded" animals (referring to those with red blood, vertebrates) with humans at the top, viviparous quadrupeds (what we would call terrestrial mammals), oviparous quadrupeds (legged reptiles and amphibians), birds, cetaceans, fishes, and then "bloodless" animals (invertebrates). He named each one of these groups a "genus."

Humans
Viviparous quadrupeds (terrestrial mammals)
Oviparous quadrupeds (reptiles and amphibians)
Birds
Cetaceans
Fish
Malacia (squids and octopuses)
Malacostraca (crustaceans)
Ostracoderma (bivalve mollusks)
Entoma (insects, spiders, etc.)
Zoophyta (jellyfishes, sponges, etc.)
Higher plants
Lower plants

Based on the "kinds" of animals and the varieties he described, we can distinguish somewhere between 550 and 600 species. Most of them he had observed directly and even dissected but others were based on tales and he warned about the accuracy of those descriptions. For example, although he mentioned information in numerous occasions provided to him by fishers, many times (but not always) he debunks some of the fallacies he heard based on his own observations, particularly when it came to reproduction.

Of what we would consider today as mammals (including cetaceans) he described about 80 and about 130 species of fishes, which, again, underlines the extensive work, he did on marine creatures, mostly while living at Lesbos. Under the genus "Cetacea" he included at least three species: (1) "dolphins" probably a combination of striped dolphin

---

[288] *b*. Stagira, Chalcidice, Macedonia, today's Greece, 384 BCE; *d*. Chalcis, Euboea, Ancient Greece, today's Greece 322 BCE
[289] Biographical information on Aristotle is largely based on Barnes (1995).
[290] We used the text available at http://classics.mit.edu/Aristotle/history_anim.html

(*Stenella coeruleoalba*, the most frequent species in the Mediterranean), the common dolphin (*Delphinus delphis*), and the bottlenose dolphin (*Tursiops truncatus*); (2) the harbor porpoise (*Phocoena phocoena*) which he described as "similar to dolphins but smaller and found in the Black Sea" ("Euxine") (*HA* 566b9)[291]; and, (3) the fin whales (*Balaenoptera physalus*) another common species in the Mediterranean at that time.

The motives behind Aristotle classification system, particularly animals, were not biological in nature but rather philosophical. For him these creatures were evidence for rational order in the universe. This approach meant that species were rigid elements of the world and, thus, he never contemplated mutability or anything close to evolution, despite the fact that earlier Greek philosophers such as Anaximander envisioned the mutability of species. Furthermore, Aristotle's motive for conducting this categorization was done in such a way that we can then identify the causes that explain why animals are organized the way they are. His investigation into those causes is carried out in other surviving biological works (e.g., *Parts of Animals*). When describing species, he adhered to his teleological doctrine of purposiveness in nature.

Aristotle was able to distinguish between homology and analogy, recognizing cetaceans as a natural group with many similarities with other mammals ("viviparous quadrupeds"). He considered cetaceans as "blooded" animals, adding, "viviparous such as man, and the horse, and all those animals that have hair; and of the aquatic animals, the whale kind as the dolphin and cartilaginous fishes" (*HA* 489a34-489b3). He also wrote: "all creatures that have a blow-hole respire and inspire, for they are provided with lungs. The dolphin has been seen asleep with its nose above water as he snores (*HA* 566b14). All animals have breasts that are internally or externally viviparous, as for instance all animals that have hair, as man and the horse; and the cetaceans, as the dolphin, porpoise and the whale -for these animals have breasts and are supplied with milk" (*HA* 521b21-25). Among the species he described were dolphins, orcas, and baleen whales, noting that "the [whale] has no teeth but does have hair that resemble hog bristle" (*HA* 519b9-15). Thus, he was the first to separate whales and dolphins from fish.

However, Aristotle placed whales and dolphins below reptiles and amphibians, because their lack of legs, despite his physiological and behavioral observations that they were related more closely to "viviparous quadrupeds" than to fish.

Aristotle followed his teacher Plato in classifying animals by progressively dividing them based on shared characters. This is an embryonic form of today's classification more fully developed by Linnaeus. The reason he ordered the different "genera" the way he did was because he considered "vital heat" (characterized by method of reproduction, respiration, state at birth, etc.) as an index of superiority placing humans at the very top. Men were superior to women because they had more "vital heat." On this he followed Hippocrates's[292] ideas, since the Greek physician thought there was an association between temperature and soul.

Yet he was not fully satisfied by this approach given that a number of "genera" had characters that were shared across groups, particularly when compared with their habitats. For example, both fishes and cetaceans had fins, but they differ markedly on other characters such as reproduction (oviparous vs. viviparous) or organs (gills vs. lungs, respectively).

Many of Aristotle's observations about cetaceans remain accurate. In terms of internal anatomy he mentioned that they have internal reproductive organs (*HA* 500a33-500b6), that dolphins, porpoises, and whales copulate and are viviparous, giving birth to between one and two offspring, having two breasts located near the genital openings that produce milk (*HA* 504b21), that dolphins reach full size at the age of 10 and their period

---

[291] These citations for *Historia Animalium* follow the Bekker' pagination.
[292] *b*. Kos, Ancient Greece, ca. 460 BCE; *d*. Larissa, Ancient Greece, ca. 370 BCE.

of gestation is 10 months, show parental care, some may live up to 30 years and this is known because fishers can individually identify them by marks on their bodies (*HA* 566b24), and that dolphins have bones (*HA* 516b11).

Regarding behavior and sensory organs, he said that dolphins have a sense of smell but he could not find the organ (*HA* 533b1), that dolphins can hear despite the lack of ears (*HA* 533b10-14), produce sounds when outside the water (*HA* 536a1), that dolphins and whales sleep with their blowhole above the surface of the water (*HA* 537a34), are carnivorous (*HA* 591b9-15), and swim fast (*HA* 591b29).

He held that cetaceans are not fishes because they have hair, lungs (*HA* 489a34), lack gills, suckle their young by means of mammae, they are viviparous (*HA* 489b4), and that their bones are analogous to the mammals, not fishes. Still, they he calls them "fishes" (*HA* 566b2-5).

These basic Aristotelian biological descriptions persisted for good and for bad until Charles Darwin's evolutionary work. On one hand his descriptions were so accurate that Darwin admired Aristotle, to the point that he said privately that the intellectual heroes of his own time "were mere schoolboys compared to old Aristotle."[293] Yet the fact that Aristotle saw the natural world as fixed in time with no room for evolution and that he kept calling cetaceans "fishes," would delay intellectual progress for many centuries when it came to the classification of these animals.

Aristotle's influence on naturalists' classification of life would extend until Darwin's times when evolutionary views replaced the fixity of species as elements in nature.

Pliny the Elder

Pliny the Elder[294] was the son of an equestrian (the lower of the two aristocratic classes in Rome) and was educated in Rome. After serving in the military, he became a lawyer and then a government bureaucrat. In these positions he travelled not only throughout what is Italy today but also what it would later become Germany, France and Spain as well as North Africa (Reynolds 1986).

He wrote a 37-volume *Naturalis Historia*[295] (ca. 77-79) in which according to himself he had compiled "20,000 important facts, extracted from about 2000 volumes by 100 authors" and was written for "the common people, the mass of peasants and artisans, and only then for those who devote themselves to their studies at leisure" (Preface 6). This is the earliest known encyclopedia of any kind, which has been interpreted as a Roman invention in order to compile information about the empire (Naas 2002, Murphy 2004). It was a rather disorganized book, whose prose has been criticized by many (Locher 1984). Pliny seemed to be more interested in what appeared to be curiosities' than what were facts. This is a big collection of facts and fictions, based mostly said on things said by others.

He devoted 9 of the 37 volumes to animals and ordered them according to where they live. Volume IX (*Historia Aquatilium*) of *Naturalis Historia* is devoted to aquatic creatures, whether living in oceans, rivers or lakes, whether vertebrate or invertebrate, real or mythical. Based on their size he categorized as "monster" anything big, whether it is a whale, a sawfish or a tuna (IX 2,3).

He grouped together all known species of cetaceans (*cete*) but constantly mixed their descriptions with those of other marine mammals such as seals as well as with cartilaginous fishes, such as some sharks (*pristis*). Pliny mentioned the three species cited

---

[293] Darwin Correspondence Database, http://www.darwinproject.ac.uk/entry-11875 accessed on 25 Feb 2012.
[294] *b*. as Gaius Plinius Secundus in what is now Como, Italy ca. 23/24 CE; *d*. near Pompeii, Italy, 25 August 79 CE.
[295] I used the version available at: http://www.perseus.tufts.edu/hopper/text?doc=Plin.+Nat.+toc&redirect=true

by Aristotle: dolphins (*delphinus*, probably a combination of striped dolphin [*Stenella coeruleoalba*] and the common dolphin [*Delphinus delphis*], IX 12-34), porpoises (*porcus marinus*, the harbor porpoise [*Phocoena phocoena*], IX 45) and whales (*ballaena*, possibly a combination of large toothless whales [mysticetes] IX 12-13). Then he added a few more: the *thursio* or *tirsio* (probably the bottlenose dolphin, *Tursiops truncatus* IX 34), the *physeter* (probably the sperm whale [*Physeter macrocephalus*] IX 8) found in the "Gallic Ocean" (probably the Bay of Biscay, IX 3, 4), the *orca* (probably the killer whale [*Orcinus orca*] IX 12-14), and the river dolphin from India (possibly *Platanista gangetica*, IX 46). He also mentioned some mythical creatures such as *Homo marinus* (Sea-Man, IX 10) and the *Scolopendra marina* (IX, 145) a mythical organism whose legend may be based on polychaeta, marine annelids characterized by the presence of many legs (Leitner 1972, p. 218).

Pliny recognized that neither whales nor dolphins have gills, that they suckle from the teats of their mothers, and that they are viviparous. In addition to these true facts copied from Aristotle, he mentioned exaggerations such as whales of four jugera (ca. 288 m) in length that because of their large size "are quite unable to move" (IX 2,3). In addition to some of the biological facts mentioned by Aristotle, Pliny adorns his narrative with all kind of casual tales about interactions between cetaceans and humans.

By lumping together all kinds of aquatic organisms it is hard to distinguish what he called "fish" and what he did not (see for example IX 44-45). His classification took a step back from Aristotle because he did not try for a comprehensive classification of animals. He failed to compare organisms based on shared or divergent characters. Many times, he ordered creatures based on size, from the largest to the smallest. Yet, his work had great influence for 1700 years, which was unfortunate because he was an uncritical compiler of other people's writings (even if they were contradictory). Pliny also created a number of unfounded impressions about the reality of nature. His only positive contribution was that he established the norm of always citing the sources of his information (in actuality 437 authors, whose works, in some cases, are no longer available).

Medieval Times

During the middle ages, little progress was made in the sciences. Students were urged to believe what they read and not to question conventional wisdom. Observation did not determined truth; logic did. Free thought was non-existent and minds were filled with mythological explanations for the unknown. Marine mammals were depicted as monsters and little new information was generated during this period of time.

The Renaissance

The Renaissance was a time of awakening and religious beliefs began to be questioned. The translations of the works of Aristotle and Pliny into Latin and the introduction of the printing press helped to spread the little knowledge accumulated until that time about natural history in the western world. For example, by 1500 about 12 editions of Aristotle's *Historia Animalium* and 39 of Pliny's *Historia Naturalis* had seen the light, which is evidence of the popularity of these works. During this age of discovery, the finding of species that were never mentioned neither by Aristotle nor the Bible, opened up scientific curiosity about new creatures around the world. Thus, people once again began to seek new knowledge. However, in these times, naturalists were more compilers of information than investigators despite the fact that they were performing more dissections that in turn uncovered new taxonomic possibilities. Still, scientists relied on environmental aspects to classify animals. Collecting was a primary activity during this era (Alves 2010, p. 54).

Pierre Belon

Belon[296] (Fig. 5.1) was the first author studying marine mammals in this historical period. Little is known about his family and early years. He traveled extensively throughout Europe and the Middle East, including the Arabian Peninsula and Egypt. Among the places he visited were Rome where he met two other ichthyologists, Rondelet and Salviani (see below). He studied medicine at the University of Paris and botany at the University of Wittenberg, Germany. He served as a doctor and apothecary for French kings, as well as a diplomat, traveler, and as a secret agent (he was murdered under strange circumstances) (Wong 1970).

His *L'Histoire Naturelle des Estranges Poissons Marins* (1551) was the first printed scholarly work about marine animals. This book was expanded and published in French in 1555 as *La Nature et diversité des poissons* including 110 species with illustrations for 103 of them.

Belon not only reproduced information from Aristotle and Pliny but also added his own observations including comparative anatomy and embryology. For him "fish" was anything living in the water. He divided "fishes" in two large groups: the first was "fish with blood" (as Aristotle had done) that included not only actual fishes but also (Fig. 5.2) cetaceans, pinnipeds, marine monsters and mythical creatures such as the "monk fish" as well as other aquatic vertebrates such as crocodiles, turtles, and the hippopotamus. He called a second group "fishes without blood" and consisted of aquatic invertebrates (see also Delaunay 1926).

Belon ordered what we know as cetaceans today in a vaguely descending order based on size: *Le balene* (mysticete whales, although in the illustration he depicted a cetacean with teeth), *Le chauderon* (sperm whale? although he mentions the sawfish), *Le daulphin* (common dolphins on which he devoted 38 pages of this 55-page book), *Le marsouin* (porpoise), and *L'Oudre* (bottlenose dolphin) (for a rationale on the identification of these species see Glardon 2011, p. 393-398). He dissected common dolphins (*D. delphis*) and porpoises (*P. phocaena*) acquired at the fish market in Paris brought in by Normandy fishers, and probably a bottlenose dolphin (*T. truncatus*) as well.

Belon described these marine mammals as having a placenta, mammae, and hair on the upper lip of their fetus. Belon wrote that apart from the presence of hind limbs, they conform to the human body plan with features such as the liver, the sternum, milk glands, lungs, heart, the skeleton in general, the brain, genitalia. He also dealt with issues of breathing and reproduction (although from the description it is clear that he never saw one of these animals giving birth, since he depicted the newborn surrounded with a membrane). He drew the embryo of a porpoise and the skull of a dolphin (Fig. 5.3). Despite all this he did not make the connection between cetaceans and "viviparous quadrupeds" and based his entire classification on environmental foundations, as he made clear in the introduction of his work.

---

[296] *b.* Soultière, near Cerans, France, 1517; *d.* Paris, France, April 1564.

Figure 5.1. Pierre Belon (public domain drawing).

Figure 5.2. Belon's "Monk Fish" (public domain illustration).

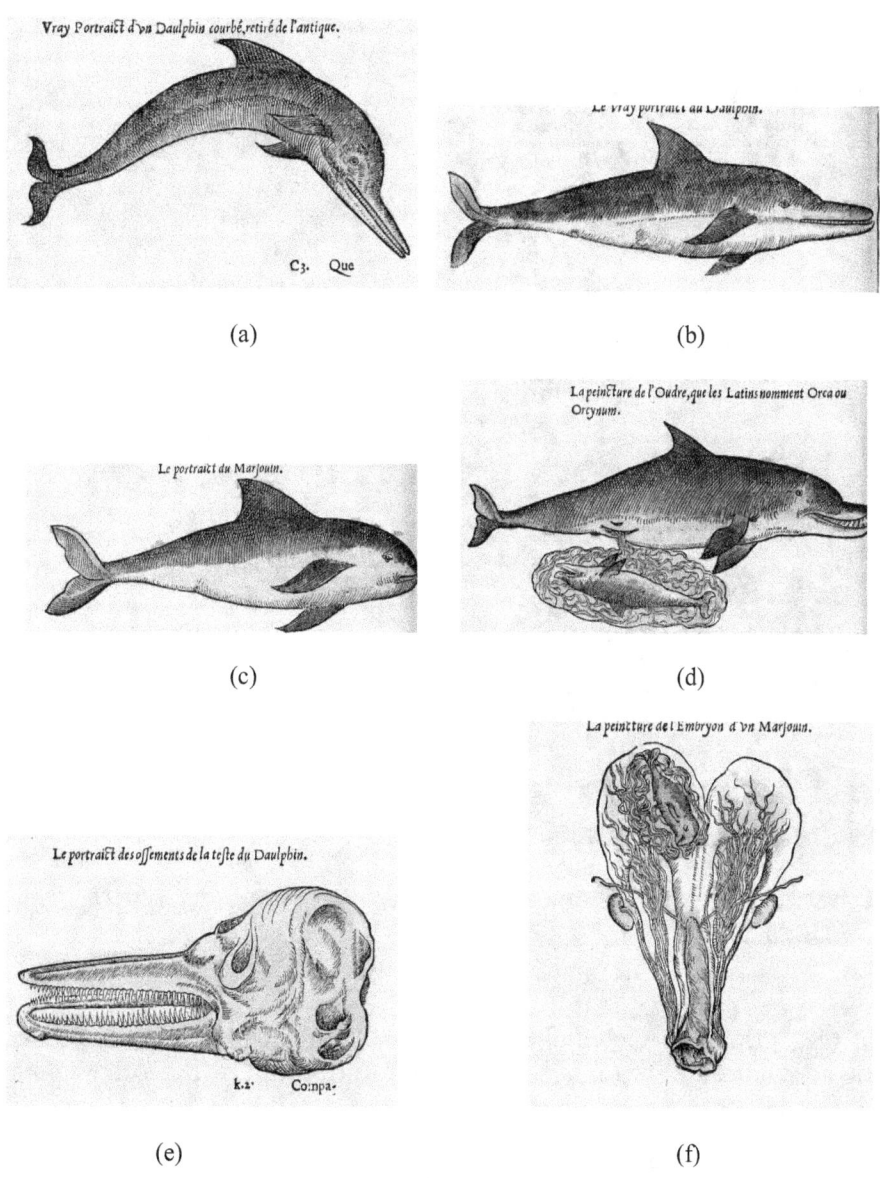

Figure 5.3. Illustrations of marine mammals by Belon (1551): (a) and (b) are representations of the common dolphin (*Delphinus delphis*); (c) a porpoise (*Phocaena phocaena*); (d) a bottlenose dolphin (*Tursiops truncatus*, although he uses the name of "Orca") presumably giving birth; (e) the skull of a dolphin; (f) a porpoise fetus in a placenta, showing that he had actually dissected these animals.

Edward Wotton

Wotton[297] (Fig. 5.4) was the son of a theologian who did general studies at Oxford and studied medicine and Greek at Padua (1524-1526). He was a practicing physician

---

[297] *b.* Oxford, England, 1492; *d.* London, England, 5 October 1555.

who published *De Differentiis Animalium Libri Decem* (1552), probably the first published book exclusively on natural history of the Renaissance. This was a 10-part ("books") treatise that followed the classification structure by Aristotle while adding some comments from Pliny. In Book 8 (pp. 171-173) he placed *Cete* together with fishes because of the medium they inhabit. Except for entomology he did not conduct any original observations on animals nor include any illustrations. His contemporaries noted his lack of originality (Nutton 1985).

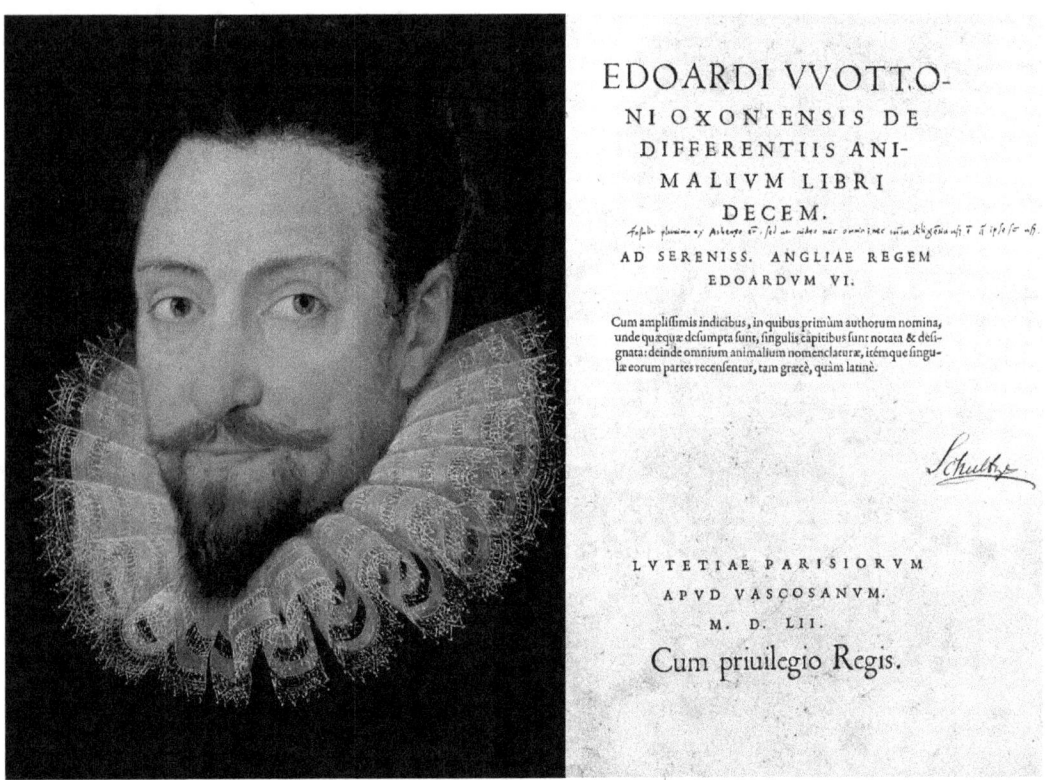

Figure 5.4. Edward Wotton and cover of one of the volumes of his *De Differentiis Animalium Libri Decem* (1552).

The list of cetacean species included *Delphino* (dolphins), *Phocaena* (porpoises), *Balaena* (mysticete whales), *Orca* (either the bottlenose dolphin or the killer whale) and *Physeter* (the sperm whale).

Guillaume Rondelet

Rondelet[298] (Fig. 5.5) was the son of a drug and spice merchant. He studied medicine at the University of Montpellier, one of the best medical schools in Europe at that time. While in Paris he studied anatomy under Johannes Guinther, who also taught Vesalius. Rondelet would later become Professor of medicine and Chancellor at Montpellier (Keller 1975). He probably acquired his interest in ichthyology at a young age while living in Montpellier (about 12 km from the coast) because his family owned a farm that was a stopping place for carts of fish from the Mediterranean (Oppenheimer 1936). During his trips as personal physician to Francois Cardinal Tournon (who was

---

[298] *b*. Montpellier, France, 27 September 1507; *d*. Réalmont, Tarn, France, 30 July 1566.

also the patron of Belon) to the Atlantic coasts of France, he became acquainted with the whaling industry. Rondelet met several contemporary ichthyologists while in Rome (1549-1550) such as Belon, Hippolyto Salviani, and Ulyssis Aldronvandi (Gudger 1934). Guillaume Pellicer, Bishop of Montpellier, who was also interested in fishes but never published on ichthyology, may have influenced Rondelet (Oppenheimer 1936, Dulieu, 1966).

He enjoyed dissecting and did so frequently for both teaching and research purposes. He published *Libri de Piscibus Marinis in quibus verae Piscium effigies expressae sunt* (1554) (Fig. 5.6) with a second part titled *Universae Aquatilium Historiae pars altera* (1555) about both marine and freshwater animals. Both were later translated into French as *L'histoire entière des poissons* (1558, 599), a monograph for teaching purposes.

After writing about food, habitat, morphology, and physiology, he described 145 freshwater and 190 marine species that included at least seven species of cetaceans: *delphino* (common dolphin), *phocaena* (porpoise), *tursione* (bottlenose dolphin, although the illustration more resembles a porpoise), *balaena vulgo* and *balaena vera* (two different species of mysticetes whose true identities are difficult to ascertain), *orca* (killer whale), and *physetere* (sperm whale) (Fig. 5.7). He also included among cetaceans the *priste* (sawfish) and mythical animals such as Pliny's *scolopendra cetacea*, the *monstruo leonino* (a lion covered with scales and with a human face), the *pisce monachi habitu* (a fish that looks like a monk), and the *pisce Episcopi habitu* (a fish that looks like a bishop) of which he was skeptic. All together his book contained more species than previous published works. Each species description included the animal's name in different languages, their morphology (external and internal), feeding habits, and use as food for humans. Species were differentiated similarly to Aristotle as blooded and non-blooded. Although Aristotle inspired the entire book, including teleological considerations in his discussions, Rondelet added some original ideas, especially concerning anatomy and descriptions of the small cetaceans he dissected. Rondelet made correlations between form, function, and environment.

Despite noting differences, he grouped marine mammals with fish based on habitat. For example, he noted that fishes with scales lack lungs and have a three-chamber heart while what we know today as marine mammals have hearts with four chambers. He compared the anatomy of a dolphin to that of the pig and humans. Based on this and his descriptions of other internal organs, he considered marine mammals to be a type of aquatic quadruped. Yet, he did not propose a system of classification. He did not advance the notion of valid classification, but because of the quality of his descriptions his work remained as the main reference for about 100 years.

Figure 5.5. Rondelet (public domain image).

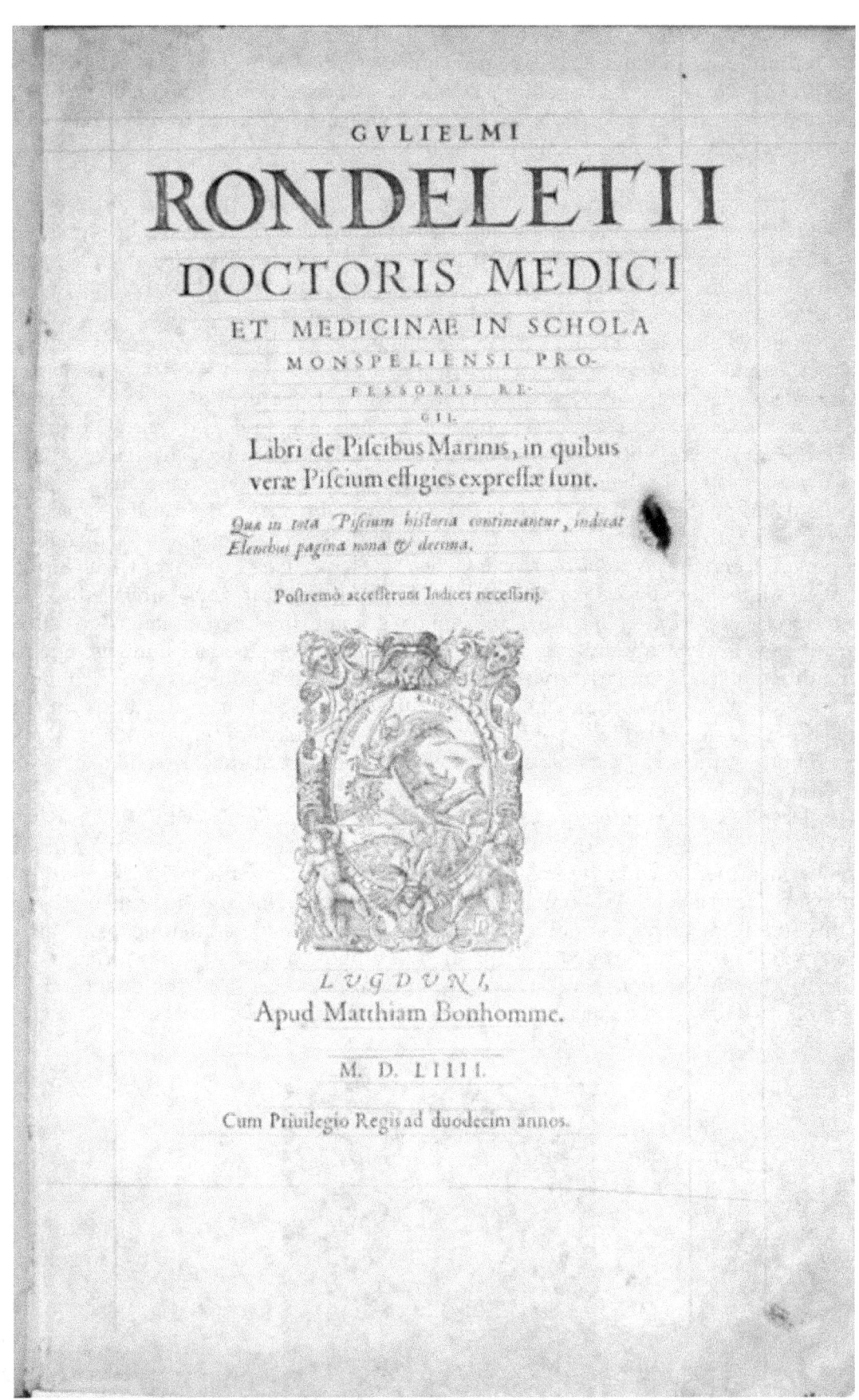

Figure 5.6. Libri de Piscibus Marinis in quibus verae Piscium effigies expressae sunt (1554) by Rondelet (public domain image).

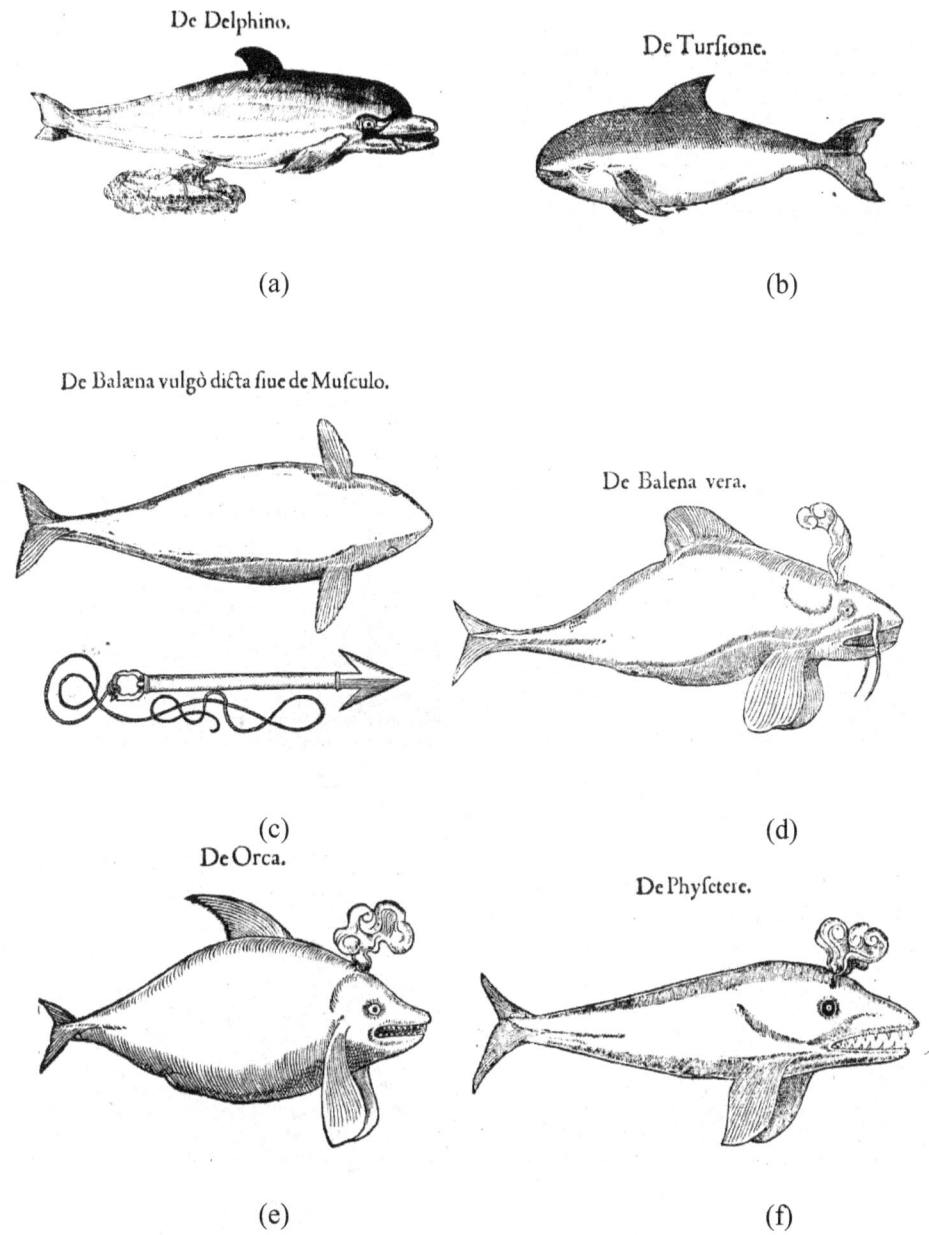

Figure 5.7. Illustrations of marine mammals by Rondelet (1554): (a) a dolphin showing a fetus surrounded by a placenta indicating it was a viviparous animal; (b) a porpoise; (c) an unidentified species of mysticete, probably a right whale because may have been observed by Rondelet during a whaling operation in the Atlantic; (d) an unidentified species of mysticete that he never saw as evidenced by the depiction of barbels above the mouth; (e) orca (*Orcinus orca*); (f) a sperm whale (*Physeter macrocephalus*).

Conrad Gessner

Gessner[299] (Fig. 5.8) probably developed an interest in zoology after seeing the carcasses of furred animals at his father's workshop where several furriers worked. He also lived with a great-uncle, an herbalist, who furthered his interest in natural history

---

[299] *b*. Zürich, Switzerland, 16 March 1516; *d*. Zürich, 13 December 1565,

(Bay 1916, Gmelig-Nijboer 1977, p. 17, Wellisch 1984, p. 1). He was an avid traveler who studied theology and medicine in Bourges, Paris, Montpellier, and Basel (Fischer 1966) and had great facility for classical languages. During his travels Gessner met with Belon and Rondelet. He is considered as the "father of bibliography" because of his work on compiling information about books (Bay 1916). Gessner himself had a very large private library of more than 400 volumes (which was a very large private collection for his time) of which 19% of the volumes were on natural history and 13 of them were on zoology (Leu et al. 2008, pp. viii, 1, 13, 21). He published *Historiae Animalium* (1551-1558), an encyclopedic (4 volumes, 4,500 pages treatise) but uncritical compilation of information and bibliography in which he intended to itemize all of God's creations. In addition to classic authors such as Aristotle and Pliny, Gessner obtained information from whomever he could correspond. He classified cetaceans among 'aquatic animals,' i.e., including fishes (Fig. 5.9). The fourth volume (*Piscium & Aquatilium*) of 1297 pages was published in 1558 and was about the aquatic animals. A fifth volume on reptiles and arthropods was not published until 1587, posthumously. *Historiae* was added to the list of prohibited books because Gessner was Protestant. Yet, the 14 editions in different languages of this book reveal its popularity.

Figure 5.8. Gesner (public domain image).

Gessner followed Aristotle's classification of animals when it came to their grouping by volume (Vol. 1: viviparous quadrupeds; Vol. 2: oviparous quadrupeds; Vol. 3: birds; Vol. 4: aquatic animals; Vol. 5: serpents). He ordered them alphabetically, like a "Dictionarium," in each volume, which did not provide a rational classification based on relationships of any kind; on the other hand, this alphabetical order facilitated its use as an encyclopedic source. Gessner's intention was to collect any piece of information ever written about each animal by any author in history, he cited nearly 250 authors including Rondelet (*Libri de Piscibus Marinis*, 1554), Belon (*De Aquatilibus*, 1553), and Salviani (*Aquatilium Animalium*, 1554). The latter only mentioned marine mammals *in passim*.

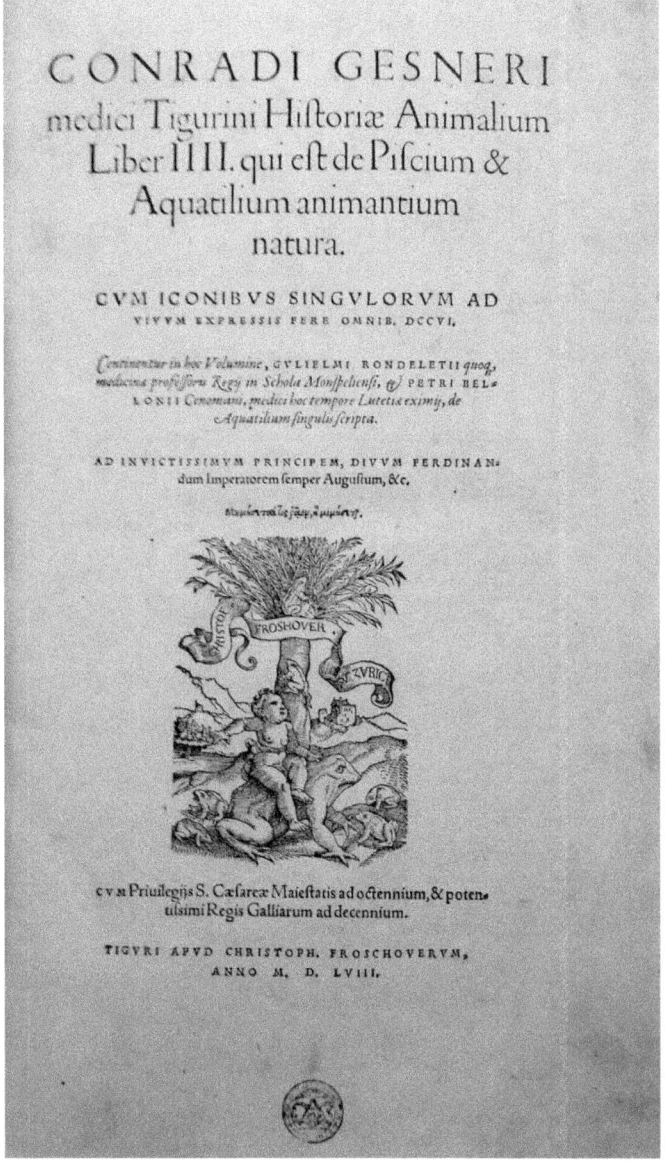

Figure 5.9. Gesner's *Aquatilium Animalium* (1563) (public domain image).

Information included names of the animals in various languages (some times more than a dozen) comprising epithets and etymology (even inventing common names in other languages when those names were not available), physical features, geographic distribution, the animal's way of living including diseases and their cures, behavior, utility towards man (e.g., for food or medical purposes), and tales. His work was full of illustrations: some were very accurate showing that he had first-hand knowledge of the animal in questions while other were bizarre or just invented, especially when dealing with mythical creatures (Fig. 5.10).

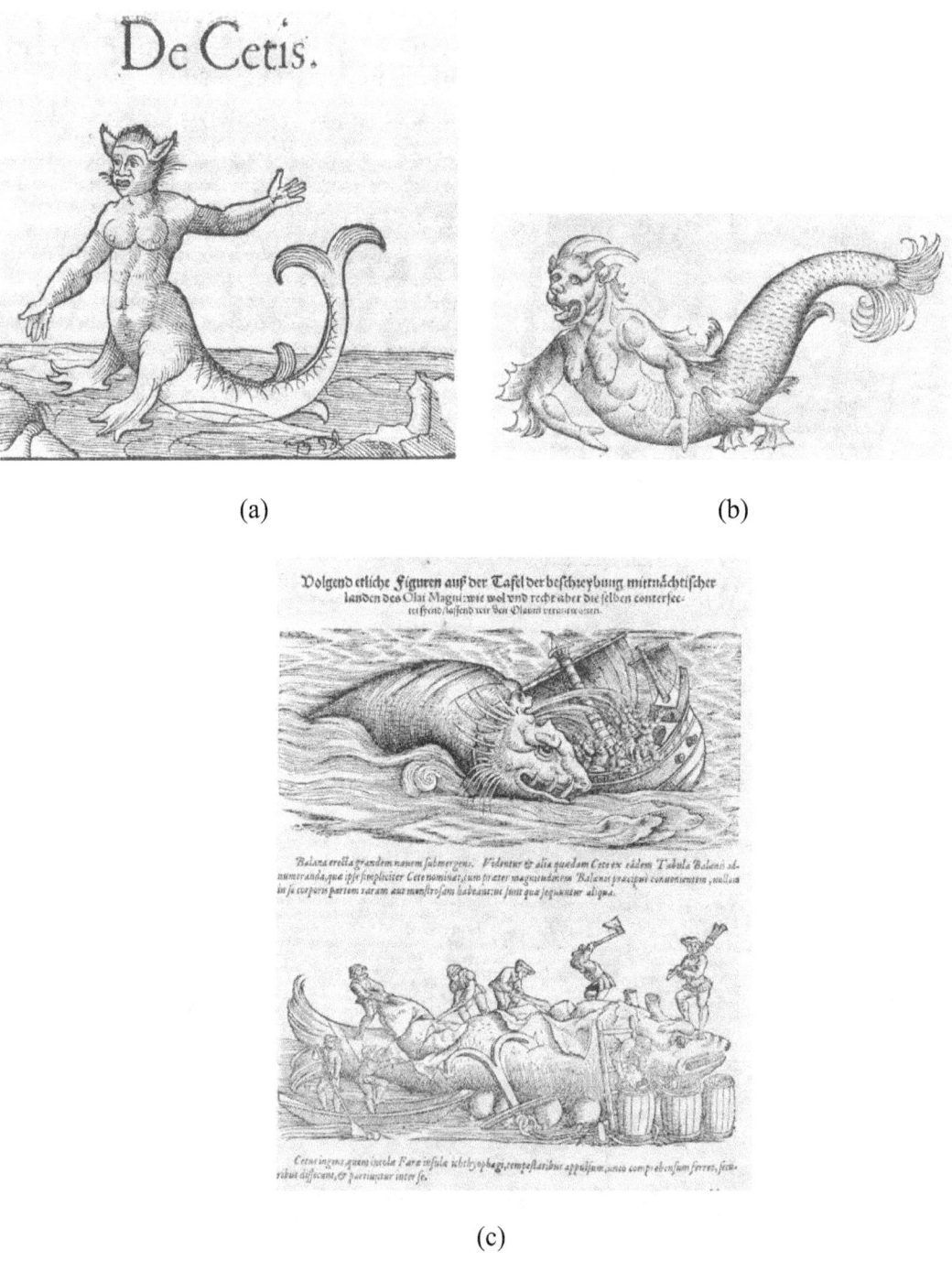

Figure 5.10. Some of the "Cetis" described by Gessner (1558): (a) and (b) two examples of marine monsters; (c) a whale attacking a ship and another being flensed during whaling operations. Both show mysteces with teeth, which indicates that Gessner never saw these animals. This exemplifies that Gessner was an uncritical compiler of information (public domain image).

Gessner included a 16-page-folio discussion about the dolphin very much along the lines of Aristotle and Pliny. As an uncritical compiler he included contradictory or totally false information such as mythical species and even "monsters." In volume 4 he relied heavily on Belon and Rondelet. For example, *Monachus marinus* (sea monk, IV, p. 519) description was copied from Rondelet who, in turn, had received the description from Marguerite, Queen of Navarre, who heard it from Emperor Charles V's ambassador, who had claimed to see the monster himself (Kusukawa 2010). He did not add much to what was already known. Among marine mammals he mentioned are the *Balaena* (mystecete whales, IV, p. 128) depicted more as sea monster than as an actual whale, *Cetis diversis* (IV, p. 207), an amalgam of marine monsters based on Olaus Magnus's descriptions of sea monsters from seas from northern Europe, *Hominis marinis* (IV, p. 438), a collection of humanoid sea monsters such as the sea-monk and the sea-bishop. To certain extent he was skeptical of accuracy of some of these descriptions by other authors.

Many of the figures were made by others and copied directly from other books including those of "cetaceous" animals as was the case of a whale which was copied from Olaus Magnus' map of the Northern Lands (IV, p. 176).

Ulysses Aldrovandi

Figure 5.11. Aldrovandi and his *De piscibus libri V, et De cetis lib. unus* (1613) (public domain images).

Aldrovandi[300] (Fig. 5.11) was the last author who published anything of significance about marine mammals during the Renaissance. He was born to a noble and wealthy family, which allowed him to initially dedicate his life to his own pursuits. He was educated in Bologna, Padua, and Rome, receiving degrees in law and medicine although he never

---

[300] *b*. Bologna, Italy, 11 September 1522; *d*. Bologna, 4 May 1605.

practiced those professions. He was appointed as the first professor of natural history in the University of Bologna. Although he was a pious Catholic, because of what he read he was charged with heresy. After producing himself in Rome, he was acquitted. While in Rome he met Rondelet and accompanied him to the fish markets where he became interested in ichthyology (which included the study of marine mammals) collecting specimens for his own museum. He traveled extensively throughout Italy and made a collection of about 11,000 animal specimens for pedagogical purposes; most of them can be found today at the Bologna Museum to which he bequeathed not only his specimens but also his library and unpublished manuscripts as well (Alves 2010, pp. 56-82). He also conducted dissections (Impey & McGregor 1985). He was a true encyclopedist following the tradition of the University of Bologna at that time (Tugnoli Pattaro 1994). He wrote extensively but the quality of his animal descriptions and illustrations were poor from the scientific viewpoint (Fig. 5.12). Aldrovandi was an uncritical compiler who included legends of mythical animals in his writings similar to the medieval bestiaries and in the tradition of Pliny.

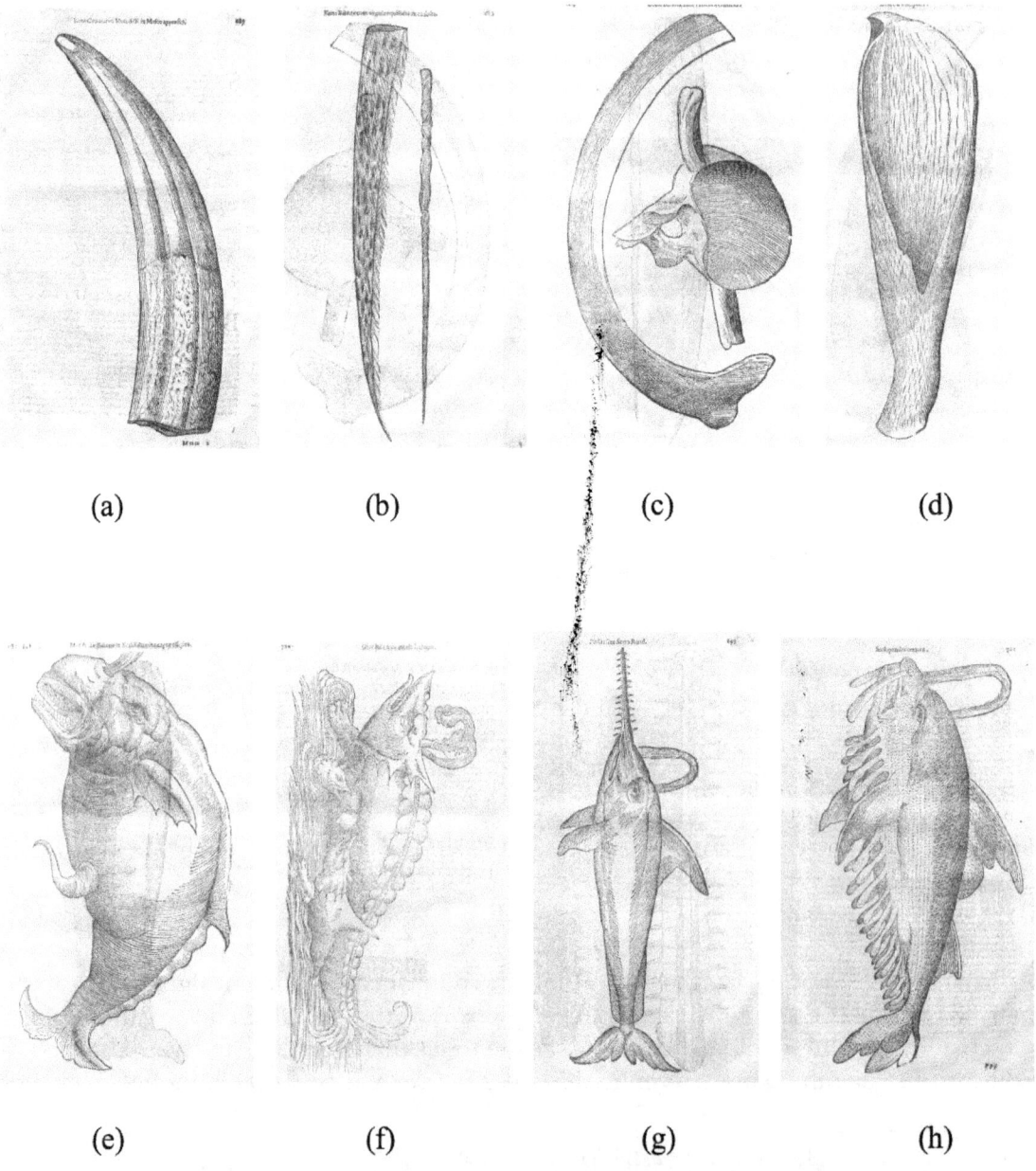

(a) (b) (c) (d)

(e) (f) (g) (h)

Figure 5.12. Aldrovandi was an uncritical compiler who included legends of mythical animals in his writings similar to the medieval bestiaries and in the tradition of Pliny. Figures e-h are highly inaccurate depictions of cetaceans with (f) being a mixed of a crocodile and a cetacean and (g) one of a swordfish and a cetacean (public domain images).

Aldrovandi published *De piscibus libri V, et De cetis lib. vnus* (1613) where he defined "Pisces" as animals covered with scales and "aquatilis" as "anything else that lives in the water" while recognizing that cetaceans are air-breathing creatures. The species that he mentioned were the ones cited by his predecessors: *Balaena, Physeter, Orca, Delphino, Phocaena,* and *Tursione,* while including the *Manate Indorum, Phoca, Pristi* (the sawfish), and the mythical *Scolopendra Cetacea*. From the illustrations (Fig. 5.11) it is clear he never saw any of these animals with the exception of some of their skeletal parts. As an uncritical compiler of information, he did not add anything new to the knowledge of these creatures and, yet, was cited by later authors.

Modern Science

In this period, observation and experimentation moved to the forefront of science. Classification was based on similarities and differences in characters. During this time English physicians travelled to Padua, Bologna and Paris to be trained in human dissection since the status of medicine in England was still poor. People involved in this kind of activities had a background in either medicine (or "physic" as it was called then) and/or theology (Kruger 2004). During this time the center of gravity of science moved from the Mediterranean world to northern Europe, mostly England.

Johann Jonston

The first researcher of the biology of marine mammals in this period was Johann Jonston[301](Fig. 5.13). Although born in Poland, Jonston's father was Scottish and his mother German. He was educated in St Andrews, Frankfurt, Cambridge, and Leiden, receiving a medical degree from the last two institutions. He traveled extensively throughout Europe teaching, and despite offers for academic positions, he decided to make a living as an independent scholar (Miller 2008). He published *Historiae naturalis de Piscibum Partem* in 1657. Jonston was another encyclopedist who when it came to natural history was more a compiler than anything else, relying heavily on Gessner and Aldrovandi while adding some new information from New World creatures from Georg Marcgrave[302]. Yet, Jonston did not offer any significant critical view to his sources although his descriptions were briefer than those of his predecessors. He gave no hint of biological classification for marine mammals and also added further mistakes and legends (even 'monsters'). He slightly modified Aldrovandi's classification of fishes by adding 'pelagic' fishes. Yet his books were widely read and translated.

---

[301] *b.* Szamotuly, Poland, 15 September 1603; *d.* Legnica, Poland, 8 June 1675.
[302] *b.* Liebstadt, Electorate of Saxony, Holy Roman Empire, today German, 1620; *d.* today Guinea, 1644.

Figure 5.14. Jonston and Tab. XLI. Whales. *Historia Naturalis, de Exanguibus Aquaticis* (ca. 1655) (public domain images).

Walter Charleton

Charleton[303] (Fig. 5.14) was the son of a church rector of modest means. He was educated at Oxford as a physician at that time when medical education in England emphasized scholastic approaches to knowledge and British colleges had inadequate anatomical staff and teaching facilities. The practical elements of practicing medicine were not acquired until after assisting a more experienced practicing medical doctor.

Charleton was a follower of epicurean atomism (materialism) (Kargon 1964) and an eclectic (Lewis 2001), whose interest in natural history was more or less theological because, as he said, men were obligated into "naming & looking into the nature of all Creatures" (Boot 2005, p. 119). In other words, just as Ray and Willoughby did later, natural science was the search a divine pattern in nature, part of the research agenda of the Royal Society – to which Charleton belonged (Rolleston 1940, Sharpe 1973). His publications showed him more as a compiler than as an innovator. His major contribution to science was the discovery that tadpoles turn into frogs (Booth 2005, p. 1).

He published two books dealing with animal classification: *Onomasticon zoicon* (1668) and *Exercitationes de Differentiis & Nominibus Animalium* (1677) works that listed the names of all known animals (including some fossils) in the western world in several languages with a somewhat taxonomy discussion, including remarks about these animals habits and habitats that contained anatomical descriptions of two animals that he had dissected. As Belon did over a century before, he divided "fishes" as either "with blood" (vertebrates) and "without blood" (invertebrates). He grouped under "Cetaceos" not only actual cetaceans but also the sawfish, seals, walruses, manatees, hippopotamus and the mythical "scolopendra cetacea." The actual cetaceans described were *Balaena vulgaris* (probably the right whale), *Physeter, & Physalus* (probably the fin whale but also other species), *Cetus dentatus* (the sperm whale), *Pustes* (indeterminate species, maybe the beluga), *Orca* (the killer whale), *Monoceros* (the narwhal), *Delphinus* (probably a composite of delphinidae), and *Phocaena* (the porpoise).

---

[303] *b.* Shepton Mallet, Somerset, England, 2 February 1619; *d.* London, England, 24 April 1707.

Figure 5.14. Charleton and his *Exercitationes de Differentiis & Nominibus Animalium* (1677) (public domain images).

Edward Tyson

Tyson[304] (Fig. 5.15) was born into an affluent merchant family. He performed numerous dissections as a college student, obtained his medical degree at Oxford University and was a lecturer of Anatomy at the Barber-Surgeons Hall in London. Tyson was the first of the comparative anatomists in the modern sense. He did extensive dissections and was the first to use a microscope as part of his anatomical studies. His description of the highly convoluted cetacean brain as well as his recognition of the many homologies with "viviparous quadrupeds", rather than the fishes that they externally resembled, constituted a major landmark contribution to the history of biology (Kruger 2003).

In *Phocaena, or, The anatomy of a porpess dissected at Gresham Colledge, with a preliminary discourse concerning anatomy and natural history of animals* (1680), he noted that

> *"What we have here is a signal Example of the same between Land-Quadrupeds and Fishes; for if we view a Porpess on the outside, there is nothing more than a fish; for if we view a Porpess on the inside, there is nothing less. (...) It is viviparous, does give suck, and hath all its Organs so contrieved according to the standard of them in Land-Quadrupeds; that one would almost think of it to be such, but it lives in the Sea, and hath but two fore-fins."* Adding later *"The structure of the viscera and inward parts have so great an Analogy [sic] and resemblance to those of*

---

[304] *b.* Clevedon, near Bristol, Somerset, England, 20 January 1651; *d.* London, England, 1 August 1708.

Quadrupeds, that we find them here almost the same. The greatest difference from them seems to be in the external shape, and wanting feet. But here too we observed that when the skin and flesh was taken off, the forefins did very well represent an Arm, there being the Scapula, an of Humeri, the Ulna, and Radius, and bone of the Carpus, the Metacarp, and 5 digiti curiously joynted. The Tayle too does very well supply the defect of feet both in swimming as also leaping in the water, as if both hinder feet were colligated into one, though it consisted not of articulated bones but rather Tendons and Cartilages."

Tyson's description of the internal anatomy of the porpoise is remarkable, particularly when it comes to its nervous system (Kruger 2003). In many ways he thought that the "porpess" was the transitional link between terrestrial mammals and fish (Fig.5.16).

In his monograph Tyson surveyed contributions from previous authors. He corresponded with John Ray (see below). Ray had also dissected a porpoise (an exercise on which he reported in a published form in 1671), nine years before Tyson but was far more superficial and added very little to what other authors such as Rondelet had done. Tyson met Ray around 1683 and the latter invited Tyson to contribute to Willughby's *De Historia Piscium* (Montagu 1943, p. 103).

Tyson was critical of encyclopedic approaches and relying on classical authors when it came to natural history. He set new standards in terms of direct observation and comparative anatomy. He also established an understanding of homology not seen since Aristotle. He proved to be a very competent observer of internal anatomy and he saw comparative anatomy as a means to explain the Great Chain of Being (or *scala naturae* or ladder of nature) as proposed by Plato and Aristotle.

Figure 5.15. Tyson (public domain image)

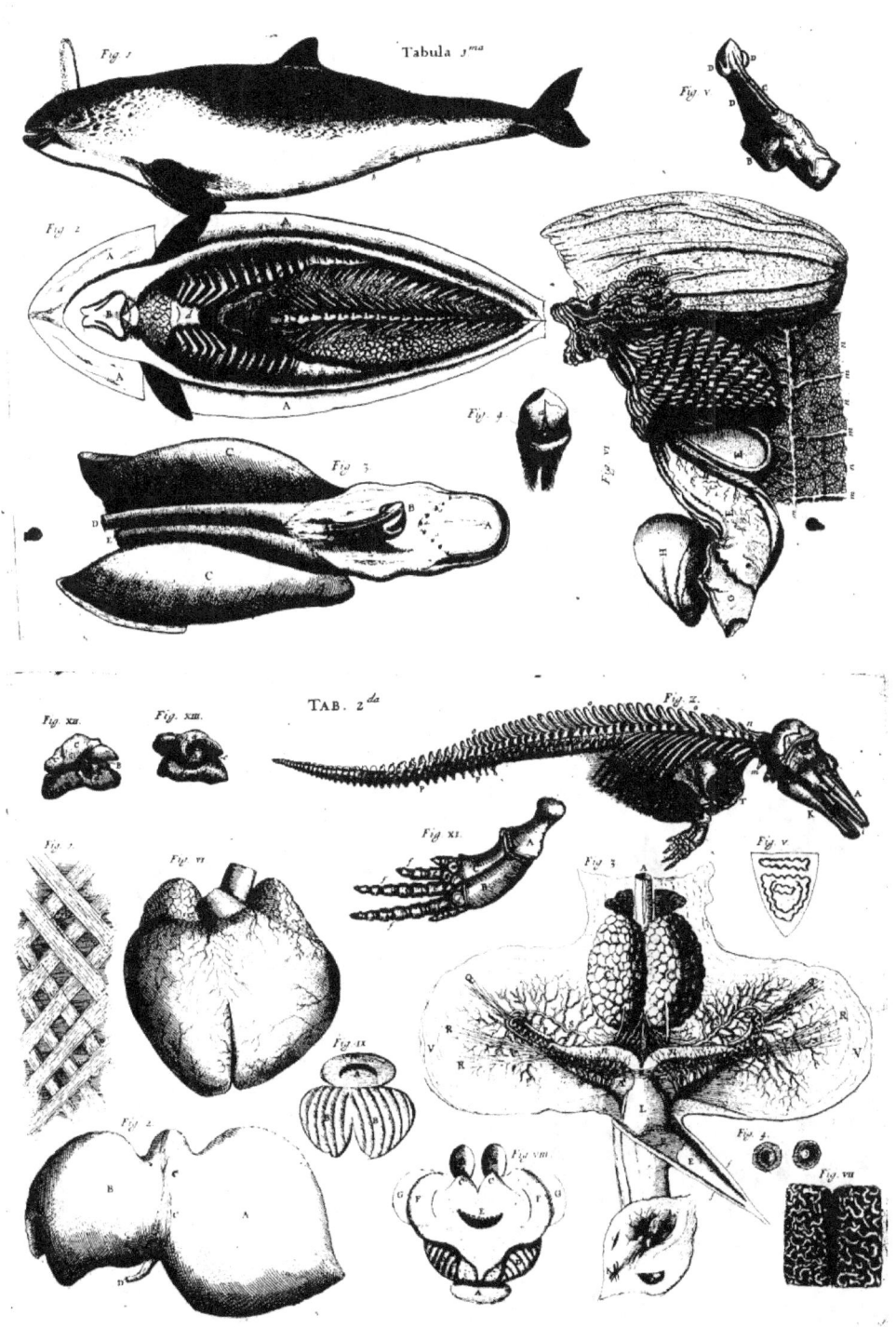

Figure 5.16. Illustrations from Tyson's (1680) description of the internal anatomy of a porpoise. Notice the remarkable accuracy of the depictions (public domain images).

Samuel Collins

A contemporary of Tyson was Samuel Collins[305](Fig. 5.17). The son of the rector of Rotherfield, Sussex, who got his education at Cambridge, Collins travelled to several

---

[305] *b.* Rotherfield, Sussex, England, 1618; *d.* Westminster, Middlesex, England, 11 April 1710,

universities in France, Italy and the Netherlands finally getting his medical degree at the University of Padua, later becoming physician of Charles II. He taught anatomy at the Royal College of Physicians[306]. Collins published *A Systeme of Anatomy* (London 1685), which was the earliest attempts to illustrate the brains of a broad variety of mammals, birds, teleosts, and elasmobranchs in a remarkable two-volume folio edition of 1,263 pages. It included 73 full-page illustrations of very high quality. There he described a female porpoise. However, it seems that he had used Tyson's previous descriptions and unfortunately says nothing about the brain of this cetacean. Had he had examined the brain of the porpoise he would have noted the great similarities of this organ with those among the "viviparous quadrupeds." Collins did not discuss the similarities between the other internal organs of the porpoise and those called mammals today either. He acknowledged Tyson' previous contributions in this matter.

In addition to Tyson, Collins's anatomy draws largely upon the works of Thomas Willis. In the opening Epistle-Dedicatory to James II he claimed that various chapters "are illustrated by the Dissection of other Animals (which I have performed with Care and Diligence, speaking the wonderous Works of the Glorious Maker) rendering the Parts of Man's Body clearer and more intelligible." In volume two of his huge work he described numerous folio copper plates containing the most extensive comparative anatomy of the brain then existing, an expansive account of the functional significance of his findings, as well as practical clinical commentary.

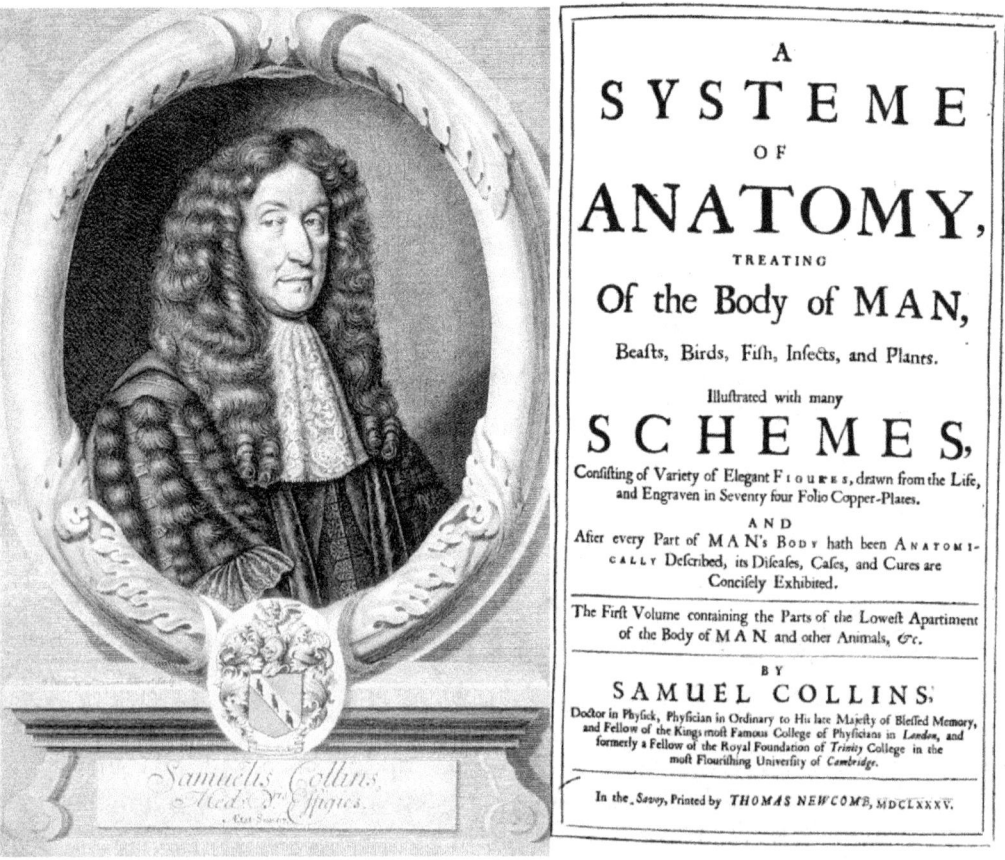

Figure 5.17. Collins and his *A Systeme of Anatomy* (1685) (public domain images).

---

[306] Biographical information obtained from: http://munksroll.rcplondon.ac.uk/Biography/Details/950 and accessed on 2 April 2012.

John Ray

Ray[307] (Fig. 5.18) was the first naturalist who truly represented this new era of careful observation. His father was a blacksmith and his mother was an herbal healer. He studied at the University of Cambridge, pursuing comparative anatomy although initially his main interest was botany. He taught Greek, mathematics and humanities at Cambridge but abandoned his teaching position after refusing to comply with the Act of Uniformity of 1662. He was a very religious person who undertook the study of nature to understand God's creation (Raven 1950). Fairly early he developed a plan with his student and patron, Francis Willughby[308] to produce a joint general natural history. To that end Ray and Willughby went on an extended tour of England and Europe (1662-1666), including the medical school at Montpellier. Although they did not always travel together both collected specimens, got involved in dissections and acquired books and illustrations (Kusukawa 2000), an endeavor bankrolled by Willughby. When Willughby died, Ray took over his parts of the general natural history. Willughby left him an annuity of £60 and Ray stayed on as tutor to Willughby's children until 1675, when Willughby's mother, also his patron, died, and the widow immediately terminated the relationship. Ray inherited a small farm that also contributed to the family's maintenance while he earned money from his productive publishing. Therefore, Ray had the financial freedom to pursue his intellectual interests.

Ray's first published work on cetaceans was *Dissection of a Porpess* (1671). He does a much better job in describing the internal anatomy of this animal when compared with Rondelet but does not get into the detail that Tyson achieved later. During the narrative of his findings, he keeps noticing that a porpoise has a lot in common with the "quadrupeds". Yet he persisted calling them "fishes."

Ray published *Historia piscium* (1686), under Willughby's name 14 years after his patron death, though Ray himself contributed the vast majority of the content. He carried out the first serious attempt to achieve a systematic arrangement, the success of which can be attributed by the fact that it served as a basis for the systematics work of the following century. His approach was based on direct observation, collaboration with other researchers, and critical reading of previous authors.

Historia Piscium is divided into two parts that were printed separatedly: the first is the narrative and the second, titled *Ichthyographia*, were the illustrations. Many libraries today have both bound together. As sources Ray used authors mentioned earlier in this chapter: Rondelet, Salviani, Gessner, Aldrovandi and Belon, among others. Yet, far from merely compiling information from them, Ray insisted in very comprehensive descriptions of species and discarded all monsters and mythical creatures mentioned by his predecessors. Ray not only removed narratives of marine invertebrates but also other aquatic animals such as the crocodile and the hippopotamus. He divided his subject matter into three groups: cetaceans, cartilaginous fishes, and bony fishes. He recognized that when it comes to reproduction and internal anatomy cetaceans are identical to the "viviparous quadrupeds." Still, he kept cetaceans within the "piscium" despite the fact that he was well aware that they were biologically distinct from fishes.

---

[307] *b.* Black Notley, near Brainton, Essex, England, 29 November 1627; *d.* Black Notley, England, 17 January 1705.
[308] *b.* Middleton, Warwickshire, England, 22 November 1635; *d.* Middleton, 3 July 1672.

Figure 5.18. John Ray and his *Synopsis Methodica Animalium Quadrupedum et Serpentini Generis* (1693) (public domain images).

In his narrative of species Ray moved away from in the practical aspects related to these animals. Aspects such as usage for medical purposes were very common among previous authors because of their medical background. Yet, Ray was very keen at compiling names on the belief that a universal language could be construct based on the knowledge of nature. As Kusukawa (2000) has argued convincingly, Ray believed that there was a need for "a construction of a universal language based on a table that properly expressed the natural order and relations between things." Hence a precise description and classification was the route to achieve that goal. The final product counted not only on the intellectual support of the Royal Society's members who provided constructive criticism and moral support but also their financial support. The cost of publishing *Historia Piscium* was not only very high, mostly because of the expense of the illustrations (187 plates), but also the 500 copies printed sold poorly. As a consequence, the Society could not print Isaac Newton's *Principia*.

Ray's third publication related to marine mammals was *Synopsis Methodica Animalium Quadrupedum et Serpentini Generis* (1693). By then he was totally convinced that cetaceans were not fishes: "For except as to the place on which they live, the external form of the body, the hairless skin and progressive swimming motion, they have almost nothing in common with fishes, but remaining characters agree with the viviparous quadrupeds." He placed today's terrestrial mammals (including the manatee) among the 'hairy animals' very close to the *Cetaceum genus* (cetaceans).

In *Synopsis* Ray included a section called *Pisces Cetacei seu Belluae marinae* where he expressed that these animals breath and give birth like the "oviparous quadrupeds." He grouped them into two categories according to the presence of teeth much as we do today to separate odontocetes from mysticetes. Ras was the first in doing so. The species he cited were *Balaena vulgaris* (Rondelet), *Balaena* (Fin-Fish), *Physeter* or *Balaena physeteris*, *Orca* (Rondelet & Belon), *Cete* (Sperm whale), Pot Walfish, *Albus piscis cetaceus* (white fish), *Monoceros cetaceo* (*Narhual islandis*), *Delphino antiquorum* (dolphin, from Rondelet), *Phocaeno* (Rondelet & Belon), dissecting a specimen of the latter in 1669.

Ray developed a division of animals characterized by having blood, breathing by lungs, two ventricles in the heart, and being viviparous. Ray subdivided this group into aquatic (cetaceans) and terrestrial or quadruped including sirenians (manatees and dugongs). He rejected tales of fabulous animals while perfecting Aristotle's classification by diving vertebrates into those having hearts with two ventricles (mammals and birds) from those with a single ventricle (reptiles, amphibians and fish). He also advanced the understanding of other groupings. He established the significance of the generic principle, defined species, and was a leading contributor to the gigantic task of classification.

Ray came close to recognizing mammals as a separate group based on "warm-blood," vivipary, and hair. He conceded the relationship of cetaceans with viviparous quadrupeds; described genera and species; established ordinal classification of mammals; systematic phrases and names; used of descriptive phrases as well as monomial names (a taxonomic name consisting of a single word); a dichotomous ("A is B or not B") classification of mammals. Yet, he lacked the vision or intellectual courage to reunite marine mammals with their terrestrial relatives and still placed the former with the fish "in accordance with common usage." Still, he was possibly the best naturalist of the seventeenth century.

Peter Artedi

Artedi[309](Fig. 5.19) was the son of a parish priest who developed an interest in fishes from an early age. He studied medicine at the University of Uppsala, devoting most of his time at studying natural history. At 29 years of age, he went to London for a year to study natural history collections and described the sighting of a whale in November 1734, probably downstream of the London Bridge. He then moved to Leiden, The Netherlands, to complete his medical studies and there he met Linnaeus, whom he knew from their native Sweden, forging a lifelong personal and professional relationship. Linnaeus introduced him to an Amsterdam chemist, Albert Seba, and Artedi started working on Seba's fish collection. Artedi died at the age of 30 by drowning in an Amsterdam canal. After his death, Linnaeus recovered his manuscripts and published *Ichthyologia* (1738) without amending Artedi's original work. Despite the fact that this was an unfinished work, it was a fundamental publication that marked the origin of ichthyology as we know it today. After a long (96 pages) introduction describing previous authorities on ichthyology the second part deals with the taxonomic terminology he used, particularly regarding the concept of genus and distinguishing between species and varieties. His system set the basis for the modern systematic classification of living organisms later established by Linnaeus. In part three he went into the classification of species including detailed description of them, some of which he had dissected himself. For this Artedi is considered the father of ichthyology (Wheeler 1962, 1987, Broberg 1987).

---

[309] *b*. Anundsjö, Västernorrland, Sweden, 10 March 1705; *d*. Amsterdam, The Netherlands, 27-28 September 1735.

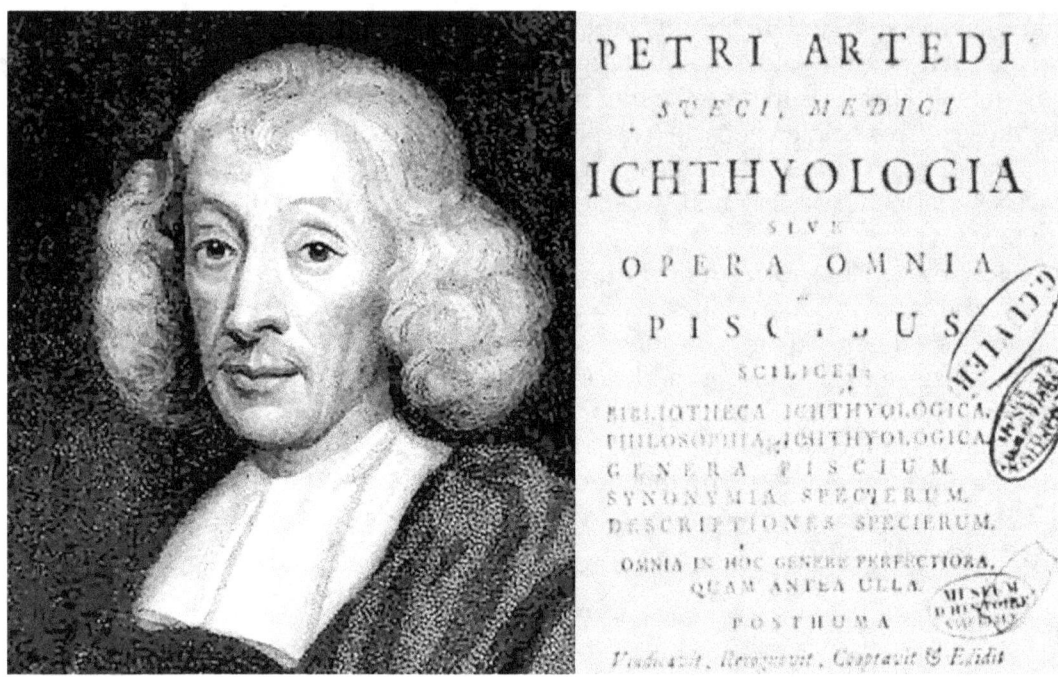

Figure 5.19. Artedi and his *Ichthyologia* (1738) (public domain images).

Artedi separated actual fishes from cetaceans (which he called "plagiuri") based on the plane of the caudal fin. He described 7 genera and 14 species including the manatee and the "siren" as follows:

Order: Plagiuri
    Physeter
        Balaena major (Ray, p. 15)
        Balaena macrocephala (Ray, p. 16)
    Delphinus
        Delphinus (*Phocaena*) (Art. Syn. 104)
        Delphinus (*Delphin*) (Art. Syn. 105)
        Delphinus (*Orca*) (Art. Syn. 106)
    Balaena
        Balaena vulgaris (Ray p. 6, 16)
        Balaena edentula Fin-Fish (Ray p. 6, 10)
        Balaena tripinnis (Ray 16)
        Balenae (*Balaena tripinnis*) (Ray 17)
    Monodon
        Monoceros pisces (Will. 42, Ray 11, Charleton 168)
    Catodon
        Balaena minor (Ray p. 15)
        Balaena major (Ray p. 17, Will. P. 41)
    Trichechus
        Manatus (Rondelet p. 490, Gessner p. 213, Charleton 169, Aldrovandi 7 28, Jonston 223)
    Siren
        Homo marinus

Artedi established the basic classification of fishes that lasted for about 200 years and separated cetaceans into a totally different order than fishes; he apparently knew that they were different, but still tradition was difficult to break and thus he included them into his ichthyological treatise. He also established the basic branching of animal groups into Class, Maniples (Families), Genera, and Species, a system that was to be closely followed by Linnaeus (Wheeler 1987, Broberg 1987). His work set the foundations for what Linnaeus would culminate as the definitely recognition of cetaceans as distinct group within mammals.

Carolus Linnaeus

Linnaeus (or Linné)[310](Fig. 5.20) had as a father a country person who loved plants. Linnaeus followed a medical career but was actually more interested in botany than in anything else. Linnaeus met Artedi in 1729 and their interests were complementary: Artedi, a zoologist interested mostly on fishes, and Linnaeus, interested in botany. He would later edit Artedi's book in ichthyology that was published in 1738. What Linnaeus learned from Artedi set the basis for a better classification not only of plants but also animals in general.

Even some of Linnaeus students were developing a better understanding of cetaceans as being really close to "viviparous quadrupeds." That was the case of Pehr Löfling[311], one of Linnaeus' students who came very close to making major contributions to the true nature of dolphins and manatees based on his observations of these animals in South America. In his description of Amazon freshwater dolphin or boto, Löfling was clear about when writing that whales and dolphins were different from fishes: ("*Pisces per pulmonibus spirantibus*"). However, his early death and the fact that his manuscripts were never published prevented him from gaining recognition in the scientific community (Romero *et al.* 1997).

With all of this background, the botanist Linnaeus was ready to revolutionize biological classification and in the 10th Edition (1758) of his fundamental work *Systema Naturae*, he introduced the term Mammalia, and included *Cete* among them (Fig. 5.21). For Linnaeus, mammals were united by having hair, being viviparous, and producing milk. He coined the term *cetacea* and separated them from fishes and grouped them with the rest of the viviparous quadrupeds based on the following characteristics: two-chamber heart, breathing by lungs, hollow ears, internal fertilization, and production of milk.

Thus, Linnaeus revolutionized the science of systematics by developing a fully natural system of classification, using consistently the binomial nomenclature, and designing species with Latinized names (genus and species). He developed a hierarchy (class, order, genus, species) as proposed by Artedi, with species defined as similar individuals bound together by reproduction, which also set the basis of the biological species concept. The use of telegraphic speech-like (very short sentences) diagnosis for species descriptions and the standardization of synonymies (same species with different names) in order to reach a taxonomic consensus made his classification even more useful since from now on one could find clarity on what a particular species was tracing its description to other authors. He also doubled the number of species described by Ray. Thus, despite the fact that he was not a zoologist *per se* nor was involved in dissection of animals, he was far from a compiler in that he applied critical thinking to the way he ordered nature.

---

[310] *b*. Stenbrohult, Småland, Sweden, 23 May 1707; *d*. Uppsala, Sweden, 10 January 1778.
[311] *b*. Valbo, Sweden, 1729; *d*. Guayana, Venezuela, 22 February 1756.

Figure 5.20. Linnaeus (public domain image).

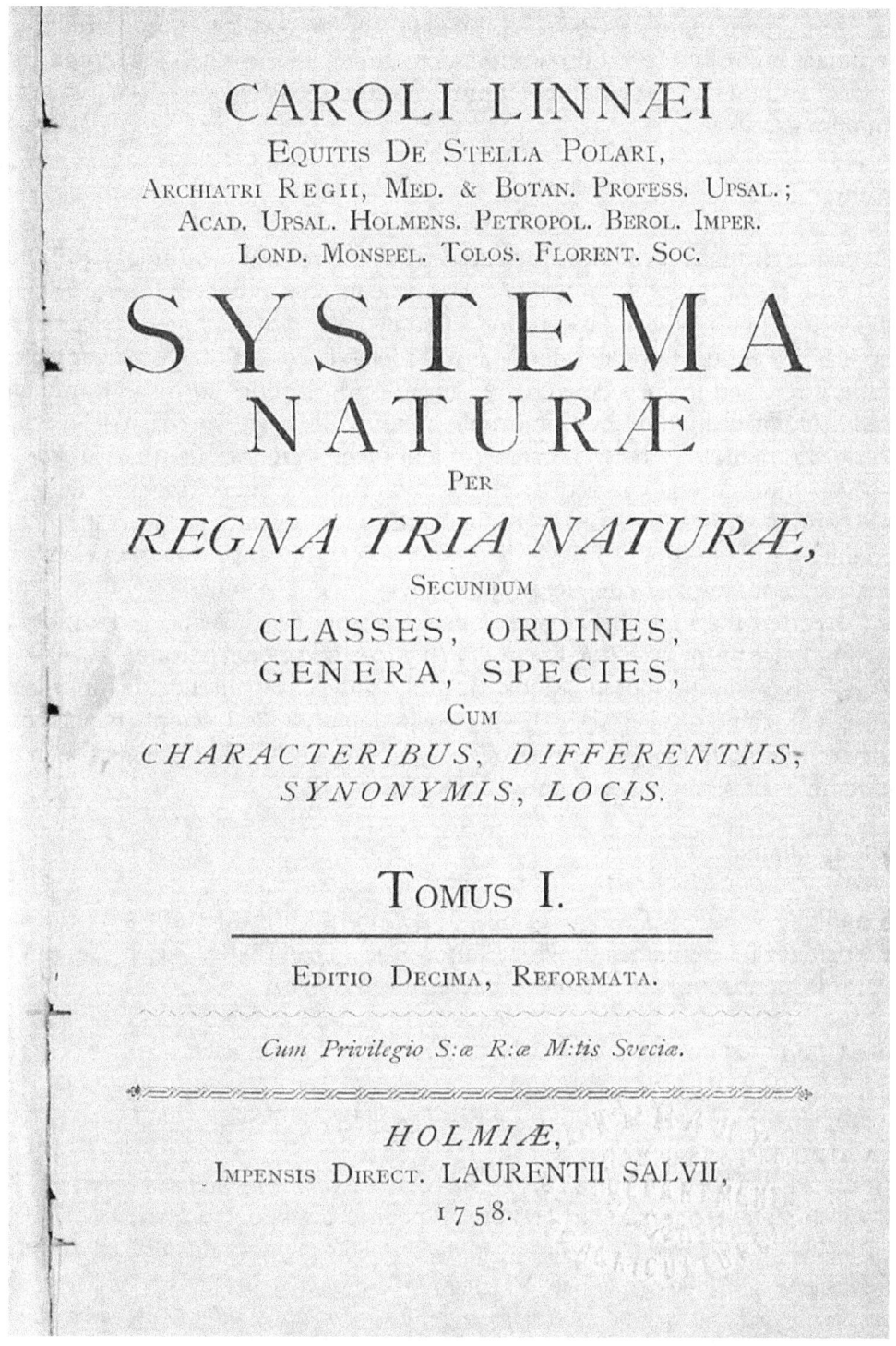

Figure 5.21. Linnaeus' 10th edition of his *Systema Naturae* (public domain image).

This progress is even more remarkable when considering that Linnaeus was far from an evolutionist. For him species were fixed except for small variations due to climatic/local conditions. Yet, Linnaeus was, without question, the founder of systematics and the one who laid the foundations for the naturalists to become specialists and, therefore, opened the door for the first group of marine mammal specialists, now that these creatures were no longer considered "fishes." It was not until Linnaeus that the science of taxonomy made the strides that have led us to where we are today in our understanding of the natural world. Linnaeus understood biological principles and placed animals in groups based on homologies rather than using environment to drive classification, and this was what allowed him to recognized cetaceans as a distinct group within mammals.

Conclusion

Pursuing at the information provided above there are a number of discernable patterns. One is the preponderance of pre-Linnean researchers interested in marine mammals who had a medical background of some sort. That is not surprising because medicine was the closest thing to science as a career existed until the eighteenth century. Also, being interested in medicine created more opportunities to dissect animals and, therefore, understanding of their internal anatomy that was particularly crucial in establishing the homology between cetaceans and the "viviparous quadrupeds." Yet, this positive influence was marred by the proliferation of encyclopedists who, for the most part, were uncritical compilers of other authors' information. However, the major impediment to any attempts to develop a natural classification for cetaceans was the insistence on classifying them by virtue of the environment in which they live, something that even diverted the thoughts of keen observers such as Ray and Artedi, despite of abundant evidence to the contrary having been collected since Aristotle.

Finally, we should not overlook the role played by intellectual inertia in the development of science. As Horder (1998) clearly demonstrated, scientists need to know the history of their field to avoid errors of the past, something that has also been argued for specific fields of biology (see Romero 2009, Chapter 1).

Acknowledgements

I thank Dr. Matthew Cashen for his advice interpreting Aristotle's writings and to Dr. Carl Springer for similar help with Pliny's work. Dr. Edward Keith read an earlier version of this manuscript and made valuable suggestions.

Literature Cited

Aldrovandi, U. 1613. *U. Aldrovandi ... de Piscibus libri V et de Cetis lib. Uncus.* Bononiae, 732 pp.

Alves, W.L. 2010. *Ulisse Aldrovandi's Opera Omnia: collecting natural wonders.* Thesis: Departmental Honors in Art History, Wheaton College, Norton, MA. 179 pp.

Artedi, P. 1738. *Ichthyologia sive opera omnia piscibus scilicet: Bibliotheca ichthyologica. Philosophia ichthyologica. Genera piscium. Synonymia specierum. Descriptiones specierum. Omnia in hoc genere perfectiora, quam antea ulla. Posthuma vindicavit, recognovit, coaptavit & edidit Carolus Linnaeus, Med. Doct. & Ac. Imper. N.C.* Wishoff, Leiden.

Barnes, J. 1995. Life and work, pp. 1-26, *In*: J. Barnes (Ed.). *The Cambridge Companion to Aristotle.* Cambridge: Cambridge University Press.

Bay, J.C. 1916. Gesner, the father of bibliography. *Papers of the Bibliographical Society of America* **10**:53-88.

Belon, P. 1551. *L'histoire naturelle des estranges poissons marins: avec la vraie peincture & description du daulphin, & de plusieurs autres de son espece.* Paris: De l'imprimerie de Regnaud Chaudiere.

Belon, P. 1551. *La nature et la diversité des poisons. Avec leurs pourtraicts representez au plus près naturel.* Paris: Ch. Estienne.

Booth, E. 2005. *'A subtle and mysterious machine'. The medical world of Walter Charleton (1619-1707)*. Dordrecht: Springer.

Broberg, G. 1987. Petrus Artedi in his Swedish context. Proc. V Congres Europeé Ichthyolofie, Stockholm 1985, pp. 11-15.

Charleton, W. 1668. *Onomasticon zoicon plerorumque Animalium Differentias et Nomina Propria pluribus Linguis exponens. Cui accedunt Mantissa Anatomica et quaedam de Variis Fossilium Generibus*. Jacobum Allestry, London. 363 pp.

Charleton, W. 1677. *Exercitationes de Differentiis & Nominibus Animalium*. Oxford. 343 pp.

Collins, S. 1685. *A Systeme of Anatomy* treating of the body of man, beasts, birds, fish, insects and plants. Illustrated with many schemes, consisting of variety of elegant figures, drawn from the life, and engraved in seventy-four Folio copper plates and after every part of man's body hath been anatomically described, its diseases, cases and cures are concisely exhibited. London, 2 vol.

Delaunay, P. 1926. *Pierre Belon naturaliste*. Le Mans: Imprimerie Mannoyer.

Dulieu, L. 1966. Guillaume Rondelet. *Clio Medica*. **1**:89-111.

Ellwood, C.A. 1938. *A history of social philosophy*. New York: Prentice Hall.

Gudger, E.W. 1934. The five great naturalists of the sixteenth century: Belon, Rondelet, Salviani, Gesner, and Aldrovandi: a chapter in the history of ichthyology. *Isis* **22**:21-40.

Fischer, H. 1966. *Conrad Gessner (1516-1565) as bibliographer and encyclopedist*. The Library 21:269-281.

Gessner, C. 1558. *Historiae Animalium*. Liber IIII qui est de Piscium & Aquatilium natura… Tiguri

Glardon, P. 2011. *L'Histoire naturelle au XVIe Siècle*. Genève: Libraire Droz S.A.

Gmelig-Nijboer, C.A. 1977. *Conrad Gessner's "Historia Animalium" an inventory of Renaissance Zoology*. Utrecht: Communicationes Biohistoricae Ultrajectinae.

Horder, T. J. 1998. Why do scientists need to be historians? *Quarterly Review of Biology* **73**:175-187.

Impey, O. & A. MacGregor. 1985. *The origins of museums. The cabinet of curiosities in sixteenth- and seventeenth-Century Europe*. Oxford: Clarendon Press.

Jonston, J. 1657. *Historiae naturalis de piscibus et cetis libri V, cum aeneis figuris, Johannes Jonstonus,… concinnavit*. J. J. Schipperi, Amsterdam. 306 pp.

Kargon, R. 1964. Walter Charleton, Robert Boyle, and the acceptance of Epicurean Atomism in England. *Isis* **55**:184-191.

Keller, A.G. 1975. Rondelet, Guillaume, pp. 527-528, *In*: C.C. Gillispie (Ed.). *Dictionary of Scientific Biography*, Vol. 9. New York: Scribner's Sons.

Kruger, L. 2003. Edward Tyson's 1680 account of the 'porpess' brain and its place in the history of comparative neurology. *Journal of the History of the Neurosciences* **12**:339-349.

Kruger, L. 2004. An early illustrated comparative anatomy of the brain: Samuel Collins' A Systeme of Anatomy (1685) and the emergence of comparative neurology in 17th century England. *Journal of the History of the Neurosciences* **13**:195-217.

Kusukawa, S. 2000. The *Historia Piscium* (1686). *Notes Rec. Royal Society London* **54**:179-197.

Kusukawa, S. 2010. The sources of Gessner's pictures for the *Historia animalium*. *Annals of Science* **67**:303-328.

Leu, U.B.; R. Keller & S. Weidmann. 2008. *Conrad Gessner's private library*. Leiden: Brill.

Leitner, H. 1927. *Zoologische terminologie beim älteren*. Hildesheim: Verlag.

Lewis, E. 2001. Walter Charleton and early Modern Eclecticism. *Journal of the History of Ideas* **62**:651-664.

Linnæus, C. 1758. *Systema naturæ per regna tria naturæ, secundum classes, ordines, genera, species, cum characteribus, differentiis, synonymis, locis*. Tomus I. Editio decima, reformata. – pp. [1–4], 1–824. Holmiæ. (Salvius).

Locher, A. 1984. The structure of Pliny the Elder's Natural History, pp. 20-29, *In: Science in the Early Roman Empire: Pliny the Elder, his sources and influence* (R. French and F. Greenaway, Eds.). London: Croom Helm.

Miller, G.L. 2008. Beasts of the New Jerusalem: John Jonston's natural history and the launching of millenarian pedagogy in the seventeenth century. *History of Science* **46**:203-243.

Montagu, A. 1943. *Edward Tyson, M. D., F. R. S., 1650-1708, and the rise of human and comparative anatomy in England; a study in the history of science.* Philadelphia: The American Philosophical Society.

Moseley, A. 2010. *Aristotle.* London: Continuum International Publishing Group.

Murphy, T. 2004. *Pliny the Elder's Natural History: The Empire in the Encyclopedia.* Oxford: Oxford University Press.

Naas, V. 2002. *Le projet encyclopédique de Pline L'Ancien.* Rome: Ecole Française de Rome.

Nutton, V. 1985. Illustrations from the Welcome Institute Library. *Medical History* **29**:93-97.

Oppenheimer, J.M. 1936. Guillaume Rondelet. *Bulletin of the Institute of the History of Medicine* **4**: 817-834.

Raven, C.E. 1950. *John Ray naturalist. His life and works.* Cambridge: University Press.

Ray, J. 1671. An account of the dissection of a porpess, promised Numb. 74, made, and communicated in a Letter of Sept. 12, 1671, by the learned M. John Ray., having therein observ'd some things omitted by Rondeletius. *Philosophical Transactions (Royal Society)* **6**:2274-2279.

Ray, J. 1686. *Franciscus Willughbei...De historia piscium libri quatuor.* Oxonii, 343+30 pp.

Ray, J. 1713. *Synopsis methodica avium & piscium; opus posthumum...*Londini, 166 pp.

Reynolds, J. 1986. The Elder Pliny and his times, pp. 1-10, *In: Science in the Early Roman Empire: Pliny the Elder, his sources and influence* (R. French and F. Greenaway, Eds.). London: Croom Helm.

Rolleston, H. 1940. Walter Charleton, D.M., F.R.C.P., F.R.S. *Bulletin of the History of Medicine* **8**:403-406.

Romero, A. 2009. *Cave Biology: Life in Darkness.* Cambridge: Cambridge University Press.

Romero, A., A.I. Agudo and S.J. Blondell. 1997. The scientific discovery of the Amazon river dolphin *Inia geoffrensis. Marine Mammal Science* **13**(3):419-426.

Rondelet, G. 1554-1555. *Libri de piscibus marinis...Universae aquatilium historiae pars altera...*Lugduni, 583 pp.

Rondelet, G. 1558. *La Premiere Partie de l'Histoire entiere des Poissons ...* Lyon, France. 599 pp.

Sharpe, L. 1973. Walter Charleton's Early Life, 1620-1659, and Relationship to Natural Philosophy in Mid-Seventeenth Century England. *Annals of Science* **30**:311-340.

Simpson, G. G., 1945. The principles of classification and a classification of mammals. *Bulletin of the American Museum of Natural History* **85**:163 – 216.

Tugnoli Pattaro, S. 1994. Scienziati pionieri all'Universita di Bologna: il caso Aldrovandi. *Forum Italicum* **8**:1130124.

Tyson, E. 1680. *Phocaena, or, The anatomy of a porpess dissected at Gresham Colledge, with a preliminary discourse concerning anatomy and natural history of animals.* London: Benj. Tooke, 48 pp.

Wellisch, H.H. 1984. *Conrad Gessner. A Bio-Bibliography.* Zug: IDC.

Wheeler, A.C. 1962. The life and work of Peter Artedi, (pp. vii-xxi) *In: Peter Artedi Ichthyologia.* New York: Wheldon & Wesley.

Wheeler, A.C. 1987. Peter Artedi, founder of modern ichthyology. *Proceedings Congress Europeé Ichthyology.*, Stockholm 1985, pp. 3-10.

Wong, M. 1970. Belon, Pierre. Pp. 595-596, *In:* C.C. Gillispie (Ed.). *Dictionary of Scientific Biography*, Vol. 1.

Wotton, E. 1552. De differentiis animalium libri decem ... cum amplissimus indicibus, in quibus primùm authorum nomina, unde quaequae desumpta sunt, singulis capitibus sunt notata & designata: deinde omnium animalium nomenclaturae, itémque singulae eorum partes recensentur, tam Graec.

Section II. Histories of Discoveries and Demythizations

This section is a collection of essays about scientific discoveries. The idea of these narratives is to contrast different approaches in scientific research and reporting. The historical conditions are given to understand better why these cases were worked out and reported the way they did.

Chapter 6 in this section ("The Scientific Discovery of The Amazon River Dolphin *Inia geoffrensis*") deals with a case of a scientist (Pehr Löfling) sent to South America by another (Carl Linnaeus) to describe new species of plants and animals. Among the latter, he found and described a South American species of freshwater dolphins. Löfling passed away before he could have his manuscripts published. Those manuscripts had a convoluted journey from the rural, colonial Venezuela to the halls of science in Eighteenth-Century Europe. Despite all this, his work was an example of scientific accuracy and diligence.

A similar case occurred with the discoveries of Cuba's first blind cavefishes ("Chapter 12: The Discovery of The First Cuban Blind Cave Fish: The Untold Story"). Despite being in a far-away colony (Cuba) Spain, the person who carried this discovery was a Cuban national (Felipe Poey). Poey had benefited from an education in Europe and while he maintained contact with the most prominent scientists and scientific institutions of his time throughout his life. This contradicts the old idea of the Latin American "Isolated genius" fable. Not only that, it was this discovery that made him change his position on the origin of species from an ambivalent creationist-evolutionist stance to a firm Darwinian position.

The other five chapters (7-11) are a different story. They deal with alleged discoveries of species that never existed. A combination of factors can explain these false narratives. First is the lack of scientific preparation by the people who claimed these discoveries. The second was the lack of scientific standards that require checking out the facts before announcing the alleged discovery. Third (and the most common one) is the human impulse to believe what you want to believe. Among the examples described along these lines is the case of Humboldt and his "volcano fish" ("Chapter 10: Humboldt's Alleged Subterranean Fish from Ecuador"). This case is striking for several reasons. First, Humboldt was a brilliant scientist whose other discoveries were well documented and respected by his colleagues at that time. Second, it is an impossibility for a first to survive inside a volcano. That defies reason. Yet, Humboldt believed he has discovered a "volcano fish" and published this finding, a claim that was repeated by other notable scientists after him.

Another similar case was Edward Drinker Cope ("Chapter 11: Myth and Reality of the Alleged Blind Cavefish from Pennsylvania"). Cope was a very well-known scientist during his time. He was also was known for creating new species galore, many of them hastily, which led to severe errors. That was the case of his alleged "cavefish from Pennsylvania."

As a whole, all these are excellent examples of how to conduct (or not conduct) scientific research and about way you should report your findings.

Chapter 6. The Scientific Discovery of The Amazon River Dolphin *Inia geoffrensis*

Summary

Analysis of little-known manuscripts revealed that there have been at least two pre-Linnean scientific descriptions of the South American freshwater dolphin *Inia geoffrensis* (Blainville, 1817). The earliest one that we found was made by Frei Cristóvâo de Lisboa in a manuscript written around 1627. The second one was by Pehr Löfling, a disciple of Linnaeus, who wrote a very detailed and accurate description of this mammal in 1755. He used the binomial system to designate this species, and his description was much more complete and more sophisticated than the ones used by Linnaeus in the 10th edition of *Systema Naturae* for other cetaceans. This and other zoological work by Löfling remain almost completely unexamined to date. Like the outcome of other outcome of other field work carried out by many Spanish scientists in America, failure to publish the findings of de expeditions resulted in scientific information being largely lost.

Figure 6.1. Alexandre Rodrigues Ferreira (public domain image).

Introduction

---

[312] Based on: Romero, A., A. I. Agudo & S. J. Blondell. 1997. The scientific discovery of the Amazon river dolphin *Inia geoffrensis*. *Marine Mammal Science* **13**(3):419-426.

Alexandre Rodrigues Ferreira[313] (Fig. 6.1) is often mentioned (e.g., Best and da Silva 1989) as the discoverer of the Amazon river dolphin ("boto" in Portuguese, "tonina de río" or "bufeo colorado" in Spanish) known today as *Inia geoffrensis* (Blainville, 1817). He collected a specimen of this species somewhere in the lower Amazon basin no later than 1790, classified it as *Delphinus delphis,* and sent it to the Museu de Ajuda in Lisbon, together with a description and a drawing (Rodrigues Ferreira 1790). The specimen was later plundered by Geoffroy St. Hilaire in 1810 on orders of Napoleon Bonaparte and sent to the *Muséum d'Histoire Naturelle de Paris* where it can be found today (de Miranda Ribeiro 1943, van Bree and Robineau 1973). Henri-Marie Ducrotay de Blainville (1817) described *Delphinus geoffrensis* based on this specimen. The genus *Inia* was created for *I. boliviensis* (d'Orbigny 1834), making this the type species, but Gervais (1856) later recognized that *D. geoffrensis* also belonged in *Inia*. We have found documentation, however, that at least two authors described chis species much earlier: one no later than 1647 and the other in 1755, i.e., well before Rodrigues Ferreira's collection. Both descriptions were unknown to Linnaeus and are still unfamiliar to cetologists today.

Figure 6.2. Frei Cristóvâo de Lisboa (public domain image).

---

[313] *b*. Bahia, Brazil, 1756; *d*. Lisbon, Portugal, 1815.

Frei Cristóvâo de Lisboa[314] (Fig. 6.2) is the author of the next mention of the boto or tonina de río that we have uncovered. A Capuchin priest, he arrived in Brazil in 1624 and stayed there until sometime between 1627 and 1631. Between 1624 and 1626 he traveled extensively throughout the Maranhao (lower Amazon) region. Apparently, by 1627 he was already working on a manuscript on the natural history of the animals and trees of the area titled *Historia dos animaes e árvores do Maranhâo* (da Fonseca 1952) and what appears to be the final draft is dated 1647, in Lisbon. The manuscript was rediscovered in 1934 when it was bought from a bookseller in Lisbon by the *Arquivo Histórico Colonial* of Portugal in Lisbon where it remains today. A facsimile version of the manuscript, including contemporary historical notes, was published in 1967 (de Lisboa 1647).

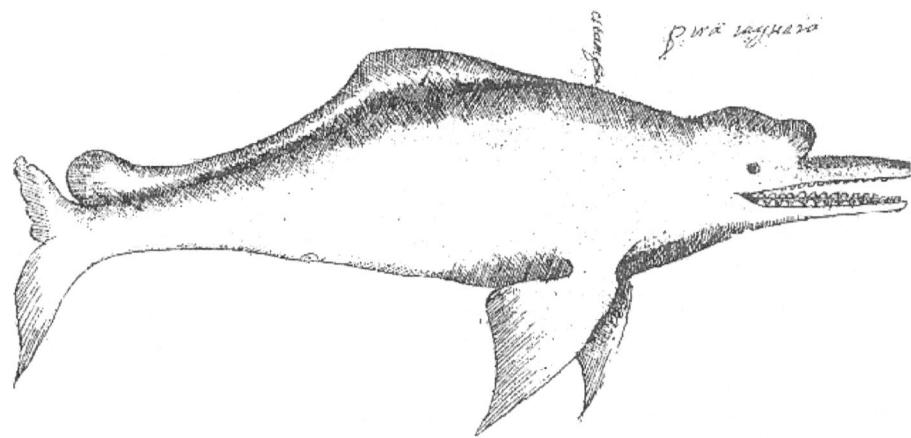

Figure 6.3. "Pyraiaguara" of Frei Cristóvâo de Lisboa (public domain image).

On page 175 of the manuscript, there is a description of the "Pyraiaguara" among the "fishes of Para." The paragraph-long description reads as follows:

"he especie de porco marinho no sabor he como porco principalmente o figado tem as partes genitais como o porco tem noue palmos de comprido e grosso nesta proporcâo face do rabo manteiga as femeias parem como os animais tem hû buraco asima do naris par onde respira e lanca algua"

("it is a species of sea-pig; with the taste of a pig, particularly the liver, it has the genital pares like a pig and has nine palms of length, and the bulk is proportional; butter can be obtained from it; the females give birth like the animals; it has a hole on top of the nose through which it breathes and throws water"). On page 54 an illustration of this dolphin (Fig. 6.3), with its long snout and short dorsal fin, leaves little doubt that de Lisboa referred to what is known today as *Inia geoffrensis*. Although this manuscript does not have the exhaustive descriptions of some of the most popular and influential natural history works of

---

[314] *b*. Lisbon, Portugal, 1583; *d*. Santo Antonio do Curral, Portugal, 1652.

its time that also describe cetaceans (for a list of books of the above-mentioned period see Allen 1881), it is a fair description of this species, especially considering that Cristóvâo de Lisboa was not a professional naturalist.

The next description that we found for this species was made by Pehr Löfling in 1755. Born in Valbo, Sweden, in 1729, he was one of Carolus Linnaeus's[315] students (for biographical details on Löfling see Rydén 1957, Pelayo and Puig-Samper 1992). Löfling went to Spain (1751- 1754) where he did valuable botanical and zoological work, some of which was incorporated into the 10th and 12th editions of Linnaeus's *Systema Naturae*.

Although a botanist by training, Löfling was also interested in animals, particularly in fish (including cetaceans as they were considered at the time, see Romero 2012). His interest was apparently prompted by Linnaeus's instructions that the Queen of Sweden wanted some specimens for her private museum. In Linnaeus's letter to him, dated 2 October 1753, Linnaeus told Löfling to collect all the fish he could find (Rydén 1957, pp. 73- 74). Among the "fishes" described by Löfling in his unpublished ichthyological work on the Bay of Cádiz ("*Piscis Gadicana, Observato Gadibus et ad Portus Sa. María 1753, Mens Nov. et Decemb.,*" MS at the Royal Botanical Garden, Madrid), which he wrote while waiting to depart for South America, there is a "*Delphinus*" (*sensu* Artedi, Syn 105 & Gen 72:2; probably a *Delphinus delphis*; Pelayo 1990, p. 121).

Löfling arrived in Cumaná, eastern Venezuela, on April 1754 as part of an expedition whose primary goal was to fix the borders of Spain's and Portugal's possessions in America. Löfling's main mission was to survey natural products that could have economic value for the Spanish Crown (Ramos Pérez 1946, Lucena Giraldo & de Pedro 1992). In May 1755 he reached the Orinoco River, where he made extensive botanical and zoological collections despite illness and the expedition's numerous logistic problems. He worked, according to his diary deposited at the Madrid's Royal Botanical Garden, until October 1755 when he fell sick. On 22 February 1756, he died at the Caroní River, near the confluence with the Orinoco. He had only a few months of time for real work in South America (Pelayo & Puig-Samper 1992, p. 14).

Among Löfling's documents known today there are more than 1,700 papers (folios), 200 drawings, and a few watercolors of plants and animals (Rydén 1957, p. 150). One of the unpublished manuscripts left by Löfling (today in Madrid's Royal Botanical Garden) was "*Ychthyologia Orinocensis*" ("Ichthyology of the Orinoco"). On sheet II,4,4,1, p. 7 of that manuscript where he lists the species, which he later described in greater detail, he writes "*Delphinus tonina (Δελφίς), Artedi p. 105 n 2, Linei System. N. Delphinus, corpore oblongo subtereti rostro longo acuto, Jonston 2b.c.2., Rondel. Li6c8., Charlton p 168, Aldrov. Cet. C7 p 103.*"

He lists the common names used by the Spaniards and by local tribes for this species as follows: Spanish-Tonina; Cabre-Muna; Maypure-Muna; Guama-Yufa; Guayana-Urinugna.

Then, in sheet 11,4 ,4 ,1, p. 56, he describes the major characteristics of the "tonina," as follows:

"*Caput subtotundum Rostro acuminato obtuso. Oris Rectum prolongatum Lingua ampla subovata (drawing) Dentes minuca in margine maxillarum. Oculi paris ad prope rictus oris. Fistula transversalis in capite loro narum? per*

---

[315] *b*. Råshult, Stenbrohult, Sweden, 1707; *d*. Uppsala, Sweden, 1778.

quam aquam recipis olfaro. Pinae pectorales cartilaginae. Pina dorsalis culta! forma acuta rel. Cauda orinzontalis bifurca angulis obtusis. Anos foramen subrotundi. Membrum femeninum prope anum muliebri similisimum. Mamae dua ad latris utrum que nacurae membre."

("Head suboval. Rostrum bluntedly pointed. Beak straight and prolonged. Tongue broad and suboval. Teeth are small and located at the sides of the maxillae. Two eyes near the base of the rictus [posterior end of the beak]. Blowhole is transverse with respect to the head in place of a nose? *[sic]* through which water is swallowed while sniffing. Pectoral fins are cartilaginous. Dorsal fin is elevated! *[sic]* with a form relatively acute. Horizontal tail is bifurcated in an obtuse angle. Anus hole is semicircular. Female membrane is in the place of the anus, as it happens similarly among females. Two breasts one at each side chat are present at birth"). (Please note that the original Latin text is defective in both grammar and spelling; his penmanship is also somewhat difficult to read).

This description, much more detailed and different in style and content than Artedi's (1738) and Linnaeus's (1748) descriptions of similar species, is accompanied by two drawings (one dorsal and one lateral), the latter being very accurate (Fig. 6.4).

The information contained in "*Ychthyologia Orinocensis*" suggests the following:

Sources of information - Löfling used Peter Artedi's[316] book (1738) and Linnaeus's Systema Naturae 6th edition (1748) for this dolphin species diagnosis and mentioned Jonston, Rondelet, Charleton, and Aldrovandi as further sources. From an inventory of his belongings made after his death, we know that he possessed (among others) Artedi 's and Linnaeus's books when he died, as well as 14 "authorless books in good shape" and 20 "in bad shape and virtually worthless" (Pelayo & Puig-Samper 1992, pp. 56-61). Thus, it is unclear whether or not he had copies of the other zoological books mentioned in the list of references for the *tonina*. We hypothesize here that one possibility is that he removed the covers of books that the Spaniards may have considered offensive at that time just because the authors were not Catholics. Book censorship started in Spain in 1502 (more stringent regulations followed in 1558) and such censorship continued until 1812. The Inquisition banned many scientific books that were standard at the time, including classical natural history references such as Conrad Gesner's books (Herr 1958, pp. 201-213; Beddall 1983). Gesner is cited by Lofting in his manuscript when he describes the manatee *(Trich Species Identification and Nomenclature* - Löfling designates this dolphin as "*Delphinus tonina* (Δελφίς)." Was he proposing a new species based on the striking differences between this freshwater species and the delphinoids known then? That is difficult to answer given that: (1) Löfling's manuscript seems to be far from polished, so we cannot know what his final thoughts on this were; (2) although he seems to be inclined to follow Linnaeus's practice of using the common name as the specific name, the fact that he adds the then monotypic species name in parentheses and in Greek (the only time he does so in his "*Ychthyologia Orinocensis*") may indicate that he was following descriptive procedures of the time for plants, which gave a generic name in Latin followed by a specific epithet in Greek characters expressing several features (Stearn 1959). Furthermore, it is even uncertain that

---

[316] *b*. Anundsjö, Sweden, 1705; *d*. Amsterdam, Holland, 1735.

Löfling intended to use a binomial designation for this (new?) species. Given that most of the "fish" species he described were totally new to science (no fish collections had ever been made in northern South America), he used mostly the common names to designate the Orinoco fishes. Linnaeus's first introduction of consistent binomial nomenclature for species dates to his *Species Plantarum* (1753) (which Löfling carried with him to America), but it is not until the 10th edition (published two years after Löfling's death) that Linnaeus first gave binomials to all the known species. Linnaeus's *Systema Naturae* 6th edition of 1748, which Löfling also carried with him to America, does not employ binomial nomenclature (Stearn 1959). Before that, the use of this system was occasional (Stearn 1971). Whenever possible, Linnaeus chose epithets that preserved an association with earlier literature (nouns in apposition). Löfling followed Linnaeus in employing the common name (often invented when necessary). In his *Critica Botanica*, aphorism 237, Linnaeus (1737) ruled that the specific name should distinguish the plant from all others of the genus. By specific name (*nomen specificum*) he then meant a diagnostic phrase. Hence when a genus had only one species, the generic name by itself was enough (Stearn 1971).
echus manatus) that he observed.

Figure 6.4. "*Tonina*" of Pehr Löfling (public domain image).

In Löfling's time there were only three differentiated species among the known delphinoids. They were grouped in the genus *Delphinus*: *D. delphis* (basically a conglomerate of what we know today as *Tursiops* and *Delphinus* itself), *D. phocaena* and *Orcinus orca*. This same classification was followed by Linnaeus in his 10th edition. Given the unpolished state of his manuscripts and that almost all of the Orinoco "fish" were new to science, Löfling might have just decided to gather as much field information about them as possible and to take care of the taxonomy later.

Higher Classification - Löfling keeps the "*tonina*" among the fishes, although, as can be read in the description, he knew that the female of this species had mammary glands. That is not surprising. Although at that time there was strong evidence that cetaceans were, at least, different from other fishes, it is not until Linnaeus's 10th edition that the class Mammalia is created; it then included today's cetaceans (Romero 2012). Furthermore, Löfling's characterization of "*tonina*" is almost exclusively on external morphology (there is no record of Löfling ever practicing a dissection), while Linnaeus's bases for classifying cetaceans as mammals are their warm, two-chambered heart, their breaching by lungs, their hollow ears, a penis that enters the female, and mammaries that exude milk (10th Edition, Vol 1, p. 17).

Löfling's notes and manuscripts were sent to Spain after his death. However, some have been lost (Rydén 1957, pp. 9, 75, 80, 120, 144). Most of the surviving documents have been preserved in the archives of the Royal Botanical Garden of Madrid since 1801. The Spanish naturalist Casimiro Gómez Ortega[317] was commissioned to put together all of Löfling's papers. He wrote some notes on the drawings of the expedition's artists Bruno Salvador Carmona[318] and Juan de Dios Castel[319]. "Ichthyologia Orinocensis" has Castel's penmanship. Castel may have written what was dictated to him by an ill Löfling, or just copied the notes when he came back to Spain and worked with Gómez Ortega on Löfling's papers. Through the chaplain of the Swedish embassy in Madrid, Daniel Scheidenburg[320], Linnaeus obtained a copy of some of Löfling's manuscripts and notes (Rydén 1957, p. 148). With that and the material that Löfling himself had sent from Spain, Linnaeus put together and published in 1758 Löfling's Iter Hispanicum ("Spanish Journey" as part of Löfling's botanical work in Spain and Venezuela) (Stockholm 1758; for the English version see Bossu 1771). Further attempts by Linnaeus to obtain the rest of Löfling's material were unsuccessful. Löfling's specimens, including his herbarium, are today lost (Pelayo & Puig-Samper 1992, p. 100).

Löfling's zoological work remains almost completely unexamined; only a mammal, an amphibian, and a reptile are mentioned in the Iter Hispanicum, and there are some references to the Bay of Cadiz fishes in the 10th and 12th editions of the Systema Naturae. Lofling's diagnosis of the Cadiz fishes found in the catalog of the Queen of Sweden published by Linnaeus as Museum Ludovica Ulricae Regina (1764) does not mention the collector's name (Pelayo & Puig-Samper 1992, p. 100).

There is no evidence that any of the animal specimens of this expedition ever reached Spain. Cetaceans, because of their size and skin, are difficult to preserve. It was not until the 1750s that the use of salt and alum allowed permanent preservation of mammal and bird skins (Farber 1977). Like most of the work by other naturalists at the service of the Spanish Crown in Spain's colonies before and after him, Löfling's work was largely lost. Basalla (1967) considers both Spain and Portugal to be special cases, not only because modern science (1450-1800) had not been extensively cultivated by either of the countries, but also because of the failure to publish the findings of the expeditions

There are other references in passim to freshwater dolphins in South America found in pre-Linnean times such as Pedro de Magalhães de Gândavo's[321] 1576 (although not published until 1922 as "The Histories of Brazil") or Jacinto de Carvajal's[322] "The discovery of the Apure River" (1647, not published until 1805). A vernacular description of this species can also be found in Filippo Salvatore Gilij's[323] "Essay of American History" (published in 1782; see Paolillo & Romero 1989 for analysis of this source). Yet it lacks the details and scientific accuracy of his predecessors Cristóvâo de Lisboa and Löfling himself. Thus, the descriptions

---

[317] b. Añover de Tajo, Spain, 1740; d. Madrid, 1818.
[318] b. Madrid?, Spain, 1737; d. Madrid, 1801.
[319] b. 1738? d.?.
[320] b. Västerhaninge, Sweden, ca. 1726, d. Spain, 1791.
[321] b. Braga, Portugal, c. 1540; d. 1579.
[322] b. Extremadura, Spain, c. 1587; d. Venezuela?, 1648?.
[323] b. Legogne, Perugia, Italy 27 July 1721; d. Rome, Italy, 10 March 1789.

by Cristóvâo de Lisboa and Löfling are to date the only known examples of pre-Linnean scientific description for any freshwater cetacean in South America.

Acknowledgments

M. Boesernan (Rijksmuseum, Leiden, Holland) provided useful information on pre-Linnean Brazilian naturalists. Maria de San Pío Aladrén (Archives Curator, Royal Botanical Garden, Madrid, Spain) gave us access to the Löfling papers. Juan Ramón Romero (State Archives, Ministry of Culture, Madrid, Spain), Ramón Moscato (Santiago de Compostela, Spain), and Alfredo López (CEMMA, A Coruña, Spain) supplied us with valuable information. Dr. Roland Moberg of Uppsala University provided valuable biographical information about Daniel Scheidenburg. Steven M. Green read an early version of the manuscript and made helpful suggestions. Two anonymous reviewers also made valuable suggestions. This is contribution number 513 of the Program of Ecology, Behavior, and Tropical Biology, University of Miami.

Literature Cited

Allen, J.A. 1881. Preliminary list of works and papers relating to the mammalian orders Cete & Sirenia (1495-1840). Bulletin of the United States Geological and Geographical Survey of the Territories **6**:399-562.

Artedi, P. 1738. *lchchyologia sive Opera Omnia de Piscibus Scilicec: Bibliotheca Ichthyologica, Philosophia Ichthyologica, Genera Piscium, Synonymia Specierum, Descriptiones Specierum. Omnia in Hoc Genere Perfectiora, quam ancea ulla. Posthuma Vinduicativ, Recognovit, Coaptavit & Edidit Carolus Linnaeus, Med. Doct. & Ac dmper. N.C. Lugduni Batavorum Apud Conradum Wishoff*

Basalla, G. 1967. The spread of western science. *Science* **156**:611-622.

Beddall, B.G. 1983. Essay review: Spanish science and the New World. *Journal of the History of Biology* **16**:433-440.

Best, R.C. and V.M.F. Da Silva. 1989. Amazon river dolphin, boto *Inia geoffrensis* (De Blainville, 1817). Pages 1-23 *In*: S. H. Ridgway and R. Harrison, eds. *Handbook of Marine Mammals*. Vol. 4. New York: Academic Press.

Bossu, J.B. 1771. *Travels through that part of North America formerly called Louisiana*. London, 2 Vols.

Bree, P.J.H. van and D. Robineau. 1973. Notes sur les holotypes de *Inia geoffrensis geoffrensis* (De Blainville, 1817) et de *lnia geoffrensis boliviensis* D'Orbigny, 1834 (Cetacea, Platanistidae). *Mammalia* **37**:658-664.

Da Fonseca, L. 1952. Frei Cristóvâo de Lisboa, O.F.M., missionary and natural historian of Brazil. *The Americas* **8**:289-303.

De Blainville, H.-M. 1817. *Delphinus geoffrensis*. Pages 151-152 *In*: A.G. Desmaret, ed. *Nouveau dictionnairc d'histoire naturelle*. Paris: Déterville.

De Carvajal, J. (1647) 1805. *Descubrimiento del río Apure*. Madrid, Spain: Historia 16.

De Lisboa, F.C. (1647). 1967. *História dos animais e árvores do Maranhâo*. Lisbon: *Arquivo Histórico Ulcramarino e Centro de Estudos Históricos Ultramarinos*.

De Magalhâes, P. (1576) 1922. 1969. *The Histories of Brazil*. Translated into English for the first time and annotated by John B. Stetson, Jr. 2 vols. New York: Kraus Reprint, Co.

De Miranda Ribeiro, A. 1943. *lnia geoffrensis* (Blainvillc). *Arquivos do Museu Nacional, Rio de Janeiro*. **37**:23-58.

D'Orbigny, A. 1834. Notice sur un nouveau genre de Cétacté des rivieres du centre de l'Amerique meridionale. *Nouvelles Annales du Museum d'Histoire Naturelle* **3**:28-38.

Farber, P.L. 1977. The development of taxidermy and the history of ornithology. *Isis* **68**:550- 566.

Gervais, P. 1856. Sur crois espêcies de Dauphins qui vivent dans la region de Haute Amazone. *Cornptes Rendus de l'Academie des Sciences de Paris* **42**:806- 808.

Herr, R. 1958. *The eighteenth-century revolution in Spain*. Princeton, NJ: Princeton University Press.

Linnaeus, C. 1737. *Critica botanica*. Leyden, Holland: Conrad Wishoff.

Linnaeus, C. 1748. *Systema naturae. Editio Sexta*. Stockholm, Sweden: G. Kiesewetter,

Linnaeus, C. 1 753. *Species plantarum*. Stockholm, Sweden: L. Salvius,

Linnaeus, C. 1 7 58. *Systema naturae, Editio Decima, reformata*. Stockholm, Sweden: L. Salvius,

Lucena Giraldo, M. and A. E. de Pedro. 1992. *La frontera caríbica: Expedición de límites al Orinoco 1754/1761*. Caracas, Venezuela: Cuadernos Lagoven,

Paolillo, A. and A. Romero. 1989. Los relatos de la fauna orinoquense hechos por Felipe Salvador Gilij, evaluados con la óptica de la zoología del Siglo XX. *Revista Montalbán* (21):159-178.

Pelayo, F., ed. 1990. Pehr *Löfling y la expedición al Orinoco*. Madrid, Spain: Real Jardín Botánico,

Pelayo, F. and M.A. Puig-Samper. 1992. *La obra científica de Lofling en Venezuela*. Caracas, Venezuela: Cuadernos Lagoven,

Ramos Pérez, D. 1946. *El tratado de límites de 1750 y la expedición de Iturriaga al Orinoco*. Madrid, Spain: Consejo Superior de lnvestigaciones Científicas, Instituto Juan Sebastian Elcano.

Rodrigues Ferreira, A. (1970) 1972. *Viagem Filosófica Pelas Capitanias do Grao Para, Rio Negro, Mato Grosso e Cuibá.*, Rio de Janeiro, Brazil: Conselho Federal de Cultura.

Romero, A. 2012. When whales became mammals: the scientific journey of cetaceans from fish to mammals in the history of science. Pp. 4-30. *In*: Romero, A. & E.O. Keith (Eds.). 2012. *New Approaches to the Study of Marine Mammals*. Rijeka, Croatia: InTech.

Rydek, S. 1957. *Pedro Loefling en Venezuela (1754-1756)*. Insula, Madrid, Spain.

Stearn, W.T. 1959. The background of Linnaeus's contributions to the nomenclature and methods of systematic biology. *Systematic Zoology* **8**:4-22.

Stearn, W.T. 1971. Linnean classification, nomenclature, and method. Pages 242- 256 *In:* W. Blunt, ed. *The complete naturalist. A life of Linnaeus*. New York: The Viking Press.

Chapter 7. Jacques Besson, Cave Eels, and Other Alleged European Cavefishes [324]

Summary

According to Shaw (1992:227), a standard reference to the history of speleology, Jacques Besson in 1569 was the first to mention in print a reference to a cavefish. This chapter is aimed at (1) analyze Besson's record of alleged cave animal; (2) summarize other early records for European cavefishes while asserting the validity of those records; and, (3) to establish the true chronology of cave fish records before the first truly confirmed scientific description of a cavefish species, *Amblyopsis spelaea* (De Kay 1842).

Jacques Besson And His "Little (Cave?) Eels"

Before a Eurocentric view of published accounts on cavefishes, we need to examine what other accounts may have taken place elsewhere in the world. The first known written account of a cavefish came from China and took place in 1540. That year, Yi Jing Xie[325], a local government wrote a never published travel report now found in the records of Luxi County, Jiangxi Province, southeast China. In 1905 Ying Huang, the local governor, encountered the document and had it engraved as an inscription on a stele (Y. Zhao, pers. comm.). In this document Xie referred to the hyaline fish (*Sinocyclocheilus hyalinus*) from the Alu caves, Yunnan, China. This fish was not collected for scientific purposes until 1991 and its scientific description was not published until 1994 (Chen *et al.* 1994).

The first known published report of an alleged cavefish in Europe took place in 1569 by Jacques Besson. He is considered one of the most important and prolific writers in engineering of the Sixteenth Century. Yet, we do not know much about him nor any portrait of him is known. The following sketch is based on the few biographical sources about him: Arnaud (1894), Droz (1976), Keller (1964, 1973).
It is believed that Besson was born around 1540 near Grenoble, France. He described himself as from Colombières, near Briançon in Le Dauphin, part of Escarton de Oulx, now in Cesana Torinese, Italy, high up in the Alps on the southeastern border of France. There are records indicating that he might have taught mathematics in Paris sometime in the 1550s. By 1557 he was working as an engineer for the city council of Lausanne for whom he designed a water-engine as part of a fountain. In 1559 he was a resident of Geneva and that year he published his first book, *De absoluta ratione extrahendi olea, & aquas e medicamentis simplicibus,* which dealt with chemical analyses and practical distilling. By then he may have been well connected since that book had a praising preface by one of the most noted natural historians of the time: Konrad Gesner[326].

By 1561, when Besson acquired Swiss citizenship, he was going through tough times. He fell seriously ill and was living in poverty. By then he may have been married and had a daughter. In addition to being a teacher, an apothecary, and a

---

[324] Originally published as Romero, A. & Z. Lomax. 2000. Jacques Besson, Cave Eels and Other Alleged European Cave Fishes. Journal of Spelean History. 34(2):72-77.
[325] Xie, Yi Jing (*b.* ?; *d.* ?).
[326] *b.* Zurich, Switzerland, 26 March 1516; *d.* Zurich 12 March 1565

mechanical engineer, he became pastor of the Reformed Church. In the town of Villeneuve-de-Berg, in the Vivarais, west of the Rhone Valley, the Protestant community was flourishing and expanding and felt the need for a preacher. Given the serious lack of preachers, almost any educated person (who could read and write) qualified for the job. But this position was far from a solution to his problems; when he arrived at his parish, he found that the two religious parties in the town were in a civil war. He and his family were forced to live in another man's house.

In 1563 he left the ministry for Lyons to work distilling oils and waters. In an attempt to avoid angering the church, which punished those who left without permission, he wrote them a letter in which he admitted to inadequacy while claiming, however, that those who appointed him (in essence, the church) were faulty in doing so. In 1565 we found him in Paris and two years later in Orleans where he taught mathematics and demonstrated his inventions to an admiring audience. In 1567 he published his second book, *Le Cosmolabe,* about a versatile and very elaborated mathematical instrument of his own invention, that could be used for almost all the purposes of navigation, surveying, cartography, and astronomy and, when not required for any of them, could double as a reading desk.

In 1569 King Charles IX went to Orleans and Besson entered the King's service as a mathematician and engineer. In that year, he published his most famous book, *Iheatrum Instrumentorum et Machinarum* which was the first printed work of mechanical inventions. Between 1578 and 1626 this book was published in four languages and had seven editions. It was also widely plagiarized and pilfered. There he introduced cams and templates (patterns used to guide the form of a piece being made) to the screw-cutting lather, thus increasing the operator's mechanical control of tool and workpiece and permitting the production of more accurate and intricate work in metal. He also improved the drive and feed mechanisms of the ornamental lathe and described a more efficient form of waterwheel, considered a prototype of the water turbine. In this book he depicts different kinds of machine tools, pumping plants, ploughs, military engines and other machines he had observed in use in different parts of Europe. In reality, it is doubtful that many of these machines were ever in operation, since they are not mechanically viable. Yet, Besson shows great ingenuity in his designs, especially in that of a screw-cutting lather.

After the Massacre of Saint Bartholomew's Day in 1572 (which actually began on August 24 and ended in October) in which 70,000 French Protestants, or Huguenots, were murdered, he fled France. Besson died in Orleans in 1573.

Besson often used completely new concepts and made major contributions as an inventor. His screw-cutting lathe, which used a cord with an attached weight instead of the earlier springy poles, was a very important invention in the development of the machine-tool industry and of scientific instrumentation. He also invented a practical fire-engine that later became common. His work shows the technological visions of the day while pointing the way to future developments (see also Besson 1571, 1573).

In his 1569 "*L'art de science de trouver les eaux et fontaines soubs terre ...,*" (Fig. 7.1) he reported "little eels" *(petites anguilles)* in a cave stream. Although Shaw (1992:227) claims that such observation took place "in a cave stream in France" the fact of the matter is that Besson did not give the locality of where he made that observation. The passage in reference reads:

> *Les entrees sont comme portaux voustez & estroits, ainsi que le tout on experimente en entrant en semblages chasteaux natureles soubs terre, la ou Ion*

trouue auec torches de fort grands lacs, & courans d'eaux viues, mesme qui bien souuet produisent des petites anguilles qui n'ont guere affaire de l'air pour leur nourriture. (Besson 1569, 1969:41, notice that we have transcribed it in its original spelling).

[The entries are like narrow arch portals, so all the people coming into these virginal natural subterranean marvels, need to use torches to see big lakes and currents of lively waters, from which one can see small eels for which there is nothing to eat but air.]

Figure 7.1. Reproduction of the cover of *L'art de science de trouver les eaux et fontaines soubs terre* ... (public domain image).

There is no indication of either the locality or of the fish itself. He does not describe the fish as being blind and/or depigmented (what would have been extraordinary characteristics to even the casual observer). Thus, is it unclear whether he observed a true cavefish, actual eels *(Anguilla anguilla)*, or a member of some European freshwater fishes with eel-like bodies that are sympatric with the areas he traveled to (France and Switzerland). Those fish families include Petromizonidae, Cobitidae, Siluridae, and Clariidae (Blanc *et al.* 1971).

Other Early Reports on European Cave Fishes

Besson may have been the first person to report in a publication of what he thought were cavefishes for Europe, but certainly he was not the only one. The second was Athanasius Kircher[327]. He was a prolific Jesuit priest polymath who published 44 books and left more than 2000 manuscripts and letters on varied topics. One of his most famous books was *Mundus subterraneus* (1665), probably the first printed work on speleology. There he wrote about the origin of subterranean water and described all kinds of alleged cave animals (including giants and dragons). On page 85 of part 2, in book 1 there is indeed a description of cave fishes.

> *So, I had much to talk about the subterranean animals, that we know. But from these, that can be in the deep and spacious holes in the earth, we cannot present much, because we do not know them; from several examples however, it appears that there are fishes and other animals, for Plinius writes that in Greece the earth, bursting open due to an earthquake, threw out a river with a large number of fishes, that without doubt had bred in a underground river. [Kircher gives no references to Plinius] There is also in the landscape of Krain close to the town Haubach a field, that every year about the Spring gives much water with fishes, so that in a few days the field changes in a Lake full of fishes. But this subject is sufficiently discussed before. [He probably meant Laibach instead of Haubach, the old name for Ljubljana in Slovenia, with its temporary lakes like Czemica - with spring caves and ponors].*

> *Cysatus confirms the same, saying: in Switzerland rivers rise from the caves of the mountains, that flow from May until September, but stop the rest of the time. He adds to that, that the rivers, as they come out of the mountains, are full of fish, and that it is clear, that they come with the waters from below the earth. [Cysatus is probably the Swiss astronomer Johann Cysat that in 1618 discovered Orion Nebula; Kircher gives no reference].*

> *We think that it is not implausible that, as under the earth all kind of fishes occur and live, also earth animals stay there, that is all kind of Mice, Snakes, Dragons, as well as others, that find their origin in rotten matter.*

These references to subterranean fishes, however, are vague, unsubstantiated, and given Kircher's reputation as an uncritical repeater of other people's tales, highly suspect (Romero 2000a). Furthermore, he makes no reference to the features that characterize true hypogean fishes: blindness and depigmentation.

The third reference to subterranean fishes in Europe was by Marc-Rene

---

[327] *b*. Geisa, Germany, 2 May 1602; *d*. Rome, 28 November 1680.

Marquis de Montalembert[328]. He was an aristocrat, military man, and an engineer known for bis design of fortifications. Montalembert (1748) reported a blind, subterranean fish in one of his properties in the Southwest of France. His description was as follows:

> *In a spring at Gabard, Angoumois, near one of (Montalembert's) estates, it is common to fish either blind or one-eyed pike; one-eyed ones always miss the right eye and among the blind ones, the right eye seems further reduced that the left eyes. This spring is a kind of bottomless pit; there are small groups of floating plants at the surface, which impede the use of fishing lines, which makes fishing a long and difficult process; however, Montalembert was fortunate enough to capture a young pike with its right eye missing; this spring drains its water into the Lissone river; despite this connection the local people say that one eyed or blind pikes are never fished in the river, while the spring contains one-eyed or blind ones only.*

Apparently, Montalembert left no drawings, much less a preserved specimen. He said that what he saw was a pike. That, by itself, is not surprising. Pikes *(Esox lucius)* are, by far, the most common freshwater fish of the Northern Hemisphere. The fact that this fish can be identified as a pike despite being blind is not surprising either. Many subterranean fishes are very much identical to their surface ("epigean") forms except for the lack of eyes and pigmentation. But Montalembert never mentioned depigmentation in his description.

Furthermore, he says that some of the fish lacked one eye and when that was the case, it was always the one on the right side. True hypogean show the same degree of reduction in both eyes. Finally, the location mentioned by Montalembert cannot be found today nor has any true blind cave fish ever been described for Europe (Romero 1999).

Another unproven report of blind cave fish for Europe is that of Scott (1866) who wrote about such a fish in Italy without specifying source or locality. Reports of blind cavefishes for North America, have been analyzed elsewhere and found to be unsubstantiated (Romero 2000b).

Conclusions

All reports of European blind cave fishes are unsupported by scientific evidence. Two of them (Besson and Kircher) do not even describe them with the features typical of hypogean fishes while the third (Montalembert) is suspect. Furthermore, no true cave fish is known for Europe at the present time.

From the chronological viewpoint, although all three records precede the mention of the first blind cavefish described in the Western Hemisphere *(Amblyopsis spelaea* by De Kay in 1842) (Romero & Bennis 1998) and are thus pre-Linnean, there is an even earlier reference to a blind cave fish that, in this case, was probably true. That is of a cavernicole fish from China *(Sinocyclocheilushyalinus)*, first reported in 1541 (Chen *et al.* 1994).

Literature Cited

Arnaud, E. 1894. Note sur Jacques Besson. *Bulletin de la Societe d'Etudes des*

---

[328] *b*. Angouleme, France, 16 July 1714; *d*. Paris, 29 March 1800

Hautes Alps. **13**:218-219.

Besson, J. 1559. *De absoluta ratione extrahendi olea & aquas e medicamentis simplicibus: accepta olim a quodam empirico, postea vero ab eodem Bessono locupletata, & rationibus experimentisque confirmata, liber*. Tiguri: Apud Andream Gesnerum juniorem, 42 p. illus. 17 cm.

Besson, J. 1567. *Le cosmolabe; ou, Instrument universel: concemant toutes obseruations qui se peuuent faire par les sciences mathematiques, tant au ciel, en la terre, cornme en la mer*. Paris: Par Ph.G. de Roville.

Besson, J. 1569. *L'art et science de trouver les eaux et fontaines cachees soubs terre: autrement que par les moyens vulgaires des agriculteurs et architects*. Orleans: E. Gibier, [6] leaves, 83 p. [We used the reprinted version by Editions Coral, Columbus, Ohio, published in 1969].

Besson, J. 1569. *Theatrvm instrvmentorvm et machinarum, Cum Franc, Beroaldi figurarum declaratiooe demonstratiua*. Lvgdvni: Apud B. Vincentium, [22] p., 60 plates, 40 cm., Plates engr. by Jacques Androuet du Cerceau and Rene Byvin.

Besson, J. 1571. *Description et vsaige du compas euclidien: contenant la plus part des obseruations qui se sont en la geometrie perspectiue, astronomie, & corographie*. Paris: Galiot du Pre, en la rue Sainct Iaques, a l'enseigne de la Galere D'Or, et au premier pillier de la grand salle du Pallays, 5 leaves; 20 cm., (4to).

Besson, J. 1573. *Art et moyen parfaict de tirer huyles et eaux, de tous Premierement receu d'un certain empirique confirme par raisons & experiences, Nouvellement corrige & augmente d'un second livre ...* Paris: Galiot du Pre, 31 [i.e., 32] 11. illus., 16 cm.

Blanc, M., P. Banarescu, J.-L. Gaudet & J. C. Bureau. 1971. *European Inland Water Fish, A multilingual catalogue*. London, Fishing News (Book) Ltd.

Chen, Y.-R, J.-X Yang and Z.-G. Zhu. 1994. A new fish of the genus *Sinocyclocheilus* from Yunnan with comments on its characteristic adaptation (Cypriniformes: Cyprinidae). *Acta Zootaxonomia Sinica* **19** (2):246--253.

De Kay, J. E. 1842. *Zoology of New York or the New-York Fauna, Part IV Fishes*. Albany: W. & A. White & J. Visscher.

Droz, E. 1976. *Chemins de L'Heresie, Testes et Documents*. Geneve: Slatkine.

Keller, A., 1964. *Theatre of Machines*. London, Chapman & Hall.

Keller, A. 1973. The Missing Years of Jacques Besson, Inventor of Machines, Teacher of Mathematics, Distiller of Oils, and Huguenot Pastor, Technol. *Culture* **14**:28-39.

Kircher, A. 1665. *Mundus subterraneus, in XII libros digestus; quo divinum subterrestris mundi opificium, mira ergasteriorum naturae in eo distributio, verbo pantamorphou Protei regnum, universae denique naturae majestas & divitiae sum.ma rerum varietate exponuntur*. Amsterdam: J. Janssonium & E. Weyerstraten, 2 vols.

Montalembert, M.-R. 1748. *Observations de Physique Generale, I, Histoire de L'Academie Royale des Sciences, Annee M. DCCXLVIII, Avec les Memoires de Mathematique & de Physique, pour la meme Anne*, pp. 27-28.

Romero, A. 1999. The Blind Cave Fish That Never Was. *National Speleological Society News* **57**(6):180-181.

Romero, A. 2000a. The Speleologist Who Wrote Too Much. *National Speleological Society* News **58**(1):4-5.

Romero, A. 2000b. Myth and Reality of the Alleged Blind Cave Fish from Pennsylvania. *Journal of Spelean History* **33**(4):67-75.

Romero, A. & L. Bennis. 1998. Threatened Fishes of The World: *Amblyopsis spelaea* De Kay, 1842 (Amblyopsidae). *Environmental Biology of Fishes* **51**(4):420.

Scott, W. 1866. Probable Existence of a Great Cavern Under Lancaster, Pa., *Scientific American* (April 7, 1866) **14**:228.

Chapter 8. The Speleologist Who Wrote Too Much [329]

Summary

To some, he is the father of speleology. His book *The Subterranean World* was a classic. And he knew how to make money. But now very few remember him. So, what went wrong for Athanasius Kircher? He paid the price for being an uncritical compiler of other people's tales. Today he is confined to a footnote in the history of science.

Introduction

Athanasius Kircher (Fig. 8.1) was born in Geisa in what is now central Germany on May 2, the feast day of St. Athanasios, 1602. He, like his five brothers, entered a religious order, probably because the family was too poor to pay for an education. But Athanasius was a bright man, and he went to a great number of Jesuit institutions, where he learned Greek and Latin and studied humanities, natural science, mathematics, philosophy, and theology. He received a doctorate in divinity, and he became a Jesuit in 1618.

Athanasius's life was full of near misses with death, something he liked to recount to anyone willing to listen. He escaped death several times during childhood. Once, while swimming in a mill pond, he was suddenly caught by the current and swept toward the wheel, where, instead of being mangled by the machinery, he passed through unharmed. Later, at a horse race, the pressure of the crowd pushed him under the feet of the oncoming horses. He crouched motionless and again emerged untouched. On another occasion, he made a two-day journey to see a play in a neighboring town. Lost on the way back and in fear of robbers, wild boars, and bears, he spent the whole night up in a tree. When he was fifteen, he got chilblains while skating, and his skin turned gangrenous. He thought he was going to die. He prayed earnestly, and the next morning he was cured. "It was a miracle," Kircher said.

He would later have to run for his life. In late 1621, Duke Christian of Brunswick, a notorious Jesuit-hater, was approaching the German town of Paderborn. where Athanasius was residing. Kircher and two others fled the city and escaped, while many other Jesuits were caught, bound, and beaten. Kircher and his fellow escapees struggled for three days through deep snow, ill-clad and penniless, begging for food, until a friendly Catholic nobleman gave them shelter. They continued on to Cologne and tried to cross the frozen Rhine. The locals had assured them that it was safe, but, when he was halfway across, a piece of ice broke loose and carried Kircher downstream. He swam to the bank through freezing water and walked for three hours until he reached safety.

During another trip through Protestant territory, Kircher obstinately refused to travel disguised in lay clothes, saying, "I would rather die in the robes of my order than travel undisturbed in worldly dress." A party of Protestant soldiers ambushed him, he was stripped and beaten, and they prepared to hang him from the nearest tree. His calm demeanor so moved one of the soldiers that he persuaded his comrades to spare the life of the young novice. Not only did they

---

[329] Based on Romero, A. 2000. The Speleologist Who Wrote Too Much. *National Speleological Society News* **58**(1)4-5

leave him with his clothes and books intact, but the compassionate soldier returned, gave him money, and urged immediate flight.

Figure 8.1. Athanasius Kircher by an unknown artist (public domain image).

Sent to Vienna in 1635, he embarked with some other brothers for the first stage of the journey, by sea from Avignon to Marseilles. They were all ill, so the captain landed them on an island "for a rest" and promptly sailed away with all their possessions. They managed to hail some fishermen, who took them the rest of the way to Marseilles. Next, on their way to Genoa, a storm blew up, and for three days they had to shelter in a cove. No sooner had they set sail again than a violent storm drove the boat back toward the coast. When at last he reached Genoa, he set out on another boat for Leghorn, ninety miles to the south. This time his ship was blown by a storm to Corsica, before managing to dock far past his destination at Civita Vecchia, the main port of Rome. Kircher could not miss the chance to see the Eternal City, so he set out on foot for the forty-mile trip. On reaching Rome, he found to his amazement that he was expected, his orders having been changed without his knowing it, and he was offered a job. He was to stay at the *Collegio Romano*, the Vatican's university, with a special commission to study hieroglyphs and to teach. His belief that God had spared him for better and greater things was fully confirmed in his mind, and he decided to become the most prolific scholar of his time.

He finally had all the leisure, assistants, and money he needed to conduct his scientific and humanistic investigations. For the next forty-five years, until his death, he dedicated his life to studying and writing in Rome. For the first eight years, he also had teaching responsibilities. He could teach an impressive range of studies, Greek, grammar, mathematics, physics, philosophy, Syrian, Hebrew, and music. He was supported by the patronage of the Pope, as well as aristocrats from all over Europe. And he knew how to be thankful. Kircher had so many supporters that he eventually began to dedicate not just his books, but individual chapters in his books, to them. Furthermore, sometime around 1660 Kircher sold exclusive rights to his books to a prominent Dutch publisher for a large sum of money. He was probably one of the first scholars able to obtain funding by the sale of his works.

Kircher was famous mainly for his study of Egyptian antiquities and hieroglyphs. He was also famous for his notable museum, probably the first public museum in history. From a wide variety of nobles and rulers, mostly German, he received extensive gifts of stuffed animals and birds from the New World. Growing increasingly pious as he grew older, Kircher discovered and restored the ruins of an ancient church, built by the emperor Constantine at the place where St. Eustace saw his vision of Christ in a stag's horns. Kircher reestablished this as a place of pilgrimage, and after he died in Rome on November 28, 1680, his heart was buried in the grounds of the church he had reconstructed.

But Kircher was a polymath, and he tried to know and publish about everything. He was called the master of a hundred arts by his contemporaries. He is believed to have been able to read sixteen languages. He published forty-four books, and more than two thousand of his manuscripts and letters have survived. Some of his books were huge, with large print and impressive illustrations. Among the discoveries that he and some of his supporters claimed he made was that bubonic plague was caused by tiny animals, which he called *contagium animalium*, he had observed under a microscope. Because of his vast knowledge of languages, he imagined he was able to translate Egyptian hieroglyphic writing, and he has been labeled by some the founder of Egyptology.

Among his mechanical inventions were a graduated aerometer and a device for measuring magnetic force with a balance. He also promulgated the use of magnetic declinations to find longitude, described a method of measuring temperature from the buoyancy of small balls, and designed and built sundials. From time to time, he surveyed and mapped, and devised an instrument for triangulation. In 1638, Kircher wrote a book to instruct knights in the solving of "the most import ant mathematical and physical problems." This involved use of a geometric calculator, Kircher's pantometer (Fig. 8.2). It has been said that he invented a counting machine, a speaking tube, and an aeolian harp. He was also credited with the inventio n of the magic lantern, a sort of slide projector. His reputation brought scholars, letters, and specimens to his study from all over the world, and he amassed a veritable museum of artifacts, curiosities of natura l history, and scientific apparatus. He wrote hundreds of letters and interviewed innumerable visitors. And he was very much fascinated by the subterranean sources of water and fire.

Kircher, the Speleologist

During a trip in 1630, Kircher witnessed a violent eruption of Mount Etna. He was so impressed by the phenomenon that after the eruption subsided, he had himself lowered into the cone for closer observation. This may have been the original inspiration for his book *Mundus Subterraneus,* probably the first printed work on speleology. He added geological observations made during his visit to Sicily in 1637 and 1638. The book, to be sure, deals with more than the geosciences, and contains chapters on other things like alchemy, chemistry, biology, and metallurgy. So, the book's title is misleading, unless Kircher meant it to only mean that he was dealing with many topics beyond human view. The book became a popular, lavishly illustrated textbook.

Figure 8.2. Kiercher's Pantometer (public domain image).

Kircher's main speleological idea in the book was his hypotheses on the origin of the water flowing from springs. He postulated that there were vast underground water reservoirs, which he called *hydrophylacia,* in many parts of the world. How did the water get there in the first place and then reemerge? According to him, sea water was sucked down in whirlpools and drawn into underground channels that conveyed into reservoir caves in the mountains. He situated some of these *hydrophylacia* hundreds of kilometers away from the sea and many hundreds of meters above sea level. He thought that "the sea, by pressure of air and wind or movement of the tide, pushes the waters through subterranean passages to the highest water-chambers of the mountains." He indicated that the presence of whirlpools such as the Maelstrom off Norway was proof of the suction mechanism. These whirlpools were places where there were connections among the oceans. This was based on his own observation in 1638 in the Strait of Messina, where, besides the noise of the surge, a dull subterranean rumble attracted his attention. He even tried to demonstrate his theory with experiments. At least part of the energy necessary to make this happen came from the "internal heat" of the earth. He was a firm believer in the idea of earth's internal heat, based on the observation that the temperature in mines increases with depth. For him, many natural phenomena, including the formation of minerals, were due to the fact that there was fire under the earth's crust. And hot water-vapor will rise in the caverns and condense in the cold rocks, producing springs. He did mention the real origin of spring water, rain, but does not give it much importance.

The preparation of *Mundus Subterraneus* began as early as 1658, and shortly afterwards a preliminary version appeared as a 171-page supplement to another book of his. The first full-fledged version of the book was published in 1665. It was *rulo* volumes totaling 892 large pages (Fig. 8.3). A new edition, published in Amsterdam in 1678, contained lengthy additions on caves in Switzerland, Austria, Italy, and the Greek islands. This edition achieved more popularity and became the standard geology text for the rest of the seventeenth century.

Kircher, Father of Speleology?

Some have advanced the idea that *Mundus Subterraneus* was the first speleology book. If, as some have proposed, Kircher is the father of geology, why not the father of speleology? Well, we all know that some of the most brilliant scientists blundered at times. Darwin did not get genetics right in his *Origin of Species,* and Einstein later regretted adding a mysterious cosmological constant to his theory of relativity. Newton and Boyle, contemporaries of Kircher, believed in alchemy. But all of these scientists are remembered and respected for theories such as Darwin's evolution by means of natural selection, Einstein's relativity, Newton's calculus and law of gravity, and Boyle's law relating the pressure and volume of gasses. Is there anything we can remember Kircher for?

Kircher not only got wrong the water cycle of the Earth and the mechanisms of underground water, but his wrong ideas had been proposed by others four centuries earlier (Fig. 8.4). Many of his statements were absurd. He believed that caves were inhabited by dragons, unicorns, and giants (Fig. 8.5). He even included drawings of such monstrosities in *Mundus Subterraneus*. According to him, "it is not implausible that, as under the earth all kinds of fish occur and live, also earth animals stay there, that is all kinds of Mice, Snakes, Dragons, as well as others, that find their origin in rotten matter." The cavefishes that he mentioned as occurring in several places in Europe must have been imagined, since no true cave fish have ever been found in the continent.

Kircher's absurd ideas were not confined to speleology. He believed in the transmutation of metals, particularly of iron into copper, in the spontaneous formation of fossils from minerals, and the spontaneous generation of living organisms. He said he had performed palingenesis, the resurrection of plants from their ashes. He also believed in astrological influences on human health and terrestrial cataclysms. He believed in mermaids and opposed the theory that the sun was the center of the solar system.

Perhaps he was just careful about espousing new, revolutionary ideas. After all, he was a Jesuit living in the theocratic Rome that had burned Giordano Bruno in 1600 and condemned Galileo 1616 for their heliocentric ideas. Deviation from the norm was heresy, and the Society of Jesus had been formed during the heat of the Counter-Reformation to convert the heathen and combat heresy. But this is, at best, a superficial explanation. He was, above all, a person who never saw things with a skeptical eye. Even many of his contemporaries criticized him for this behavior, and twentieth-century scholars have been fierce in their criticisms. He has been accused of gullibility, error, ignorance, and filching ideas. Someone called him a scholarly windbag. Even his alleged discoveries and inventions have come under fire. Medical historians do not credit him with authentic observations on microbiology or infectious diseases. He established no useful general laws and made no stimulating suggestions for research. The invention of the magic lantern was not his, but a Danish physicist's. Real translation of the Egyptian hieroglyphs was not achieved until after the Rosetta Stone was discovered in 1799. Even the belief that he might have actually visited the caves in his native Rhön cannot be confirmed.

Maybe his fault really was in believing everything he heard. Kircher enjoyed the privilege of living in Rome, the center of a worldwide network of Jesuit missionaries and others who reported to there on their journeys. No matter how fantastic a tale may have been, he recorded and published it. As the editor of one of our present-day tabloids might say, "If you believe it, we will publish it." That is why Kircher is seen today as more a historical curiosity than a real scientist. And that is why few take seriously the idea that he might be considered the father of speleology.

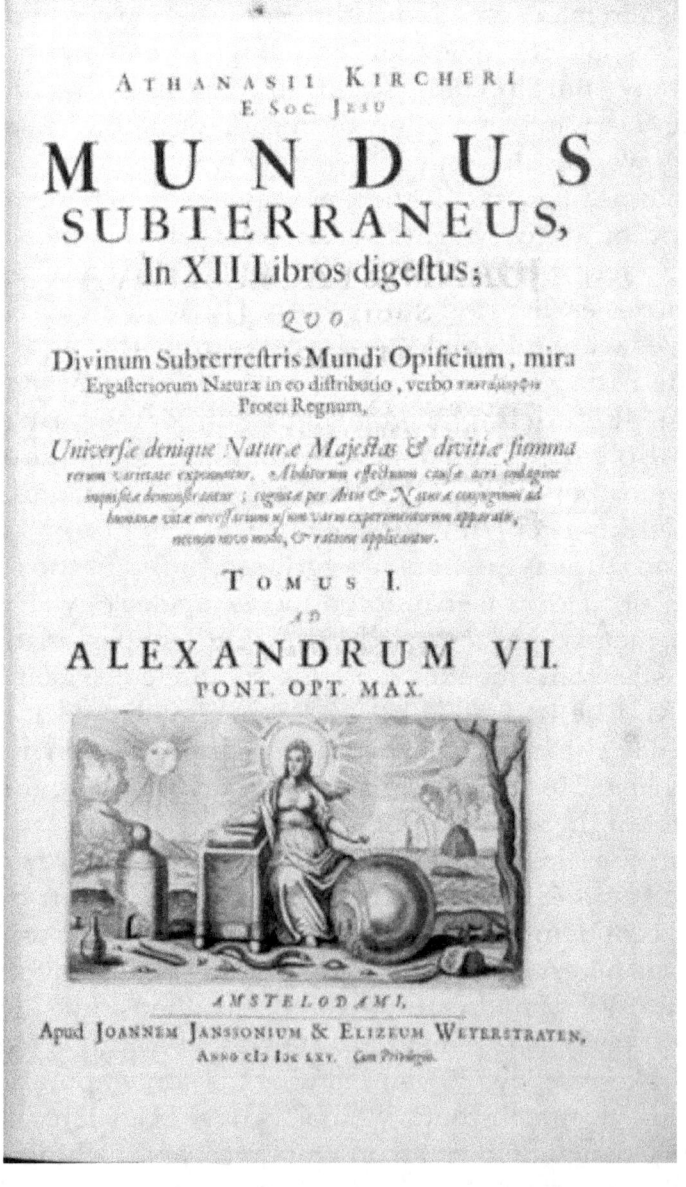

Figure 8.3. Cover of Kircher's Mundus Subterraneous (public domain image).

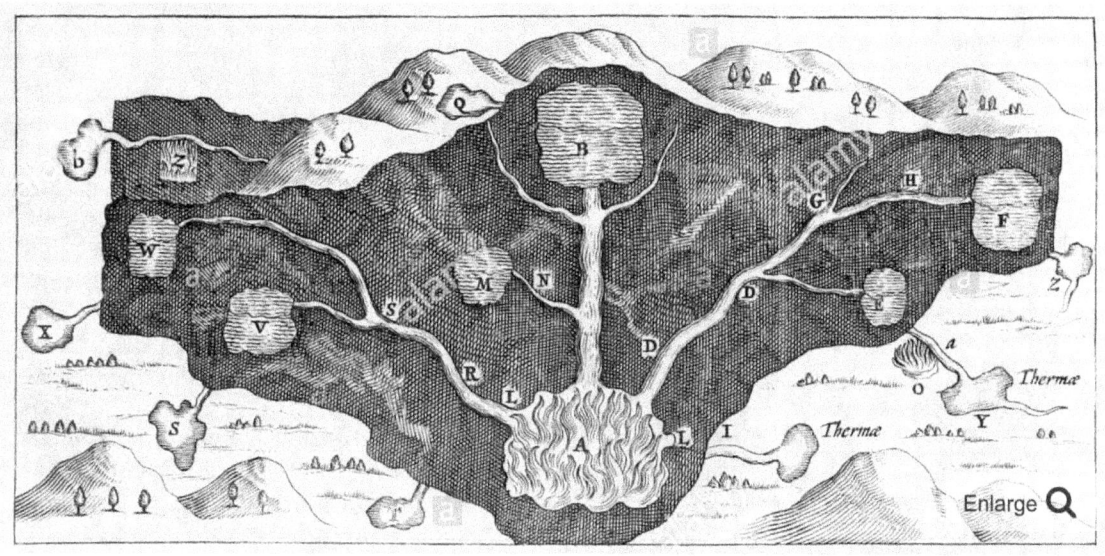

Figure 8.4. The underground world according to Kircher (public domain image).

Figure 8.5. One of the many beasts that Kircher believed lived in caves (public domain image).

Acknowledgements

Professor Herman de Swart located the Dutch translation of Kircher's *Mundus Subterraneus* in the library of the National Museum for the History of Science and Medicine in Boerhaave, Leiden, and sent the author a translation of a portion of it. Andrea Romero read the manuscript and made valuable suggestions.

Literature Consulted

Brauen, F. 1982. Athanasius Kircher (1602-1680). *Journal of the History of Ideas* **43**:129- 134.
Godwin, J. 1979. *Athanasius Kircher: A Renaissance Man and the Quest for Lost Knowledge.* London: Thames and Hudson.
Kangro, **H.** 1973. Kircher, Athanasius. Pp. 374-378, volume 7, *In*: C. C. Gillispie (ed.), Dictionary *of Scientific Biography.* New York: Scribners.
Shaw, T. R. 1992. *History of Cave Science: The Exploration and Study of Limestone Caves, to* 1900. Second edition, Sydney Speleological Society.
Strasser, G.F. 1996. Science and Pseudoscience: Athanasius Kircher's Mundus *Subterraneus* and His *Scrutinum . . . Pestis.* Pp. 219-240 *In*: G. S. Williams and K. Schindler (eds.), *Knowledge, Science, and Literature in Early Modem Germany.* Chapel Hill: The University of North Carolina Press.

Chapter 9. The Blind Cavefish that Never Was [330]

Summary

In 1748 a French aristocrat thought he had discovered the first blind, subterranean fish ever. His discovery was never acknowledged... and it may never be.

The Man

Marc-René Marquis de Montalembert had a rather odd background for a man who was about to claim to be the discoverer of the first blind, subterranean fish. Born in Angouleme on July 16, 1714, he belonged to a distinguished family with an illustrious military tradition whose origins can be traced to the "chevaliers" of St. Louis' Crusade of 1249. He received an education typical of his times and social status, entering military service at the age of 18. He distinguished himself in the War of the Polish Succession (1733-1738), the War of the Austrian Succession (1740-1748), and the Seven Years' War (1756-1763). Despite his brilliant military career, Montalembert's heart was in engineering. By the time he was admitted to the prestigious Paris Academy of Sciences in 1747, he had already written about water evaporation and he had developed a good professional relationship with the famous French physicist Lazare Carnot.

After the Seven Years' War, he distinguished himself in the field of fortification, which became the passion for the rest of his life. He wrote an eleven-volume treatise (*La Fortification Perpendiculaire*), a very controversial but influential work where he proposed the polygonal method of fortification. His design replaced the complex star-shaped fortresses sponsored by Sebastien de Vauban with a simplified polygonal structure that became the standard European fortification system of the early 19th century. His private life was also filled with excitement. In 1770 he married Marie de Comarieu, a French socialite and amateur theater actress, for whom general Montalembert wrote a dramatic piece and later even wrote some operatic works. Despite his aristocratic roots, he embraced the principles of the French Revolution.

Alarmed by the turn of political events, he moved to England between 1789 and 1790 and divorced. He remarried to Rosa lie Louise Cadet with whom he had a daughter born in July 1796. He was dispossessed of his iron forges and obliged to sell his estate in Angoumo is for very little money. With changes in the French political environment, he regained his pension, became a consultant in military affairs, and was promoted to "general de division" in 1792 becoming the oldest general in the French army. In 1797 he tried to become a member of the Class of Physical and Mathematical Sciences of the Institute of France (which was the reconstituted Academy of Sciences abolished during the Revolution), but withdrew his candidacy in favor of General Bonaparte. He died in Paris, on March 29, 1800 at the age of 86.

It was in 1748, a year after his admission to the *Academie* that he reported the discovery of a blind, subterranean, fish in one of his properties in the Southwest of France.

---

[330] Based on Romero, A. 1999. The Blind Cave Fish that Never Was *National Speleological News* **57**(6):180-181.

Figure 9.1. Portrait of René de Montalembert (courtesy of J. Langis).

The Times

The XVIII Century was a time of flourishing and expanding scientific activity. The historian Roger Hahn has written "science was the passion of the century at all literate levels of society, in every urban center of France, and even among the progressively minded gentlemen-farmers. Under these circumstances the Paris Academy of Sciences reached its peak. Founded in 1666, it was definitively organized in 1699. From then on, the Academy developed as one of the most prestigious institution s of Europe until the French Revolution suppressed it in 1793. Two years later it was reorganized as a branch of the *Institut de France*. In 1816 it was again reconstituted as *Academie Royale de Sciences*. The first half of the 1700 s is also an interesting time for the natural sciences. Impressed by the expeditions to the New World and the less-than-reliable tales of explorers, the Old World was extremely interested in rare creatures. In the l 720s, for example, anatomy, and especially grotesque embryonic monsters, were in vogue. Prior to the French Revolution (1789), "cabinets" (collections) of natural hist or y were extremely popular; in Paris there were more than 200 private collections alone.

The Fish

There is no wonder that these times were the right ones to report such odd things as blind and one-eyed fishes. And that was what Montalembert did. In his report of 1748 to the Academy (which was not printed until 1752) he wrote: In a spring at Gabard Angoumois, near one of (Montalembert's) estates, it is common to fish either blind or one-eyed pike; one-eyed ones always miss the right eye and among the blind ones, the right eye seems further reduced than the left eyes. This spring is a kind of bottomless pit. There are small groups of floating plants at the surface, which impede the use of fishing lines, and makes fishing a long and difficult process.

However, Montalembert was fortunate enough to capture a young pike with its right eye missing. This spring drains its water into the Lissone river, but despite this connection, the local people say that one-eyed or blind pikes are never fished in the river, while the spring contains one-eyed or blind ones only.

What Was It?

Montlembert's alleged discovery is more interesting and surprising than most people may think. While there have been some reports around the world of blind depigmented fish that come out of spring and wells, the fact of the matter is that to this day, no blind, subterranean fish has ever been reported from Europe- without question the continent whose caves have been most intensely explored. Although Jacques Besson, another French Engineer, claimed in his 1569 book to have observed "small eels" living in subterranean waters in France he never said they were blind or depigmented. Montalembert said that what he saw was a pike. That, by itself, is not surprising. The pike (*Esox lucius*) is, by far, the most common freshwater fish of the Northern Hemisphere. The fact that this fish can be identified as a pike despite being blind, is not surprising either. Many subterranean fishes are very much identical to their surface (what speleologists call "epigean") forms, except for the lack of eyes and pigmentation.

But here comes the other puzzling fact from Montalembert's description: reduction of eyes and pigmentation always goes hand in hand among subterranean species of animals. The more reduced the eyes are the less the pigmentation. Yet, Montalembert never mentioned the latter, which should have been a very peculiar and unusual feature. Furthermore, he says that some of the fish lacked one eye and when that was the case, it was always the one on the right side. Which leads us to wonder: were these pikes really blind because of their adaptation to the subterranean environment or due to some other phenomenon? The only other way that a fish may become blind due to natural causes is because of some disease. A number of parasites which damage fish eyes have been described.

However, there are two problems with this explanation. The first one is that, as far as I know such a type of disease has never been reported for the pike. But let's concede that what Montalembert observed was a peculiar disease for a particular location.

The second objection, however, seems more formidable. Why was it always the right eye that was missing? When parasites attack the eyes of fishes (or other animals), they tend to infect both eyes and certainly have no preferences for a particular side.

So, what was it? All of these questions lead us to two possible explanations: First, Montalembert observed a real but yet very strange phenomenon that nobody else has been able to observe again. Second, the Marquis was, indeed a sloppy observer and a very amateurish naturalist (even by the standards of his times: after all, his only natural history article is on this fish). His description leaves us with more questions than answers. Following Ockham's Razor (the simplest explanation should be the correct one), we should assume that what he left us was the product of his careless observations. Yet, it would be worthwhile to try to find Montalembert's fish and reconstruct what he really saw. After all this may be the only blind, subterranean fish of Europe and the only one-eyed specimen ever reported in the history of science!

The first problem is that, apparently, he left no drawings, much less a preserved specimen. Then, why not go to the same location and try to collect new ones? And now comes the second problem: nobody knows where that locality, Gabard, is! Neither the countless hours I spent examining old and modem maps of the area, nor the studies that Prof. Janis Langins of the University of Toronto (and the expert on Montalembert) however yielded that location or any mention of it. What may have been a great historical discovery will remain, at least for the time being, as a simple curiosity. As Montalembert's compatriots would say, *c'est la vie!*

Acknowledgments

Prof. Janis Langins of the University of Toronto provided me with Montalembert's portrait. Prof. Langins and Phoebe Vanselow read the manuscript and made valuable, suggestions.

Literature Consulted

Academie des Sciences. 1968. *Index Biographique des Membres et Correspondants de l'Academie des Sciences du 22 December 1666 au 15 December 1667*. Paris: Gauthier-Villars, 574 pp.

Bertrand, J. 1869. *L'Academie des Sciences et les Academiciens de 1666 á 1793*. Paris: J. Hetzel, 434 pp.

Besson, J. 1569. *L'Art et science de trouver les eaux et fontaines cachees soubs terre*. Orleans: Pierre Trepperel, 83 pp.

Hahn, R. 1971. *The Anatomy of a Scientific Institution: The Paris Academy of Sciences, 1666-1803*. Berkeley: University of California Press, 433 pp.

Lloyd, E.M. 1887. *Vauban, Montalembert, Carnot; Engineering Studies*. London: Chapman and Hall, 239 pp.

Maindron, E. 1888 ·. *L'Academie des Sciences*. Paris: Bailliere; 344 pp.

Maury, L.F.A. 1864. *L'Anciennne Académie des Sciences*. Paris: Didier et Cie., 395pp.

Chapter 10. Humboldt's Alleged Subterranean Fish from Ecuador [331]

Humboldt and His Report

Friedrich Wilhelm Heinrich Alexander von Humboldt[332] was one of the most respected scientists of his time. Humboldt made enormous contributions to geography and natural history and could be considered the last great polymath of the natural sciences. His travels of exploration through the American continent and central Asia are among the most famous scientific explorations ever. Even Charles Darwin felt greatly influenced by him (Brent 1981:98).

Figure 10.1. Alexander von Humboldt. Unknown artist.

Among the many contributions made by Humboldt there is a paper titled "Dissertation on a new species of pimelodid thrown out by the volcanoes of the Kingdom of Quito"[333]. There, after describing volcanoes in general, he claimed that among the things they spew forth are an "innumerable quantity of fish" *(une innombrable quantite de poissons)*. Further, he says that although he had not witnessed this phenomenon himself, during the year he spent in Quito

---
[331] Originally published as Romero, A. & K. Paulson. 2001. Humboldt's Alleged Subterranean Fish from Ecuador. *Journal of Spelean History* **35**(2):56-59.
[332] *b*. Berlin, Germany, 14 September 1769: *d*. Berlin. 6 May 1859.
[333] Humboldt, F.H.A. von. 1805. Mémoire sur une nouvelle espèce de pimelode, jetée par les volcans du Royaume de Quito. *In*: Voyage de Humboldt et Bonpland, Deuxième partie. Observations de Zoologie et d'Anatomie comparée (F.H.A. Humboldt & A. Bonpland, eds.), [Paris: Unspecified Publisher]: **1**:21-25.

"volcanoes vomiting fish is such a common phenomenon, and so well-known among all the local in habitants that there cannot be the slightest doubt of its authenticity" *(Les poissons vomis par les volcans sont un phenomene si commun et si generalement connu de tous les habitans de ce pays, quil ne peut rester le moindre doute sur son authenticite).*

His sources were the archives of small villages around Cotopaxi, where he sometimes came upon notes regarding fish that came from the great depths of the earth. Some went into greater detail: the rotting fish strewn across the earth created a great stench; fish were enveloped in volcanic mud. Some natives assured him that sometimes the fish would still be alive after their trip through the hot core of the volcano, through the air, and then onto the ground.

He goes on to describe a new fish species: *Pimelodus cyclopum* known today as *Astroblepus cyclops.* Other synonyms include *Cyclopium chimborazoi* and *Astroblepus chimborazoi.* This species is known in Ecuador as Prenadilla ("the little pregnant one").

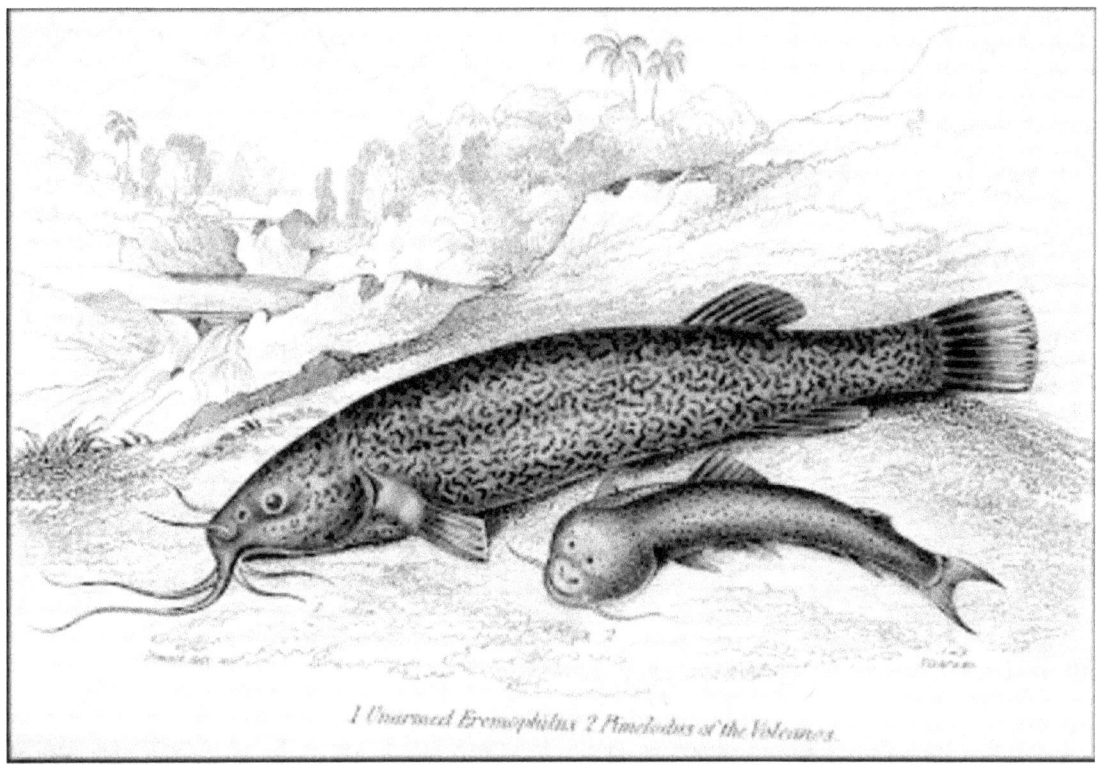

Figure 10.2. Original drawing of Humboldt's "*Pimelodus cyclopum*".

Schomburgk's Popularization of the Tale

Humboldt's publications, which he financed himself, were very limited in the number of copies printed. Thus, his report of this fish did not achieve full notoriety until it was summarized in the English literature by Robert Hermann Schomburgk[334] Although German born, he would later become British and

---

[334] b. Freyburg, Germany, 5 June 1804; d. Schöneberg, Germany, 11 March 1865.

famous for his explorations of the then British Guiana from 1841 to 1895 as well as those of the British Virgin Islands, Santo Domingo (today the Dominican Republic) and Southeast Asia. His most famous ichthyological work was his Natural History of the Fishes of Guiana (Schomburgk 1841-1843). There, he gives Humboldt's fish an English common name: "The Pimelodus of the Volcanoes" and summarized Humboldt's tale by stating that:

> *The singular fact in the history of this fish is, that from the volcanos in the vicinity it is, during the periodical eruptions, discharged in thousands; and in a state so perfect, as to show little mutilation either from scorching or from the effects of the hot water with which it is discharged. Baron Humboldt states, that in turning over the records kept by the small villages in the vicinity of Cotopaxi, he found mentioned, that, on the lands of the Marquis Selvalege, so large a quantity was thrown, that a putrid odour was spread over the country. The almost extinct volcano of Imbaburu, in 1691, discharged thousands over the plains surrounding the village of Ibarra, and the miasmata which occurred from them, fevers were attributed: and from another volcano, in 1698, thousands were also thrown, encased in algamaceous balls. Humboldt is of opinion that these volcanos contain subterranean lakes, from whence the supply is afforded, the numbers in the little rivulets around being comparatively small; he adds, many of these rivulets communicate with these subterranean caverns; and that the first Pimelodi which have stocked them must have ascended against the stream".*

Myth and Reality

Obviously, no fish can be "vomited" from volcanos. The temperature alone (between 700 and 1,300° C (1,292 to 2,192 °F) would disintegrate any living organism. As Schomburgk states in his narrative. Humboldt went into great pains in trying to explain this phenomenon. His explanation may have been influenced by stories made popular by Athanasius Kircher whose highly speculative -and often wrong- interpretations about subterranean waters were very popular into the eighteenth century (Romero 2000).

Having said that, though, the fish that Humboldt described from the rivers around Quito is not only real but it has also been found in caves. At the British Museum of Natural History in London, there is a specimen of *A. cyclopus* catalogued as BMNH 1977.5.2.4 .13 collected in a cave north of Puyo, quite far away from the localities mentioned by Humboldt. Cave individuals of the genus *Astroblepus*, whose species has yet to be determined, have been cited as occurring in Peru (Vilchez Murga 1968, Ribera & Bellés 1984). Also, two troglobitic species of the genus *Astroblepus* have been formally described: *A. pholeter* by Collette (1962) in Latas, 4 km north of Archidona, Napo Province, eastern Ecuador (again, far away from Humboldt's localities) and *A. riberae* by Cardona & Guerao (1994) for Peru. *A. cyclopus* has also been reported in rivers of Colombia.

Therefore, Humboldt's account of these subterranean fishes being spewed from volcanoes can be considered more a myth than a fact based on local legends. These types of accounts are not unusual in the literature (see Romero

1999a,b, Romero 2000; Romero & Lomax 2001) and represent styles of scientific reporting typical of times when facts were not necessarily checked before being published.

Literature Cited

Brent, P. 1981. *Charles Darwin, A Man of Enlarged Curiosity*. New York: Harper Row Publishers, 356 pp.

Burgess W.E. 1989. *An Atlas of Fresh water and Marine Catfishes: A Preliminary Survey of the Siluriformes*. Neptune City: T.F.H. Publications, 784 pp.

Cardona, L. & G. Guerao. 1994. *Astroblepus riberae*, una nueva especie de Siluriforme cavernícola del Peru (Osteichthyes Astroblepidae). *Memoires de Biospeleologie* **11**:21-24.

Collette, B.B. 1962. *Astroblepus pholeter*, a new species of cave dwelling catfish from Eastern Ecuador. *Proceedings of the Biological Society of Washington*. **75**:311-314.

Eschmeyer, W.N. (Ed.). 1998. *Catalogue of Fishes*. San Francisco: California Academy of Sciences. 3 vols.

Humboldt, F.H.A. von. 1805. Mémoire sur une nouvelle espèce de pimelode, jetée par les volcans du Royaume de Quito. *In: Voyage de Humboldt et Bonpland, Deuxième partie. Observations de Zoologie et d'Anatomie comparée* (F.H.A. Humboldt & A. Bonpland, ed.), [Paris: Unspecified Publisher]: **1**:21-25.

Ribera, C. & X. Bellés. 1984. Perou, pp. 569, In: C. Juberthie & V. Decu (Eds.), *Encyclopaedia Biospeologica*. Moulis: Societé de Biospeleologie, Vol 1.

Romero, A. 1999a. The blind cave fish that never was. *National Speleological Society News* **57**(6):180-181.

Romero, A. 1999. Myth and reality of the alleged blind cave fish from Pennsylvania. *Journal of Spelean History* **33**(4):67-75.

Romero, A. 2000. The speleologist who wrote too much. *National Speleological Society News* **58**(1):4-5.

Romero, A. & Z. Lomax. 2000. Jacques Besson, Cave eels and other alleged European fishes. *Journal of Spelean History* **34**(2):72-77.

Schomburgk, R.H. 1841-1843. *The Natural History of Fishes of Guiana*. Edinburgh: W.H. Lizars, 2 Vols.

Vilchez Murga, S. 1968. *Parques Nacionales del Peru*. Lima, Peru: Ediciones Cajamarca, 128 pp.

Chapter 11. Myth and Reality of the Alleged Blind Cavefish from Pennsylvania [335]

Introduction

It is not uncommon to find unsubstantiated reports of blind cavefishes in the speleological literature. Examples include accounts of hypogean fishes in France (Montalembert 1752), Italy (Scott 1866), and North Africa (Anonymous 1879) (for analysis of some of those reports, see Romero 1999, 2000). Although several species of cave fishes have been described for the United States (Romero 1998 a,b,c; Romero and Bennis 1998), we have found persistent, but yet unconfirmed, reports of a blind cavefish for Pennsylvania.

Cope (1864) published a paper in which he claimed the discovery of a new species of troglobitic (blind, depigmented, and obligatory cavernicole) fish from Pennsylvania. Despite the fact that the validity of such species is not recognized today, numerous other reports of cavefishes for that state have continued to be published until the present time (e.g., Kranzel 1986).

This paper is aimed to analyze those reports and to establish the facts that may have given credence to the belief of (as yet) the unconfirmed presence of a blind cave fish species for Pennsylvania.

Materials and Methods

I reviewed as much published literature on vertebrate cave fauna for the state of Pennsylvania as I could find. I conducted extensive searches not only in the speleological and ichthyological literature but also all the bibliographic material pertaining to these topics at the library of the Pennsylvania State University, College Park, PA. I also reviewed all the pertinent literature on the people mentioned in this article.

Results

The First Claim

Edward Drinker Cope[336] was one of the most prolific American naturalists. He published about 1400 articles and books in many different areas of knowledge. Most of them were on vertebrates, both extinct and living, but his list of publications also includes works on invertebrates, geology, anthropology, evolution, behavior, sociology, education, philosophy, religion, and history, as well as on issues of public interest of his times (Osborn 1931). Cope published 27 papers pertaining to speleology which makes him one of the pioneers of this science in the U.S. (Grady 1987, 1992; Romero & Romero 1999).

In his 1864 paper, Cope described three new species of fish. One of them, according to him, was a blind subterranean fish. In that paper he stated (correctly) that many aquatic organisms found in subterranean environments are blind and that

---

[335] Based on Romero, A. 1999. Myth and Reality of the Alleged Blind Cave Fish from Pennsylvania. *Journal of Spelean History* **33**(4):67-75.
[336] *b*. Philadelphia, PA, 28 July 1840; *d*. Philadelphia, 12 April 1897.

they belong to diverse taxonomic groups. He briefly reviewed examples of blind, subterranean fishes from around the world. He also explained that among silurid fish, there are many examples with reduced or sunken eyes.

Then, he went on to describe a new genus and species of "blind silurid" which he called "*Gronias nigrilabris*" based on two specimens collected by Jacob Stauffer, Secretary of the Linnean Society of Lancaster, Pennsylvania, which he had received a year earlier.

It is occasionally caught by fishermen, and is supposed to issue from a subterranean stream, said to traverse the Silurian limestone in that part of the Lancaster county, and discharge into the Conestoga.

Two specimens of this fish present an interesting condition of the rudimental eyes. On the left side of both a small perforation exists in the corium, which is closed by the epidermis, representing a rudimental cornea; on the other the corium is complete. Here the eyeball exists as a very small cartilagenous sphere with thick walls, concealed by the muscles and fibers tissue attached, and filled by a minute nucleus of pigment. On the other the sphere is large and thinner walled, the thinnest portion adherent to the corneal spot above mentioned; there is a lining of pigment. It is scarcely collapsed in one, in the other so closely as to give a tripodal section. Here we have lapsed an interesting transitional condition in one and the same animal, with regard to a peculiarity which has at the same time physiological and systematic significance, and is one of the comparatively few cases where the physiological appropriateness of a generic modification can be demonstrated. It is therefore not subject to the difficulty under which the advocates of natural selection labor, when necessitated to explain a structure as being a step in the advance towards, or in the recession from, any unknown [italics in the original] modification needful to the existence of the species. In the present case observation of the species in a state of nature may furnish interesting results. In no specimen has a trace of representing the lens been found.

The two syntypes described by Cope are deposited today in the Academy of Natural Sciences of Philadelphia and catalogued as ANSP 22082 and ANSP 22083. The label in the jar says: "Pennsylvania: Conestoga Creek, tributary of the Susquehanna; Coll. Jacob Stauffer." Fig. 1 is a drawing of the first of those syntypes.

There are two issues here. The first one is the true location where these fish were found. The second is if the fish he examined were true blind cave fish or the result of a biological phenomenon.

For the first one, Cope provides no evidence whatsoever that the fish was of underground origin. Furthermore, his appreciation that the specimens that he had examined really represented individuals not only of a new species but even a new genus of true blind cave fish, has been countered by those who have examined them.

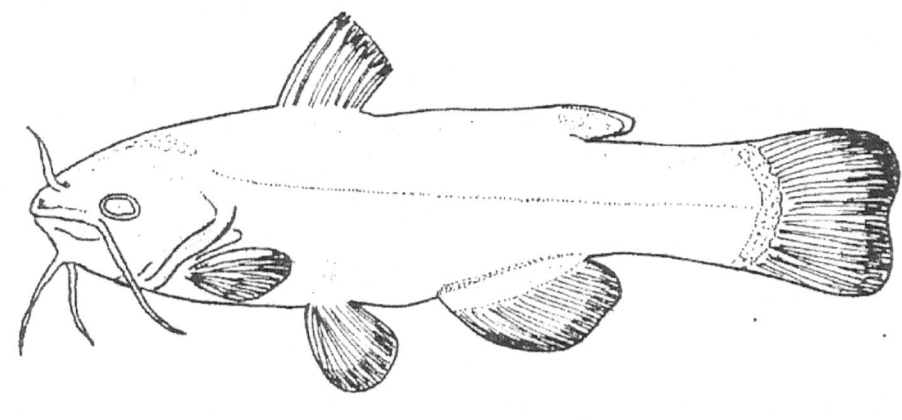

Figure 11.1. Specimen of *"Gronias Nigrilabris"* (Cope 1864). Drawing by Olga Mayayo.

The Counter Claims

Henry Weed Fowler[337], was the first curator of "cold-blooded"(every vertebrate but mammals and birds) at the Academy of Natural Sciences of Philadelphia, where the above-mentioned specimens were deposited (Conant 1966, Phillips 1964). Cope had donated most of his ichthyological collection (specimens of 341 out of the 424 nominal species) to the Academy (Smith-Vaniz & Peck 1991). Fowler, who had met Cope in 1894 (Fowler 1963), oversaw the transfer of Cope's entire alcoholic collection to the museum in 1898 and spent the next several years deciphering Cope's notes and organizing the collection (Smith-Vaniz & Peck 1991).

As early as 1915, Fowler challenged the validity of Cope's original interpretation of classifying these specimens as a new genus. In his paper on the nematognathous fishes contained in the collections of the Academy of Natural Sciences in Philadelphia, he named Cope's specimens as *Ameiurus nigrilabris* and placed this species next to *A. nebulosus* as indicating the close relationship between the two nominal species (Fowler 1915).

By 1945, Fowler was fully convinced that Cope's specimens are nothing more than unusual individuals of the brown bullhead (*Ameiurus nebulosus nebulosus*), as noted in his study of the fishes of the southern Piedmont and coastal plain of the United States (Fowler 1945:55). Unfortunately, Fowler never left any published rationale for dismissing Cope's original description as just a synonym of a fish that had been originally described by French ichthyologist Charles Alexandre LeSeur[338] (Hamy 1968) in 1819, based on a specimen collected also in Pennsylvania.

Fowler's initial reinterpretation has been confirmed by other ichthyologists who have examined Cope's specimens. Thus, Hubbs and Bailey (1947), in their analysis of the blind catfishes from the artesian waters of Texas, wrote:

> *Gronias nigrilabris Cope (1864:231-232) has also been regarded as a blind cave catfish related to Ameiurus. The specimens, however, were caught in Conestoga Creek, Pennsylvania, and were assumed to have issued from limestone caves merely because the specimens had defective eyes. They were*

---

[337] *b*. Holmesburg, PA, 23 March 1878; *d*. Philadelphia, 21 June 1965.
[338] *b*. Le Havre, France, January 1, 1778; *d*. Le Havre, December 12, 1846.

well pigmented and had an air bladder. We see no reason for thinking that the types of Gronias nigrilabris were other than specimens of Ameiurus nebulosus nebulosus (LeSeur) with eyes defective due to injury or some other cause. Such fish are not infrequently encountered. The serrated pectoral spine as well as the locality precludes the relationship with Ameiurus melas postulated by Jordan and Evermann (1896:142). A figure of the type of Gronias ni.grilabris, recently published by Fowler (1945:55, Fig. 160), confirms our reference of Gronias nigrilabris to the synonymy of Ameiurus nebulosus nebulosus.

Later, Taylor (1969), in his monograph of the catfish genus *Noturus* wrote:

The two syntypes (ANSP 22082-3) of Gronias nigrilabris Cope (1864) were examined in the course of this study. They do not represent a distinct genus, but are specimens of Ictalurus nebulosus (LeSeur) as maintained by Hubbs and Bailey (1947, p. 12). Fowler (1915a, p. 208) regarded them as a distinct species of Ameiurus [=Ictalurus]. Both have eight rays in each pelvic fin; the anal fins have 19 and 20 rays; the pectoral spines are long and serrated posteriorly. Contrary to report the eyes are present, but are asymmetrically developed - undoubtedly a teratological condition.

These analyses seem to have settled the issue. As a matter of fact, in modem ichthyological literature, Cope's specimens are no longer considered a distinct species (e.g., Eschmeyer 1998). That, together with the fact that Cope never provided any hard evidence that these specimens originated from hypogean (subterranean) environment, should have been sufficient as to dismiss any evidence of blind, subterranean fishes for Pennsylvania.

Yet, numerous authors have continued to publish accounts of an alleged blind cavefish for Pennsylvania.

The Persistence of the Myth

Walter Scott wrote on March 20, 1866 (two years after Cope's original description) an article published on April 7, 1866, in *Scientific American*, which reads as follows:

MESSRS. EDITORS. - It is well settled belief among many of our most intelligent residents, that underneath the city of Lancaster and vicinity there exists a vast cavern. Many facts are recited giving extreme plausibility of this theory, the most important of which may be briefly stated, as follows: -

The city is located within the great limestone belt extending across the southeastern part of the State, and of all the geological formations limestone the most abounds in caverns, many of which are known to be of vast extent. In sinking wells in certain parts of the city, the bottom crust breaks through before reaching water, and the pump is suspended from above by chains.

There have been several well authenticated cases in the vicinity of the city, of the crust of the earth breaking and engulfing farm animals. In two instances men engaged in plowing, saw their teams disappear beneath the surface and only a funnel-shaped cavity remained to mark the spot.

The earthquake of Sept. 29, as well as several lighter shocks, may be very reasonably accounted for this theory. Huge masses of rock breaking from the roof of the cavern and falling into the depths beneath may cause such a quaking of the upper crust and dull rumbling noise as that which astonished the inhabitants on that day.

One of the most convincing proofs of the existence of this subterranean cavity is the discovery of an eyeless catfish in the waters of the Conestoga, a stream flowing past the city and supposed to connect with the hidden waters beneath. This fish is entirely destitute of organs of sight, having only small spots in place thereof.

In a celebrated grotto of Italy eyeless fish have been found, and it is inferred that the eyeless catfish of the Conestoga must originate in a similar underground locality and escape through the fissures of the rocks. I have endeavored to present as concisely as possible the principal facts bearing on the theory, and leave it for others to elaborate.

Walter Scott. Columbia, Pa. March 20, 1866.

Note that the penultimate paragraph very likely makes reference to Cope's specimens since it refers to the same locality. The last paragraph is, however, even more surprising because there are no records whatsoever of blind cavefishes in Italy. Scott (on whom I have not been able to find any information) may have been referring to a citation by Athanasius Kircher[339], a polymath, notorious for his unsubstantiated assertions about underground creatures that included dragons and other mythical animals (Romero 2000). Kircher (1665) writes about cave fishes and cites a couple of localities in Italy, as well as others in Greece and Switzerland. However, there is no evidence whatsoever of blind cave fishes for Europe.

Thus, Scott seems to be referring to Cope's specimens and offers no evidence that those fishes are actually hypogean but that they "must originate in a similar underground locality and escape through the fissures of the rocks."

Five years later we find another article, also in *Scientific American* that, again, refers to blind cavefishes in Pennsylvania (Anonymous 1871).

It is well known that great trouble and expence [sic] have been caused by the sinking of a portion of the track of the new Jefferson Railroad, where it crosses a swamp in Ararat township, Pa. It has been found, say the Montrose Republican, that under the swamp is a subterranean pond of several acres in extent and of considerable depth. This pond, of several acres in extent and of considerable depth. This pond is covered by about six feet in depth of black earth, which supports a heavy growth of woods. The trees are mostly soft maple, pine, hemlock and birch, many of them ranging from six inches to three fe.et in diameter. Last fall it was discovered that the subterranean pond contained many fish, of the kind usually found in ponds in this part of the country - pickerel and "shiners" among others - but all without eyes! In the darkness of their subterranean abode, they have no use for the organ of vision. The Ball Pond, about a mile and a half distant, is now "growing over." A considerable part of it has become subterranean within the last twenty years, and, probably, before many years it will be entirely covered like the other. This pond is about

---

[339] *b.* Geisa, Germany, 2 May 1602; *d.* Rome, Italy, 28 November 1680.

twenty acres in extent. For some distance from the shore, it is filled with a dense growth of water-lilies, and these, no doubt, furnish the foundation on which the superstructure of earth is commenced.

This article is unsigned, but given the nature of the topic, the locality (Pennsylvania), and the publication *(Scientific American),* we cannot rule out that Walter Scott was its author. Since then, reports of blind cave fishes for Pennsylvania have been dismissed. Mohr (1953), for example, writes that:

There are no blind salamanders or eyeless fish in Pennsylvania caves, no white crayfish or blind beetles (p. 15). Blind, white fishes have been reported at intervals but always have proved to be half-starved, pale, eyed fishes, often washed into caves during floods and trapped there as the waters subsided.
A single blind fish is known from Pennsylvania. It was a blind Catfish sent to Edward Drinker Cope in 1864, from the Conestoga Creek, Lancaster County. While some thought that the fish came from a subterranean stream feeding into the Conestoga, ichthyologists say that blindness is by no means extraordinary in this group of fishes (p.18).

Holsinger (1976), in his extensive monograph on the cave fauna of Pennsylvania, wrote:

There are apparently no established records for fishes from Pennsylvania caves. True cave fishes (troglobites) are unknown from the Appalachian Valley and most of the Appalachian Plateau, although accidentals are sometimes observed in cave systems of Virginia, West Virginia, and eastern Tennessee. The northern muddler, Cottus b. bairdii Girard (sculpin family Cottidae) is reported from a few caves in the central and southern Appalachians but has not been found in caves as far north as Pennsylvania (p. 87).

Despite this lack of evidence supporting the true blind cave fishes for Pennsylvania, Kranzel (1986) thinks blind cavefish for that state are real. After summarizing the Scott (1866) paper, he wrote:

Troglobitic species are scarce; in fact, troblobitic vertebrates are unknown in PA. There are apparently no established record for fishes from PA caves, yet this single 1864 specimen exists. Believing that the blind catfish came from a subterranean stream feeding the Conestoga, the discoverer sent it to the cave exploring naturalist, Edward Drinker Cope in 1864. Cope was engaged in the world-wide examination and reclassification of fishes. Ultimately, he published 125 papers on fishes, describing over 220 new species.

The fact that no other eyeless catfish specimens have been observed either in the creek or in local caves would not totally confirm the fishes nonexistence. Perhaps it inhabits inaccessible crevices and fissures in the limestone or caves that have no entrances. The author was not trying to defend the blind species existence, but merely using the specimen to substantiate his theory of a vast underground cavern in the vicinity of Lancaster.

Note that Kranzel, (1) does not provide any hard or new evidence for the existence of blind cavefish species for Pennsylvania; (2) his speculations are the same in kind as Cope's conjecture for the existence of this type of creature; and, (3) he ignores all other sources that, based on the examination of specimens, are not consistent with his explanation (Fowler 1945, Hubbs and Bailey 1947, Taylor 1969). However, part of his reasoning is that "The fact that no other eyeless catfish specimens have been observed either in the creek or in local caves would not totally confirm the fishes nonexistence." In other words, based on the fact tha nobody can prove a negative (e.g., to prove that "Santa Claus does not exist"), Kranzel perpetuates this myth.

Conclusions

There is no evidence supporting the reports of blind cave fishes for Pennsylvania. The original description of alleged troglobitic fishes for that state was based on an erroneous identification. It has been well documented that because of his constant rush to publish, many of Cope's writings were either superficial or contained numerous errors, a fault that was even recognized by his most ardent supporters which gave him a reputation for sloppiness (Davidson 1997, Romero & Romero 1999). Later accounts of troglobitic fishes for Pennsylvania are little more than repetitions of Cope's early assertions not substantiated by either the examination of the fish he described nor by extensive cave explorations.

Unless an actual troglobitic fish is captured in Pennsylvania, we must assume that such fish do not exist and attempts to challenge this assertion by asking to prove a negative should be dismissed as non-congruential thinking.

Acknowledgments

William G. Saul of the Academy of Natural Sciences provided useful information on the specimens originally classified as "Gronias nigrilabris" and deposited at the Academy's collection. Olga Mayayo drew Fig. l. Andy Miller read the manuscript and made valuable suggestions. This is contribution no. 1 of the Cape's Papers Project.

Literature Cited

Anonymous. 1871. A Subterranean Pond - Eyeless Fish. *Scientific American* April 1, **24**(14):209.

Conant, R. 1966. Henry Weed Fowler 1878 – 1965. *Copeia* **1966**(3):628-629.

Cope, E.D. 1864. On a Blind Silurid, From Pennsylvania, *Proceedings of the Academy of Natural Sciences of Philadelphia*, **16**:231-233.

Davidson, J.P. 1997. *The Bone Sharp. The Life of Edward Drinker Cope*. Philadelphia: The Academy of Natural Sciences of Philadelphia, 237 pp.

Eschmeyer, W.N. (Ed.). 1998, *Catalog of Fishes, San* Francisco: California Academy of Sciences. 3 vols.

Fowler, H.W. 1915. Notes on Nematognathous Fishes, *Proceedings of the Academy of Natural Sciences of Philadelphia*, **1915**, pp. 203-243.

Fowler, H.W. 1945, *A Study of the Fishes of the Southern Piedmont and Coastal Plain*, Academy of Natural Sciences of Philadelphia, Monograph 7, 408 pp.

Fowler, H.W. 1963. Cope in Retrospect. *Copeia*. **1963**(1):195-198.

Grady, F.V. 1987. Edward Drinker Cope's Contributions to Speleology. *Journal of Spelean History* **21**(3-4):35-37.

Grady, F.W. 1992. More on Edward Drinker Cope's Caving. *Journal of Spelean History* **26**(1):20-23.

Hamy, E.-T. 1968, *The Travels of the Naturalist Charles A. Lesueur in North America, 1815 – 1837*. The Kent State University Press.

Holsinger, J.R. 1976. The Cave Fauna of Pennsylvania, pp. 72-87, *In*: W.B. White (Ed.). *Geology and Biology of Pennsylvania Caves*, Harrisburg, Pa., 87 pp.

Hubbs, C.L. & Bailey, R.M. 1947, Blind Catfishes from the Artesian Waters of Texas, Occasional Papers Museum of Zoology, University of Michigan (499):1-15.

Jordan, D.S. & B.W. Evermann. 1896. The Fishes of North and Middle America, *U.S. National Museum Bulletin,* **47**(1):1-1240.

Kircher, A., 1665, *Mundus Subterraneus,* Amsterdam, 2 vols.

Kranzel, R., 1986. Eyeless Catfish of Conestoga Creek, Lancaster County, Pennsylvania, *Speleo Digest,* p. 317.

Mohr, C.E., 1953. Animals That Live in Pennsylvania Caves, The American Caver. *Bulletin of the National Speleological Society* (15):15-23.

Montalembert, M.-R. de 1752. *Observations de Physique Generale, Histoire de L'Academie Royale des Sciences avec Les Memoires de Mathematiques & de la Physique pour la meme Anne, 1748,* pp. 27-28.

Osborn, H.F. 1931. *Cope: Master Naturalist.* Princeton University Press, Princeton, NJ, 740 pp.

Phillips, M.E. 1964. Henry Weed Fowler, 1878, *Quarterly Journal of the Taiwan Museum,* **17**(3/4):128-133.

Romero, A. 1998a. Threatened Fishes of the World: *Amblyopsis rosae* (Eigenmann, 1842) (Amblyopsidae). *Environmental Biology of Fishes* **52**(4):434.

Romero, A. 1998b. Threatened Fishes of the World: *Typhlichthys subterraneus* (Girard, 1860) (Amblyopsidae). *Environmental Biology of Fishes* **53**(1):74.

Romero, A. 1998c. Threatened Fishes of the World: *Speoplatyrhinus poulsoni* Cooper and Kuehne, 1974 (Amblyopsidae). *Environmental Biology of Fishes* **53**(3):293-294.

Romero, A. 1999. The Blind Cave Fish That Never Was. *National Speleological Society News* **57**(6):180-181.

Romero, A. 2000. The Speleologist Who Wrote Too Much. *National Speleological Society News* **58**(1):4-5.

Romero, A. & Bennis, L., 1998. Threatened Fishes of The World: *Amblyopsis spelaea* De Kay, 1842 (Amblyopsidae). *Environmental Biology of Fishes* **51**(4):420.

Romero, A. & Romero, A. 1999. Cope, Caves and Skeletons in the Closet. *National Speleological Society News* **57**(11):341-343.

Scott, W. 1866. Probable Existence of a Great Cavern Under Lancaster, Pa. *Scientific American,* April 7, 1866. 14:228.

Smith-Vaniz, W.F. & Peck, R.M. 1991. Contributions of Henry Weed Fowler (1878-1965), with a Brief Early History of Ichthyology at The Academy of Natural Sciences of Philadelphia. *Proceedings of the Academy of Natural Sciences of Philadelphia* 143:173-191.

Taylor, W.R. 1969, A Revision of the Catfish genus *Noturus Rafinesque,* With an Analysis of Higher Groups in the Ictaluridae, *Bulletin U. S. National Museum* (282):1-315.

Chapter 12. The Discovery of The First Cuban Blind Cave Fish: The Untold Story [340]

Introduction

Although there have been numerous published reports on blind cavefishes dating as far back as 1541 (Romero 2001), the first scientific description of one species was not published until 1842. That description was of the northern cavefish *Amblyopsis spelaea* from Mammoth Cave, Kentucky (Romero 2002). That scientific description took place shortly after it had first been sighted (Romero & Woodward 2005).

With the exception of the northern cavefish, the history of the discovery of cave fish species and how their discovery impacted the views of speleologists and other scientists at the time they were described has rarely been told.

The two first species of blind cave fishes scientifically described from outside the United States were found in Cuba and their description was published in 1858. They are the Cuban cusk-eel *Lucifuga* (*Lucifuga*) *subterranea* and the toothed Cuban cusk-eel, *Lucifuga* (*Stygicola*) *dentata*.

I have found the original documents that relate the discovery of these species. These sources are in hard-to-find Cuban publications and in Spanish. Therefore, I will condense the information I have been able to find and discuss the impact of these discoveries at the time they were made.

Felipe Poey

One of the two central characters in the saga of the discovery of the first blind cave fishes from Cuba was Felipe Poey y Aloy (hence Poey) (Fig. 12.1).

Poey (po'-ay)[341] was the son of a Frenchman who had been involved in the slave trade and a Cuban-born mother who was half Spanish, half Cuban. At the age of five, Poey went to France with his family where his father died two years later. While there, he was struck by polio, which paralyzed the right side of his body, after which he returned to Cuba. To follow his mother's wishes, he went to Madrid, Spain, where he obtained a law degree in 1822. He began his lawyer's career in Spain and became involved in politics (he was a liberal in the parlance of the time) but soon became disenchanted, returned to Cuba, married, and decided to become a full-time naturalist.

In 1826 he traveled to Paris, France, where he worked under the famous naturalist Georges Cuvier[342]. At that time Cuvier and Achile Valenciennes[343] were working on the encyclopedic *Histoire naturelle des poisons* (The Natural History of Fishes), which would end up being a 22-volume publication.

Poey arrived to Paris with 35 specimens of Cuban fishes in a barrel of brandy plus 85 drawings of fishes from that island. The information from these fishes was incorporated into Cuvier and Valencinnes's gigantic work. From early on Poey was very interested in working with fishes, visiting the fish market of Havana almost every day. Yet, he was a polymath with areas of interests ranging from anthropology to poetry. He was a very prolific author who corresponded with virtually every noted naturalist of his

---

[340] Originally published as Romero, A. 2007. The discovery of the first Cuban blind cave fish: the untold story. *Journal of Spelean History* **41**(131):16-22.
[341] b. Havana, Cuba, 26 May 1799; d. Havana, 28 January 1891.
[342] b. Montbéliard, France, 23 August 1769; d. Paris, France, 13 May 1832.
[343] b. Paris, France, 9 August 1794; d. Paris, 13 April 1865.

time and went on to occupy important academic posts at the University of Havana. He was a member of almost every major scientific society in the U.S. and Europe, and many of his new specimens and life-size drawings are found in the collections of the United States National Museum (Smithsonian), the Museum of Comparative Zoology (Harvard), the Natural History Museum of Madrid and the National Museum of Natural History in Paris.

Figure 12.1. Felipe Poey portrait and signature.

Because of the dates of his birth and death, he was technically a Spaniard, but he confessed that "As a naturalist I have never been a Spaniard, I have been cosmopolitan."

Despite cataracts at old age, he never stopped writing (for more biographical information about Poey see Jordan 1899, Sánchez Roig 1937, Vivanco & Díaz 1951, Cruz 1979, González López 1999. Romero 2014).

Tranquilino Sandalio De Noda

The other character in this story is Tranquilino Sandalio de Noda y Martínez (hence de Noda)[344] (Fig. 12.2). Despite having been born in a rural area, his neighbors were French planters that owned good libraries. He received personal tutoring from his mother, a primary school teacher, and another local teacher José María Dau. From the latter he not only acquired basic knowledge in the sciences and humanities but also the habit of educating himself in any field, including the learning of several languages (for more biographical information on Noda, see Guerra 1924 and Sánchez Roig 1942).

The Discovery

In 1823, the *Capitán General* (governor designated by the Spanish government) of Cuba, Francisco Dionisio Vives, asked Dau, de Noda's tutor, to conduct a geological survey of Santa Cruz de los Pinos (Pinar del Río Province, western Cuba). Dau asked de Noda, then only 15, to accompany him. That is when his interest in caves surged and provided him the opportunity to observe the hypogean fauna and to find many fossils in caves.

In 1831 de Noda learned of the Cuevas del Cajío (Cajío caves), at Güira de Melena, in the southern portion of the Havana Province where there were rumored to be blind fishes. After finding the cave and crawling into a very hot, bat-crowded hall, he and his companions reached a pool with "white" fishes in crystal clear waters between 60 and 90 feet underground. His companions captured one fish using a basket. De Noda wrote that the fish were easily disturbed by splashing on the surface of the water.

The captured individual died within few hours. De Noda drew the specimen and preserved it in a bottle with rum. The fates of the illustration and the specimen are unknown but that he sent them to Poey is mentioned in Poey's correspondence. Poey told him that he classified this fish as a new genus and species and gave it the scientific name of *Lucifuga subterraneus* [name later changed to *Lucifuga (Lucifuga) subterranea* for Latin grammatical reasons as well as for classification ones].

In 1876 Poey published in the Havana newspaper *El Mercurio* the correspondence between him and Tranquilino Sandalio de Noda. A copy of the original article containing such correspondence cannot be found, but it was reprinted in Poey (1888) and Carbonell y Rivero (1928). Three letters from Noda to Poey and one from Poey to Noda are reprinted in Poey (1888), and those are the primary sources I am using for this article.

The scientific description by Poey of cave fishes from Cuba was published in different venues. The first was "*Memorias sobre la historia natural de la isla de Cuba*" (Memoirs about the natural history of the island of Cuba) (Fig. 12.3). This work is complicated to cite, particularly when it comes to dates. First of all, this is a collection of papers in two volumes. The first set of papers was published between 1851 and 1854 and the second set between 1858 and 1861.

The paper containing the description of the cavefishes from Cuba was published under the title "*Peces ciegos de la isla de Cuba, comparados con algunas especies de distinto genero*" (Blind fishes from the Island of Cuba, compared with some species of a different

---

[344] *b.* Las Cañas, Guanajay, Pinar del Río, Cuba, 3 September 1808; *d.* San Antonio de los Baños, Havana, Cuba, 27 May 1866.

genus) as the paper "xivi" of volume 2, in pages 95-114 of the whole opus. Therefore, the correct date of publication should be 1858.

In that publication Poey creates a new genus and species (*Lucifuga subterraneus*) based on 12 specimens and goes on with a very detailed and accurate description of both external and internal morphology, including some minor differences among the specimens he studied such as variability in eye development from rudimentary to totally blind.

Poey cited the species as from five localities in Cuba: la Cueva del Cajío (near Güiria de Melena), Cueva del Cafetal La Industria (between Alquízar and Guanímar), la Cueva de Ashton (in San Andres), Cueva del Dragón (in San Isidro) and Cueva del Cafetal La Concordia (near Alquízar). Poey does not say if he visited these caves, but despite the fact that he was a man of fair complexion, medium height, and heavy build, polio had paralyzed the right side of his body, so chances are he never visited these caves. Furthermore, he cited as collectors of specimens from the different localities different people: de Noda, Antonio Dubrocá, Juan Antonio Fabre, and Fernando Layunta. He does give priority to de Noda for visiting the Cajío Cave in 1831.

Figure 20.2. Tranquilino Sandalio de Noda (public domain photo).

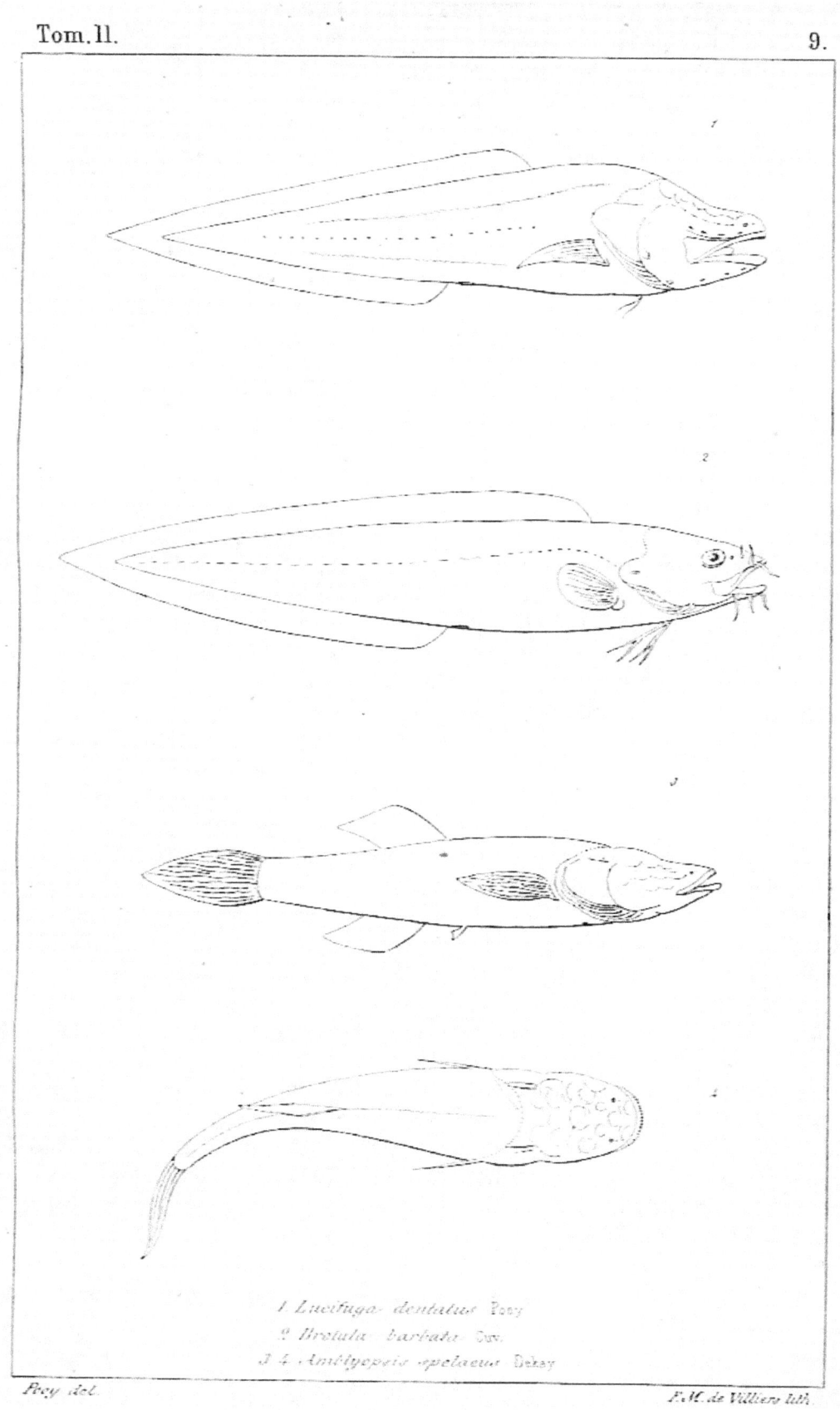

Figure 12.3. Illustration of Poey's *Memorias* where he compared the Cuba cave fish *Lucifuga dentata*, the deep-sea *Brotula barbata*, and the northern cavefish *Amblyopsis spelaea* (public domain image).

Poey then described a second new species: *Lucifuga dentatus* (now known as *Lucifuga* (*Stygicola*) *dentata* also for Latin grammatical and classification reasons). This one was described as totally depigmented and found in the three first localities cited above. He also comments on the bearded brotula *Brotula barbata*, a deep-sea fish that sometimes shows a fatty tissue covering its eyes, the northern cavefish *Amblyopsis spelaea*, and the pirate perch *Aphredoderus sayanus*.

The reason he included the bearded brotula was because it is the closest relative to the Cuban cave fishes that he described. He used the northern cavefish for comparison purposes because it was the only other known species of cavefish in the world, and he included the pirate perch because that fish is a close relative to the northern cavefish.

Later, between pages 108 and 114, he discussed whether these fishes were the product of special creation or the product of evolution. In this article published a year before Darwin's *Origin of Species*, Poey compares the two schools of thought (creationism and evolution), the first one defended by his former teacher Cuvier, and the second by Geoffroy Saint-Hilaire[345].

After saying that he does not belong to either school, Poey says that on one hand he has a great deal of inclination for Cuvier's creationism, but that on the other hand he has found very convincing arguments for evolution. The first statement seems to be the product of respect for his former (an only) ichthyology teacher. But, then, given the variability that he observed among the Cuban cavefishes (the biological characteristic that later Darwin would use to explain how natural selection works), he concluded that these species are the product of "transformation" (older term for evolution). He further rejects the use of the term "degeneration" to describe the loss of eyes and other features among cave organisms but rather prefers "modification".

This is the first time that Poey makes any mention of evolution in his writings. Although ambivalent at the beginning (Pruna, 1994), he went on to become an evolutionist based on his studies of blind cavefishes. Thus, when the works of Charles Darwin[346], Thomas Henry Huxley[347], and Herbert Spencer[348] were published, he embraced them enthusiastically. He seemed to be particularly impressed with Spencer (Jordan 1899) probably because of Spencer's progressionist ideas and his belief that cave colonization was not the result of "accidents" but rather an active process. This intellectual development took place rather smoothly. After all, the introduction of Darwinian ideas in Latin America was progressive and without much resistance at least among academic circles (for an overview of this issue see Pruna & García González, 1989).

Poey wrote that he had had extensive correspondence on the Cuban blind cavefishes with Charles Frédéric Girard[349] with whom he agreed that to better understand the issue of "transformation" of blind cavefishes, studies on the development of these fishes had to be carried out. Actually, this was an idea that Girard's teacher, Jean Louis Rodolphe Agassiz[350], had proposed previously (Romero 2001). Girard would go on to describe a new species, the southern cavefish *Typhlichthys subterraneus* in 1859, one year after Poey's description of the Cuban species.

Poey went on publishing two other pieces on the blind cave fishes of Cuba. One was in 1865 in his "Repertorio Físico-Natural de la isla de Cuba" (Physical-natural repertoire of the Island of Cuba) in which he updated some of the classification of these

---

[345] *b*. Etampes, France, 15 April 1772; *d*. Paris, France, 19 June 1844.
[346] *b*. Shrewsbury, England, 12 February 1809; *d*. Downe, Kent, England, 19 April 1882.
[347] *b*. Ealing, west London, England, 4 May 1825; *d*. Eastbourne, England, 29 June 1895.
[348] *b*. Derby, England, 27 April 1820; *d*. Brighton, England, 8 December 1903.
[349] *b*. Mulhouse, France, 8 March 1822; *d*. Neulilly-sur-Seine, France, 29 March 1895.
[350] *b*. Motier-en-Vuly, Switzerland, 28 May 1807; *d*. Cambridge, Massachusetts, 14 December 1873.

fishes based on some comments from the ichthyologist Theodore Nicholas Gill[351]. In this he also added a few new localities for these fishes and reprinted a letter from de Noda that included the fact that the "negros" usually go to the Cajío Cave to fish these animals to eat them. This seems to be the only reported case of cave fishes being consumed by humans. Cave fishes in general are not appreciated as a source of food for humans: they are generally small, found in small numbers, and many of them feed on bat guano.

He would finally cite these Cuban blind cavefishes again briefly in 1868 in his "*Synopsis Piscicum Cubensium*" or "*Catálogo razonado de los Peces Cubanos*" (An annotated catalog of Cuban fishes), an atlas of 10 volumes with more than 1,000 illustrations drawn by himself, with the descriptions of about 800 tropical American fishes. This work was purchased by the Spanish government, placed in the "*Biblioteca de Ciencias Naturales*" (Natural History Library) at Madrid, Spain, and exhibited by the Spanish government at the International Exhibit of Amsterdam in 1883, where it received a gold medal and honorable mention.

Where are the Specimens Used by Poey?

There is one question worth exploring about Poey's work on the Cuban cave fishes. That is, what was the fate of the specimens that he used to describe these new species (type specimens or holotypes)? Given that there is no trace of those specimens in the Natural History Museum of Havana, which he founded, and the fact that he usually sent specimens to colleagues and museums around the world, it is worth looking at existing specimens to see which one could be his holotypes.

The list of museums that have specimens of Cuban cave fishes collected in the nineteenth century and that have kept correspondence from Poey are in Table 1.

Table 1. Museums outside Cuba that have specimens of Cuban cave fishes collected in the nineteenth century and that keep correspondence from Poey.

Correspondent(s)	Museum	Museum Acronym
Georges Cuvier, Achile Valenciennes	Museé Nationale d'Histoire Naturelle	MNHN
Charles Frédéric Girard, Theodore Nicholas Gill	U.S. Museum of Natural History	USNM
?	British Museum of Natural History	BMNH
?	American Museum of Natural History	AMNH
?	Museum of Comparative Zoology (Harvard University)	MCZ
?	Museo Nacional de Historia Natural (Madrid, Spain)	MNCN

The list of specimens known to exist in museums for the two species in question are in tables 2 and 3.

Table 2. Known specimens of *Lucifuga* (*Lucifuga*) *subterranea* (in chronological order of collection) in museums around the world.

---

[351] *b*. New York City, U.S.A., 21 March 1837; *d*. Washington, D.C., 25 September 1914.

Catalogue #	Locality	Collector	Collection Date
USNM 00001739	Cuba: Cuevas de Alquizas	Poey	?
MNHN a-5234	Cuba	Valenciennes	1865
SU 8510	Cuba: Hawey	C.H. Eigenmann	1893 or before
CAS 30438	Cuba	C.H. Eigenmann & Riddle	1902
AMNH 18463	Cuba		ca. 1904
AMNH 18712	Cuba: Tranquilidad	C.H. Eigenman	March 1902
AMNH 18714	Cuba	C.H. Eigenmann	March 1902
FMNH 3934	Cuba: Cañas	Dr. C.H. Eigenmann & O. Riddle	March 1902
FMNH 33090	Cuba	Eigenmann?	?
FMNH 52631	Cuba	C.H. Eigenmann	?
SU 8509	Cuba: Jaiguan	C.H. Eigenmann	?
BMNH 1904.1.28.137	Cuba: Cañas	?	1904
BMNH 1904.1.28.135-136	Cuba: Cueva Tranquilidad	?	1904
MCZ 29902	Cuba: Matanzas: Cañas	C.H. Eigenmann	1910
USNM 00204452	Cuba		1936?
MCZ 31221	Cuba: Guira de Melena	Carlos de la Torre	?

In Table 2, the list includes one specimen in the U.S. National Museum of Natural History (Smithsonian) that was collected by Poey but without a date. Another is in the Museum of Natural History of Paris labeled as collected by Valenciennes in 1865. This specimen could not have been collected by Valenciennes for the simple reason that he never visited Cuba. It is most likely that he received it from Poey in 1865 (six years after the species was described). The same can be said about similar labeling for the *L. dentata* in Paris in Table 3. All the other specimens were collected by someone else or on dates much after the actual discovery. Thus, the holotype for *L. (L.) subterranea* is either the one in Washington, D.C. or the one in Paris.

Table 3. Known specimens of *Lucifuga* (*Stygicola*) *dentata* (in chronological order of collection) in museums around the world.

Catalogue #	Locality	Collector	Collection Date
MCZ 32329	Cuba: Cuevas en San Antonio	F. Poey et al.	?
MCZ 12415	Cuba	F. Poey	1861
MNHN A-5232	Cuba	Valenciennes	1865
MNHN A-5233	Cuba	Valenciennes	1865
FMNH 3933	Cuba: Cave near Alacranes	Dr. C.H. Eigenmann & O. Riddle	March 1902
CAS 6714	Cuba: Cañas	C.H. Eigenmann	March 1902
CAS 30437	Cuba	C.H. Eigenmann & Riddle	1902

FMNH 112219	Cuba	C.H. Eigenmann	Ca. 1902
CAS/SU 8511	Cuba: Cave near Pedregales	C.H. Eigenmann	ca. 1902
FMNH 96223	Cuba	H.B. Ward	25 August 1904
BMNH 1904.1.28.132-134	Cuba: Cueva Tranquilidad	?	1904
MCZ 29903	Cuba: Matanzas: Cañas	C.H. Eigenmann	1910
FMNH 33084	Cuba	Eigenmann?	2 November 1936
MCZ 36166	Cuba: Pinar del Rio	L.R. Rivas	1940
BMNH 1981.10.27.1-4	Cuba	?	1981
FMNH 52595	Cuba: Cañas	?	?
AMNH 1537	Cuba: Pinar del Río: Cueva Modesta	B. Dean	?
AMNH 10143	Cuba: Pinar del Rio: Rio Taco Taco	?	?
MCZ 30615	Cuba: Finca La Carbonera	Thomas Barbaour	?
MCZ 30616	Cuba: Matanzas: Alacranes: Cueva del Mar	T. Barbour	?
MCZ 30617	Cuba: Matanzas: Alacranes: Cueva del Mar	T. Barbour	?
MCZ 31212	Cuba: Matanzas: Alacranes: Cueva del Mar	Carlos de la Torre	?
MCZ 31213	Cuba: Guira de Melena	Carlos de la Torre	?

In the case of *L. dentata*, there is a specimen in the Museum of Comparative Zoology at Harvard University (MCZ 32329) whose collector's information is labeled as "Poey et al." Since in his original description Poey acknowledges the fact that the specimens of these fishes were collected by others than him, that specimen could well be the holotype for that species. Howell y Rivero (1938) analyzed the vouchers with which this and other specimens were deposited at that museum and reached the same conclusion. Another potential holotype is MCZ 12415, but I lacked sufficient information about that one to make any determination. This could also be a syntype (one of two or more specimens simultaneously selected as types by the original author of a name of a species).

Conclusions

Poey relied on others to collect the blind cave fishes in Cuba. His anatomical and taxonomic analyses of these specimens were highly accurate, and these fishes helped to convince him to embrace the idea of evolution. He kept ample correspondence with contemporary colleagues from the U.S. and Europe and most likely sent the specimens he used for describing the two species of Cuban blind cave fishes to museums abroad, particularly the National Museum of Natural History in Washington, D.C. and the Museum of Comparative Anatomy at Harvard University.

Acknowledgements

Joy Trauth read an earlier version of this article and made valuable suggestions

Literature Cited

Carbonell y Rivero, J.M. 1928. *La ciencia en Cuba*. La Habana, Cuba: Montalvo y Cárdenas.
Cruz, M. 1979. *El ingenioso naturalist Don Felipe de La Habana*. La Habana, Cuba: Editorial Gente Nueva.
González López, R.M. 1999. *Felipe Poey y Aloy. Obras*. La Habana: Imágen Contemporánea.
Guerra, A. 1924. *Un procer humilde (Tranquilino Sandalio de Noda)*. La Habana, Cuba: Imprenta La Moderna Poesía.
Howell y Rivero, L. 1938. List of fishes, types of Poey, in the Museum of Comparative Zoölogy. *Bulletin of the Museum of Comparative Zoölogy* **82**:169-227.
Jordan, D.S. 1899. *Science Sketches*. Chicago: A.C. McClu.
Poey, F. 1858. Peces ciegos de la isla de Cuba, comparados con algunas especies de distinto genero, pp. 94-114, *In: Memorias sobre la Historia Natural de la isla de Cuba, acompanadas de sumarios latinos y extractos en frances, etc.* La Habana, Cuba: Barcina.
Poey, F. 1865. Peces ciegos, pp. 113-116, *In: Repertorio físico natural de la Isla de Cuba*. La Habana, Cuba: Gobierno y Capitanía.
Poey, F. 1868. Synopsis Piscium Cubensium, pp. 279-484, Vol. 2, *In: Reporte Físico-Natural de la Isla de Cuba*. La Habana, Cuba: Gobierno y Capitanía.3
Poey y Aloy, F. 1888. *Obras literarias de Felipe Poey*. La Habana, Cuba: La Propaganda Literaria.
Pruna, P.M. 1994. National Science in a colonial context. The Royal Academy of Sciences of Havana, 1861-1898. *Isis* **85**:412-426.
Pruna, P.M. & A. García González. 1989. *Darwinismo y sociedad en Cuba. Siglo XIX*. Madrid: Consejo Superior de Investigaciones Científicas.
Romero, A. 2001. Scientists prefer them blind: The history of hypogean fish research. *Environmental Biology of Fishes* **62**(1-3):43-71.
Romero, A. 2002. Between the first blind cave fish and the Last of the Mohicans: the scientific romanticism of James E. DeKay. *Journal of Spelean History* **36**(1):19-29.
Romero, A. 2014. Felipe Poey, hyperbole, and the myth of the "isolated genius" among Spanish and Latin American scientists. *Polymath* **4**(4):85-95.
Romero, A. & J. S. Woodward. 2005. On white fish and black men: did Stephen Bishop really discover the blind cave fish of Mammoth Cave? *Journal of Spelean History* **39**(1):23-32.
Sánchez Roig, M. 1937. *Felipe Poey, el máximo naturalista de Hispanoamérica*. La Habana, Cuba: Imprenta Molina y Cia.
Sánchez Roig, M. 1942. *Un sabio olvidado: Tranquilino Sandalio de Noda y Martínez*. La Habana, Cuba: Compañía Editora de Libros y Folletos.
Vivanco Y Díaz, J. 1951. *Don Felipe. Su vida y su obra*. La Habana, Cuba: Imprenta "El Siglo XX".

Section III. Biographies

This section is about biographical essays in different fashions. The first one ("On White Fish and Black Men: Stephen Bishop and the Discovery of the Blind Cavefish of Mammoth Cave") is about Stephen Bishop, a slave in Kentucky during antebellum, who reached notoriety for being the first famous cave explorer and guide of Mammoth Cave. It is believed that he was the first person who observed and collected the blind cavefishes in that locality. Because of the color of his skin, the whole narrative about his life has been complicated.

The second essay ("Between the First Blind Cavefish and the Last of The Mohicans: The Scientific Romanticism of James E. DeKay") deals with the life of an amateur naturalist noted for his scientific description of many species of plants and animals of the state of New York. He did not hesitate to include species no native to the Empire State, such as the blind cavefish of Kentucky or the Florida Manatee. The contemporaries romantic writers of his circle influenced his decision to include those exotic species in the list of New York Fauna.

The third one ("The Life and Work of a Little Known Biospeleologist: Theodor Tellkampf") deals with an obscure but excellent scientist on which no biography had been written before.

The fourth ("Felipe Poey and the myth of the "Isolated Genius") is about how unnecessary hyperbole, far from helping, obscure the real achievement of a very brilliant naturalist in Cuba and how this kind of hagiography has been so common among Spanish and Latin American scientific luminaries.

Next ("Cope, Caves, and Skeletons in the Closet") is a chapter about one of the more colorful American naturalists in the Nineteenth Century: Edward Drinker Cope. This is a case study on how personality can heavily influence science and scientist to the point of extravagance.

The following essay ("He Wanted to Know Them All. Eigenmann and His Blind Vertebrates") is about a German-born scientist who developed his entire career in the United States. Thanks to his perseverance, discipline, and unbound energy became one of the most productive naturalists in American history.

Chapter 19 ("The Unsung Heroes of Speleology") deals with an epitome of how obscure individuals can make a difference by bringing to the attention of professional scientists species who were unique. Unfortunately, we know almost nothing about these contributors to science.

The next two chapters ("Charles Marcus Breder, Jr. 1897-1983" and "Herbert L. Needleman 1927-2017 ") are contrasting obituaries I wrote about two scientists whose careers were remarkable. Breder, despite his lack of formal education in biology, achieved great things in the study of blind cavefishes. Needleman made outstanding contributions to public health but whose scientific integrity was furiously attacked by commercial interests.

The last two essays ("Columbus, Christopher" and "Vespucci, Amerigo") are encyclopedic entries about those two individuals who played a significant role in the history of the American continent.

Chapter 13. On White Fish and Black Men: Stephen Bishop and the Discovery of the Blind Cavefish of Mammoth Cave [352]

Introduction

Printed references in the popular literature on blind cavefishes have a long history. From the sixteenth throughout the nineteenth centuries a number of blind cavefish observations (some certifiable, others not) had been published in China, Europe, and the American continent (Romero 2001). Because none of them met the criteria established by the Rules of Zoological Nomenclature, they are not considered valid by the scientific community. However, in 1842, James DeKay described the northern cavefish *(Amblyopsis spelaea),* making it the first species of blind cavefish recognized in the scientific literature. When DeKay described *A. spelaea,* he cited the River Styx in Mammoth Cave as the type locality, that is, where the specimen used for its description (holotype) was collected. Yet, DeKay never visited Mammoth Cave and the specimen, along with the information of when it was collected, and by whom, is lost (Romero 2002).

Stephen Bishop, a slave, and the most famous Mammoth Cave guide, has always been credited with the discovery of River Styx and the first sighting of a blind cave fish in that cave (e.g., Brucker & Watson 1987). However, the first published reference to a blind cave fish (*"white fish"*) in Mammoth Cave appears to be by Robert Davidson in 1840 (p. 55). According to Davidson, he visited Mammoth Cave in October or November 1836 and said that River Styx had been discovered the year before (1835). This chronology challenges the conventional wisdom of Bishop being the discoverer of a blind cavefish for Mammoth Cave and its type locality. Given that Mammoth Cave is inhabited by two different species of blind cavefishes, there is also the question of which of the two species was first sighted.

This chapter presents the results archival and field research aimed at clarifying the question of who might have been the first person who saw (and probably collected) this fish. To that end, we reviewed all the available printed references to Mammoth Cave that mentioned its fauna previous to DeKay's 1842 publication. We investigated all possible primary literature related to Robert Davidson in order to clarify the chronology of his visit. We visited Mammoth Cave and tried to retrace the route likely taken by whoever might have been the first person to see a blind cave fish in that cave. We researched all the known specimens of the two species of blind cave fishes ever found at Mammoth Cave to see if that information could provide evidence of which of the two species was seen first. We conclude that: (1) Davidson's chronology, as presented in his book, is probably wrong and that he did not visit the cave until 1838 or 1839; (2) it is possible that Bishop was the first person to sighted the fish, but others cannot be definitely excluded from having been involved in this discovery; and (3) that although there are two species of blind cave fish that inhabit the waters of Mammoth Cave, the first one sighted was likely *Amblyopsis spelaea,* which

---

[352] Based on Romero, A. & J. S. Woodward. 2005a. On white fish and black men: did Stephen Bishop really discover the blind cave fish of Mammoth Cave? *Journal of Spelean History* **39**(1):23-32.

was also the first one to be recognized in the scientific literature. Finally, we conclude that the facts surrounding Stephen Bishop's fame need to be further investigated under the perspective of the romantic movement of the mid-nineteenth century that gave rise to the " noble savage" mythology as well as on the perspective of race in the United States prior to the Civil War.

The Conventional Narrative and its Challenges

Arguably, Stephen Bishop is the most frequently mentioned person in the history of Mammoth Cave. He was born into slavery probably around 1821 and died in 1857, a year after obtaining his freedom. Bishop was acquired by Franklin Gorin (1798-1877), a lawyer from the nearby town of Glasgow, Kentucky, when he was about 13 years old. Gorin purchased Mammoth Cave in 1838. Bishop soon became a guide and explorer of the cave. Although by that time the most accessible parts of Mammoth Cave had been visited, explored, and mapped, a major obstacle remained to continued exploration: Bottomless Pit. Bishop is consistently credited with having suspended either a cedar pole or a log pole ladder across Bottomless Pit and thus, was able to significantly expand the known area of the cave (Anonymous 1981, Barr 1986, Anonymous 1992). More significantly, Bishop and others could now visit River Styx and Echo River and observe the blind cave fishes found there (for a good summary of the history of Mammoth Cave see Brucker & Watson 1987). Bishop is also specifically credited with discovering the fish (e.g., Anonymous 1992).

However, conventional wisdom on Bishop's accomplishments in this regard has been challenged (see for example Meloy 1977). First, it has been stated that the first crossing of Bottomless Pit was carried out by Bishop and a visitor, Hiram C. Stevenson, not by Bishop alone (Brucker & Watson 1987, pp. 266-267). Did they together (or perhaps others, later) keep exploring until arriving at the River Styx and/or Echo River and see the blind cavefish? Also, it should be noted that reaching the far side of Bottomless Pit can be accomplished by crawling and an eight-foot climb, something that any caver would have checked prior to risking a crossing on a cedar pole or ladder. A short ladder would have helped, of course. Was Bishop doing the ladder crossing to add an exciting aura to his tour for the more adventuresome visitors? (Brucker & Watson 1987, pp. 268, and Roger W. Brucker *pers. comm.*).

In 1840 Robert Davidson published an account of his visit to Mammoth Cave that seems to support the contention made by some that Bishop was not the first to cross Bottomless Pit and/or to see the blind cavefish. He wrote:

> *This is a stream of water twenty feet wide, and they said as many deep. It was discovered only about a year ago. Its current is very sluggish, as has been proved by launching a piece of wood bearing a lighted candle, on its bosom. We were informed that a species of white fish (in italics in the original] were found here without eyes, and the keeper of the hotel assured us he himself bad seen them, but that their other senses were so acute, the slightest touch of water overhead was sufficient to alarm them, and make them dan off like lightning. There had been a canoe here; but the day before it had got loose from its mooring and floated away. In this visiters [sic] would row down the stream two hundred yards, still stop ped by a ledge of rock Two of my acquaintances a week afterwards, obtained a new skiff, and resolved to pass the barrier. Accordingly, lifting the skiff over the rock, they launched it on the other side, and rowed, as thought, for two miles. They beheld a great many new scenes and*

chambers never explored before[353]. *They also saw some of the white fish. As for us, on our visit, we were not favoured with a sight of these natural curiosities.*

Davidson's book is titled *An Excursion to the Mammoth Cave and The Barrens of Kentucky read before the Society of Adelphi of Transylvanian University. January 16, 1840.* (Fig. 13.1). We can therefore assume that his visit to the cave took place at the time of Davidson's visit, tours were already being given at river level and it is unlikely that Stephen (or others) would have been deterred from continuing by a few rocks. It is possible that Davidson's acquaintances might have thought they were discovering a new cave, but probably not (Rick Olson Pers. Comm.) which would have been as interesting a spectacle as Prince Bonbobbin's white mice with green eyes, for which he ransacked the world. All we found was a poor miserable mudfish, caught with the hand by the guide, near the shore, blinded by the light It was certainly a wonderful thought, that such a body of water should have been flowing here a furlong at least underground, in the silence and gloom of centuries. (Davidson, 1840, pp. 54-56). "Before 1840" but when, exactly? Davidson, while describing his visit, says: "It was the early part of October, 1836, that we first set foot in this interesting region" referring to Henderson, Henderson County, Kentucky. From there, he and his companions hired a barouche and headed south to Hopkinsville where they spent one week after having traveled for two days. From there they went to Elkton and stopped for a night at Russellville. The next day they passed through Shakertown "Just midway on the road between Russellville and Bowling-Green". They apparently reached the latter town that very same day. Davidson then started describing Mammoth Cave "twenty-four miles from Bowling-Green" and immediately begins the account of his visit.

Based on this description one might conclude that Davidson's visit to Mammoth Cave took place late in October or early November 1836. Furthermore, Davidson mentioned in his narrative that the river where the fish lives "was discovered only about a year ago," i.e., in 1835. If that were the case, the "*white fish*" might have been discovered almost three years before Bishop became a guide to Mammoth Cave and thus, contradicting the accepted story that Bishop (and/or his companions) discovered the fish in question.

However, Davidson's narrative is confusing. First, he did not mention by name the body of water where the fish was seen except for calling it "*The River*" (p. 54). Second, he mentioned visiting "the Dome, called Gorin's Dome, for its discoverer and later owner" (p. 56). Because we do not know exactly when Davidson wrote his book it is hard to determine whether he is referring Gorin as owner at the time of his visit or well after the fact. Gorin purchased Mammoth Cave on April 17, 1838. Does this date mean that Gorin actually explored the cave at least three years before he purchased it? But if Gorin explored the cave at least three years before purchasing it, could he have been accompanied by Bishop himself? Furthermore, Davidson wrote about crossing Bottomless Pit using a ladder *after* seeing the "*white fish.*" Was he confused by the passage of time between the occasion of his visit and that of writing his account that he switched the order of events? Davidson probably wrote his account sometime in 1839 because he states that "It was with a feeling like regret that I

---

[353] At the time of Davidson's visit, tours were already being given at river level and it is unlikely that Stephen (or others) would have been deterred from continuing by a few rocks. It is possible that Davidson's acquaintances might have thought they were discovering new cave, but probably not (Rick Olson Pers. Comm.) which would have been to the full as interesting a spectacle as Prince Bonbobbin's white mice with green eyes, for which he ransacked the world. All we found was a poor miserable mudfish, caught with the hand by the guide, near the shore, blinded by the light It was certainly a wonderful thought, that such a body of water should have been flowing here a furlong at least underground, in the silence and gloom of centuries. (Davidson, 1840, pp. 54-56).

heard that the present owner, Dr. Croghan, of Louisville, who has just purchased the estate for ten thousand dollars" (pp. 62-63). John Croghan (1790-1849), a physician, did not buy Mammoth Cave until 1839.

There is a smoking gun favoring the hypothesis that Davidson got not only the order of events of his trip wrong, but the chronology as well. On page 66, he states that after leaving the cave that night while at the hotel "the landlord afterwards averred, that some of his guests had made a terrible noise in the night, and called out lustily for *Stephen, the guide!*" Because there are no other guides on record with the first name of Stephen for that period, we must conclude that he was referring to Stephen Bishop.

Although Davidson does mention that his party was accompanied by "guides" without specifying names, number, or ethnicity, it is most likely that such guides were, in addition to Stephen Bishop, the brothers Materson and Nicholas Bransford, two other slaves owed by Thomas Bransford of Nashville and hired by Gorin for $100 each per year to serve as guides in the increasingly popular cave tours (Brucker & Watson 1987, pp. 269).

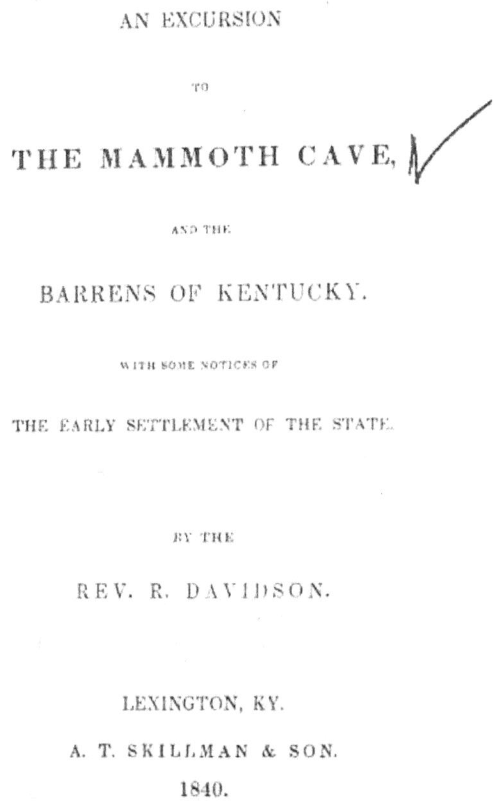

AN EXCURSION

TO

THE MAMMOTH CAVE,

AND THE

BARRENS OF KENTUCKY.

WITH SOME NOTICES OF

THE EARLY SETTLEMENT OF THE STATE.

BY THE

REV. R. DAVIDSON.

LEXINGTON, KY.

A. T. SKILLMAN & SON.

1840.

Figure 19.1. Cover page of Davidson's book.

This being so, Davidson's excursion most likely took place in late October or early November 1838 or 1839 because Bishop became guide at Mammoth Cave after April 17, 1838, when Gorin bought the estate. Furthermore, this is consistent with the conventional chronology that Bishop crossed Boomless Pit for the first time on September 20, 1838. Therefore, Davidson and his party must have been among the first outsiders visiting River Styx. If Davidson's phrase that the river had been discovered "only about a year ago" before his visit is correct, then it is quite possible that his visit occurred in October 1839.

Can Robert Davidson's Chronology be Authenticated?

Of course, despite the contradictions, it is possible that Davidson (or others) visited River Styx before Bishop. Is there any documentation on Davidson's life that could substantiate such a hypothesis? Davidson was a clergyman, born in Carlisle, Pennsylvania, on February 23, 1808. He was the son of Robert Davidson, an educator, theologian, and also a Presbyterian clergyman of some notoriety. Robert Jr. was a graduate of Dickinson College (1828) and Princeton Theological Seminary (1831). He was pastor of the second Presbyterian Church in Lexington, Kentucky (1832-1840) and in the latter year became president of Transylvania University, a post he held until 1842. He later held pastorates in New Brunswick, New Jersey, New York City, and Huntington, Long Island. He went co Philadelphia in 1868, where he died on April 6, 1876.

We contacted Dickinson College, Princeton Theological Seminary, Transylvania University, and the Second Presbyterian Church of Lexington, KY, to see if Davidson had donated his papers to any of those institutions in the hope that such papers might contain some information clarifying the chronology of his visit to Mammoth Cave. Unfortunately, there is no evidence that he did so. Interestingly, all of them had only scant information about Davidson, on the basis of which we have written the above biographical sketch. We could not therefore find any corroborative document on the chronology of Davidson's visit to Mammoth Cave.

What about others? We checked all relevant published descriptions of Mammoth Cave prior to the publication of the scientific discovery of the blind cavefish in 1842. For example, John Hay Farnham (1791-1833) published a letter in 1820 describing Mammoth Cave in which he makes no mention of any fauna whatsoever. In 1832 Constantine Samuel Rafinesque (1783-1840), a noted naturalist, published an article, "The Caves of Kentucky," in which he described Mammoth Cave, mentioning its bats and rats. He was a professor at Transylvania College (1819-1826) and explored Kentucky thoroughly (Warren 2004). He also had discovered the cave salamander *Euycea lucifuga*. There is no question that if he had had knowledge of (or seen) a blind cavefish in Mammoth Cave he would have mentioned it.

In 1835 Edmund F. Lee, who extensively explored Mammoth Cave between 1834 and 1835, published a very precise description of the cave where he cited bats, rats, and a "nearly transparent" cricket, but no fish. Dr. Robert Montgomery Bird (1806-1854) of Philadelphia, published extensive accounts of the cave in 1837 and 1838 based on his 1833 and 1835 summer explorations of it in which he mentions the bats, rats, and insects of the cave- but no fish. Another 1838 article, this time anonymous, written by someone who obviously had visited the cave and explored it at length, makes no mention of the fish either (Anonymous 1838).

Therefore, it seems that the fish in question was not seen before 1838, which would be consistent with our assertion that Davidson's visit took place in 1839 and hence, his reference that the river where the "*white fish*" was found was discovered a year earlier, is consistent with the conventional wisdom that Bishop and/or others did not cross Bottomless Pit, finding the fish, until the autumn of 1838.

But Which "White Fish"?

Blind cavefish from Mammoth Cave were routinely captured by Bishop and others and exhibited at the lodge built by Gorin to accommodate visitors (Brucker & Watson 1987, p. 272). Two species of blind cave fish have been reported for Mammoth Cave: the northern cavefish *(Amblyopsis spelaea)* and the southern cavefish *(Typhlichthys subterraneus)*. Which one was seen first? We surveyed major museum collections of specimens for both species and the results are shown in Tables 1 and 2, we found 29 confirmed records for the northern cavefish and eight for the southern cavefish, having Mammoth Cave as the collecting locality. The data in these tables show that records for the northern cavefish in Mammoth Cave date back to at least 1842, the year in which the species was scientifically described by DeKay (Romero 2002), while records for southern cavefish do not appear until 1879. Therefore, because of the number of records and their chronology, it seems that most probably the first blind cave fish seen at Mammoth Cave was *A. spelaea*. Furthermore, *A. spelaea* is larger and easier to see, which would be important when using dim lanterns. Additionally, the base level habitat of River Styx is dominated by this species; *T. subterraneus* is more prominent in streams with currents above base level (Rick Olson, *pers. comm.*).

Unfortunately, many of the registers of the hotel and cave were apparently destroyed in the December 9, 1916 fire that consumed the old Mammoth Cave hotel (Goode 1986, p. 14). Although some of the post-1842 registers are available (Robert H. Thompson, *pers. comm.*), none contemporary with the discovery of the blind cave fish are known to exist. With the destruction of those records, we have lost any chance of knowing who came to Mammoth Cave and if they could have traveled with Stephen Bishop. However, the specimens at museum collections do indicate that many of the specimens were collected (or brought by) notable people from New England. For example, the first time that a blind cavefish from Mammoth Cave was mentioned in the scientific literature was in a short note in the *Proceedings of the Academy of Natural Sciences of Philadelphia* (Anonymous 1842). There it was reported that a W.T. Craigie donated to the Academy at the May 24, 1842 meeting a specimen of "a small white fish, also eyeless (presumed to belong to a subgenus of *Silurus*), taken from a small stream called the 'River Styx' in the Mammoth Cave, Kentucky, about two and one-half miles from the entrance." In the Academy's collections today, there are four specimens of *Amblyopsis spelaea* in alcohol that appear linked to this donation. Two are catalogued as ANSP 7964 collected by W.T. Craigie, another is ANSP 7963 collected by "Mrs. C.H. Graff, Messrs. Craige & Lambert," and a fourth is ANSP 7966 collected by J. Lambert. Because of the consecutive numerals, we also suspect that ANSP 7961 and ANSP 7962 (one specimen each) were also collected at the same visit to Mammoth Cave. The names of these and other collectors belong to distinguished people of cities such as Philadelphia and New York.

Conclusions and Further Research

Two major factors influence historical research: one is access to primary sources, i.e., documentation that is contemporary with the facts we are trying to discern; the other is the surrounding mythologies of the facts that create barriers to understanding the real story. The case of Stephen Bishop epitomizes both. Firstly, there are very few contemporary sources on Bishop and his accomplishments. One is a letter of Franklin Gorin's, an account of Bishop after his former slave had died (Forwood 1870). Another is Marianne Finch's (1853), an Englishwoman who visited the pre-Civil War South 20 years or so after slavery had been abolished throughout the British Empire. A third is by Nathaniel Parker Willis (1806-1867) (in Randolph 1853,

p. 53-54), a noted romantic and idealist of American nature. All these writings took place in the midst of the Romantic Movement which worldwide, was especially profound in the United States, influencing even the first describer of the first blind cave fish species, James DeKay (Romero 2002). That movement created the now famous myth of the "noble savage," the man of another race who was essentially a good person and whose value as a human being was to be admired and publicized. These characters figured prominently in popular travel accounts. Facts were made to fit (or at least portrayed within) these romantic values. Therefore, in order to understand who Bishop really was and how stories and history interact in this case, we need to look into these facts in the context of nineteenth century romanticism and race during antebellum. Or maybe we deluded in pretending that the real story will really be known.

Acknowledgements

The following people provided help with our archival work: Robert Benedetto, Librarian for Archives and Special Collections, Princeton Theological Seminary Libraries; James Duane Bolin, Murray State University; Kaiyi Chen, University of Pennsylvania; Jim Gerencser, Dickinson College; BJ. Gooch, Transylvania University Library; Elliot Greenwald, University Archives and Records Center of the University of Pennsylvania; DiAnna Hemsath, University of Pennsylvania; Nancy M. Shader, Mudd Manuscript Library, Princeton Theological Seminary. Rick Olson, Park Ecologist, and Vickie Carson, Public Information Officer of Mammoth Cave National Park, accompanied us in our visit to the cave, and provided very useful information. Mr. Olson read an earlier version of this article and made valuable suggestions. Bob Thompson of Maineville, OH, provided us with stimulating insights and read an earlier version of this article and made valuable suggestions. Roger W. Brucker, of Beavercreek, OH, provided some insights and read an earlier version of the MS and also provided useful suggestions. This research was partially funded by a NSF DBJ-0243765 grant to Robyn Hannigan and Jerry Farris.

Table 1. Known specimens of the northern cavefish (*Amblyopsis spelaea*) collected in Mammoth Cave (for museum acronyms see legend in Table 2).

Date	Collector	Catalog Number
b. 1843 (?)	J.E. Mitchill	ANSP 7961
b. 1843 (?)	J. Darley	ANSP 7962
b. 1843 (?)	Mrs. C.H. Graff, Messrs. Craige & Lambert	ANSP 7963
b. 1843 (?)	W.T. Craig	ANSP 7964
b. 1843 (?)	J. Lambert	ANSP 7966
1844 May	J.A. Granger	NYSM11464[1]
1851	?	BMNH 1851.11.20.1
1853 June 18	?	NRM 8380
1866	Vatble	MNHN 0000-4184
1876 December 24	J. Lindahl	NRM 8000
1890?	Tison	MNHN 1890-0043
1893	H.C. Ganter	USNM 00044435
1896(?)	?	BMNH 1896.9.30.13
1901 September 1	W.P. Hay	USNM 00127056[2]
1905 May 15	C.H. Eigenmann	USNM 00127055
1909 October 9	Columbia University, Dept. Comp. Anat. Coll.	AMNH 879
1909 November 16	Füllhorn	ZMH 13174
1918 April 9	E.O. Hovey	AMNH 12112
?	Finn	AMNH 1156
?	Holmes	AMNH 1646
?	Mrs. Frederick	ANSP 20373
?	?	MNHN 0000-2762
?	Claine	MNHN 1890-0042
?	Bromer, Fr.	NRM 8001
?	J. Sloan	ROM 08046
?	?	USNM 00005863
?	?	USNM 00048867
?	?	USNM 00237001
?	?	USNM 00237004[3]
?	Baum u. K. Hoffmann	ZMH 13175

[1] Specific location: River Styx
[2] Roaring River
[3] Echo River

NOTE: The holotype or specimen used to describe the species was collected at River Styx and was deposited at the collection of the Lyceum of Natural History of New York but is today lost (Romero 2002).

Table 2. Known specimens of the southern cavefish *(Typhlichthys subterraneus)* collected in Mammoth Cave.

Date	Collector	Catalog Number
20 December 1879	?	ZMUC * 2
20 December 1879	?	ZMUC * 3
1884	Swain Gilbert	USNM 00036632
1884	?	USNM 00036806
1903	C.H. Eigenmann	AMNH 18715
?	?	AMNH 8103
?	?	CAS 125283[1]
?	W.P. Hay	USNM 00101172[2]

[1] River Styx
[2] Roaring River

Acronyms for museum collections: AMNH: American Museum of Natural History (New York); ANSP: Academy of Natural Sciences of Philadelphia; BMNL: British Museum of Natural History (London); CAS: California Academy of Sciences (San Francisco); MNHN: Musée Nationale d' Histoire Naturelle (Paris); NMR: Swedish Museum of Natural History (Stockholm); NYSM: The New York Survey Museum; ROM: Royal Ontario Museum (Ontario, Canada); USNM: United States National Museum (Washington, DC); ZMH: Universitat Hamburg, Zoologisches Institut und Museum (Hamburg, Germany); ZMUC: Zoological Museum of the University of Copenhagen.

Literature Cited

Anonymous, 1838. The Mammoth Cave of Kentucky. *Chamber's Edinburgh Journal* **6**:234-235.

Anonymous, 1981. Stephen L. Bishop. 1821 - 1857. Explorer and Guide. Mammoth Cave. *Journal of Spelean History* **15**(1):11.

Anonymous, 1992. Bishop, Stephen, pp. 82-83, In: *Kentucky Encyclopedia* (John E. Kleber, ed.). Lexington: University Press of Kentucky.

Barr, T.C., 1986. Mammoth Cave in the years 1836-1855. *Journal of Spelean History* **20**(2):39-40.

Bird, R.M., 1837. The Mammoth Cave of Kentucky. *The Plaindealer* **1**(24):379-381.

Bird, R. M., 1838. *Peter Pilgrim: or rambler's recollections*. Philadelphia: Lea & Blanchard.

Brucker, R.W. & R.A. Watson, 1987. *The longest cave*. Carbondale: Southern Illinois University Press.

Davidson, R., 1840. *An excursion to Mammoth Cave, and the barrens of Kentucky with some notices of the early settlement of the state*. Lexington: A.T. Skillman & Son.

Farnham, J.H., 1820. Extract of a letter from John H. Farnham, Esq. a member of the American Antiquarian Society, describing Mammoth Cave, in Kentucky. *Transactions and Collections American. Antiquarian Society* **1**:355-361.

Finch, M., 1853. *An Englishwoman's Experience in America*. London: Richard Bentley.

Forwood, W.S., 1870. *An historical and descriptive narrative of the Mammoth Cave of Kentucky*. Philadelphia: J.B. Lippincott.

Goode, C.E., 1986. *World Wonder Save. How Mammoth Cave Became a National Park*. Mammoth Cave: Mammoth Cave National Park Association.

Lee, E.F., 1835. *Notes on the Mammoth Cave to accompany a map.* Cincinnati: James & Gazlay.

Meloy, H., 1977. The legend of Stephen Bishop. *Journal of Spelean History* **10**(1): 5-7.

Randolph, H.F., 1924. *Mammoth Cave and the cave region of Kentucky.* Louisville: Standard Print. Co.

Romero, A., 200 I. Scientists prefer them blind: the history of hypogean fish research. *Environmental Biology of Fishes* **62**(1-3):43-71.

Romero, A., 2002. Between the first blind cave fish and the last of the Mohicans: The scientific romanticism of James E. DeKay. *Journal of Spelean History* **36** (1):19-29.

Warren, L., 2004. *Constantine Samuel Rafinesque. A voice in the American wilderness.* Lexington: The University Press of Kentucky.

Chapter 14. Between the First Blind Cavefish and the Last of The Mohicans: The Scientific Romanticism of James E. DeKay [354]

Summary

James DeKay, the man who first described a species of blind cavefish for science, was an unlikely hero in the history of biospeleology: he was a physician by training, not a natural historian; he was closer to the romantic literary writers of his time than to any group of scientists; there is no evidence that he ever visited any cave, and the cavefish he described was collected by someone else in Kentucky, far away from New York State, his area of research. Yet his name is indelibly tied to the beginning of biospeleology in the United States. This article represents an attempt to understand DeKay's scientific approach to his work, particularly to the first scientific description of a blind cavefish.

The Beginnings

James Ellsworth DeKay (he himself spelled it sometimes as Dekay or De Kay) (Fig. 17.1) was born in Lisbon, Portugal, on 12 October 1792. He was the eldest son of George and Catherine (Coleman) DeKay. George was a descendant from a Dutch family that settled in America in the seventeenth century (Fisher 1973) while Catherine was from Cork, Ireland. George was a sea captain sent from the American colonies to Europe in 1775. He and Catherine met in Lisbon at a Dance. James was brought back to Scarsdale, New York, when he was 2 years old. His mother died when he was 6 and his father when he was 10. Apparently, his father left him with a pension of $3,000 a year[355], a sum with which he would live comfortably for the rest of his life.

Apparently, he also traveled to Paris and Germany pursuing his medical studies (Anonymous 1852, Fisher 1973). He may have used previous schooling in order to shorten his stay at Edinburgh. From the dedications in his thesis, it can be inferred that he may have studied with Samuel Latham Mitchill whom he described as a professor of natural history in *Academia Novaeboracensi* (New York City). The other dedication is to Aemelio (Emil) Osann[356], M.D., whom he described as professor of "*materia medica*" in the *Academia Literarum Regia Berolinensi* (Berlin). Mitchill, as DeKay, had graduated from the University of Edinburgh, and it is possible that the former played a role in getting James into that University. Also, Mitchill switched from medicine to the natural sciences and was the founder of the New York Lyceum of Natural History, an association in which DeKay participated actively; therefore, it is reasonable to think that Mitchell acted as both mentor and role model to the young DeKay. We know much less of Dr. Osann as to speculate on his influence on the young American.

After returning from Europe, DeKay became very close to Henry Eckford[357], the eminent marine architect and shipbuilder, who built in 1822 the *Robert Fulton* which made the first successful trip by a steam boat from New York to New Orleans to Havana (Eckford & Huxley 1988). DeKay would marry Henry's daughter Janet Eckford (1802-1854) on 31 July 1821. He traveled briefly to Quebec with Fitz-Greene Halleck and later

---

[354] Based on Romero, A. 2002. Between the first blind cave fish and the Last of the Mohicans: the scientific romanticism of James E. DeKay. *Journal of Spelean History* **36**(1):19-29.
[355] About $260,000 in 2020 dollars.
[356] b. 25 May 1787, Weimar, Germany; d. 11 January 1842, Berlin, Germany.
[357] b. Kilwining, near Irvine, Scotland, 12 March 1775; d. Constantinople, Turkey, 1832.

sailed with his father-in-law as surgeon in the frigate built for Constantinople's Sultan's navy. Eckford was to take charge (as superintendent) of the navy yard at that Turkish city but died the year after his arrival. In 1833 DeKay published (anonymously) his impressions of Turkey in a volume called *Sketches of Turkey in 1831 and 1832 by an American*, in which he gave a favorable view of the country and its institutions; yet, Hellenists of the day were incensed that an American should appear as a defender of the oppressors of Greece.

Figure 17.1. Only known portrait of James Ellsworth DeKay.

DeKay's father-in-law at one time had a controlling interest in *The National Advocate*, a New York political journal and toyed with the idea of installing DeKay as editor. DeKay also wanted to start a literary magazine with Halleck as editor, but nothing came of that initiative either (Dictionary of American Biography, Vol. 3:203-204).

While in Turkey, DeKay made a special study of the Asiatic cholera, about which little was known in America. After his return to New York, he had the opportunity to put in practice what he had learned on this disease: in 1832 he became famous because he promoted the use of port wine as a cholera remedy. Despite its uncertain health benefits, the advice was so highly regarded that "Dr. DeKay" became one of the bars pours of New York's cholera days while he was being nicknamed "Dr. Port." Yet, none of the city's doctors had any idea what caused Asiatic cholera (Koeppel 2000). This is the last time we know he practiced as a physician, a practice that he found repugnant (Wilson & Fiske 1888) at a time when anesthesia did not exist and medical treatments were usually more harmful than beneficial.

Shortly after his return from Europe he settled permanently in Oyster Bay, Long Island, devoting himself to cultivate friends in literary circles, studying natural history, and contributing to the New York press. Among the literary men he befriended were the Washington Irving[358], the author of *The Legend of Sleepy Hollow* and *Rip Van Winkle*; Joseph Rodman Drake[359] (who would marry Sarah Eckford, sister of DeKay's wife) a noted poet and physician, James Fennimore Cooper[360] who wrote *The Last of the Mohicans* (1826), and Fitz-Greene Halleck[361], a famous poet. In 1837 they started the Authors Club (Washington Irving president, Halleck vice-president), with all the members being part of America's romantic literary movement.

The main characteristics of the American romantic literary movement were the sense of frontier philosophy (a vast country with the ideas of freedom with no geographic limitations), optimism (greater than in Europe because of the presence of vast frontier lands), experimentation (in both science and institutions), the mingling of races (epitomized by the arrival of immigrants in large numbers to the US), and the growth of industrialization (with the subsequent polarization of North and South; where North becomes industrialized while the South remains agricultural). We will see how DeKay transferred some of those values into his scientific writings.

Scientific Career

DeKay's first scientific paper was published in 1821, just two years after his return from Europe. He soon joined the major scientific associations of New York. For example, in 1825 we find him as Curator of the Literary and Philosophical Society of New York. Despite the name of this group, founded in 1814, virtually all its officers were naturalists. This association disappeared by the end the 1820's when most of its members, including DeKay, joined the Lyceum of Natural History of New York (today New York Academy of Sciences), founded in 1819 by Mitchill. James was one of the most active members of that association where he acted as a Librarian (1826-1827), Editor of the *Annals* (1819-1830), editing volumes1 and 2, Corresponding Secretary (1824-1836), Recording Secretary (1834-1836), and First Vice-president (1840-1846). He was also largely responsible for the development of the Lyceum's collection. He was a member of the American Association for the Advancement of Science (1848-1851, Kohlstedt 1976). He also published in *The American Journal of Science and Arts, Transactions of the Albany Institute, Monthly American*

---

[358] *b*. New York City, 3 April 1783; *d*. Tarrytown, New York, 28 November 1859.
[359] *b*. New York City, 7 August 1795; *d*. New York City, 21 September 1820.
[360] *b*. Burlington, New Jersey, 15 September 1789; *d*. Cooperstown, New York, 14 September 1851.
[361] *b*. Guilford, Connecticut, 8 July 1790; *d*. Guilford, 19 November 1867.

Journal of Geology and Natural Science (Philadelphia). Although some claim that he was one of the founders of the Academy of Medicine (Wilson & Fiske 1968), archival papers from that institution do not support such contention (Shaner, pers. comm.).

However, it was a new government-sponsored initiative that placed him in the position of generating his main scientific opus while contributing to the advance of the study of biospeleology in the U.S. On 18 April 1835 the New York State Legislature approved the Geological Survey of New York, which was to include the preservation of specimens of "zoological productions" (Dix 1836). The legislature was responding to lobbying from the Lyceum of Natural History and the Albany Institute, among others, that were seeking a statewide survey of natural resources. That, and the need for coal, convinced the State to pursue this initiative (Sterling 1999). This can also be framed within the movement that started in the 1840's when several states of the United States inaugurated natural history surveys and published catalogues of the local faunas (Coe 1918). The Survey was established in Albany in 1836, which makes it the oldest continuously functioning geological (and biological) survey in the New World (Fakundiny & Albanese 1988). The Survey hired DeKay as its zoologist in July 1836 with an annual salary of $1,500 (Anonymous 1837).

In the wake of his literary friends' vision of an expanding America, DeKay soon began to include as fauna of New York, virtually everything he could think of in the North American continent. He justified it by saying that "The State of New-York is connected on its southern border with the ocean, and its numerous products; at the north will be found many inhabitants of the arctic regions; while the rivers on its south-western frontier will be found to connect it with the great valley of the Mississippi. From its magnitude and geographical position, it will therefore be found to comprise in all probability, more than two-thirds of all animal species existing within the limits of the United States." (DeKay 1838).

Yet, most of the citations to non-New-York species were rather brief. Although DeKay made extensive use of correspondence in order to acquire both information and specimens from farmers, hunters, and fishermen, he also embarked on extensive fieldwork, including a water-borne tour of the Adirondacks. He helped to establish what would become the major elaboration of the story of the Adirondacks as a romantic landscape and setting the pattern for increasingly popular camping trips seeking to recapture the vigor of body and soul weakened by the stresses of modern life. Native Americans were romanticized in those times, now that they had been placed in reservations and, as far as the northeast Americans were concerned, were no longer an obstacle to American expansionism (Terrie 1997). This work took him eight years (1836-1844), and the results were published between 1842-1844 in the form of five quarto volumes titled *Zoology of New-York; or, the New York Fauna, comprising detailed descriptions etc..* (Fig. 17.2). It encompassed both recent and fossil organisms, although most the latter were mentioned only briefly.

Additionally, a list of mammals, birds, reptiles, and amphibia, drafted by DeKay prior his death in 1851, were published in the *Catalogue of the Cabinet of Natural History of the State of New York and of the Historical and Antiquarian Collection Annexed Thereto*. For other groups of animals, he wrote "The Fishes, Insects, Shells, etc. are for the present omitted, in the hope that they may soon be increased in number, and duly arranged and named" (DeKay 1853).

This contribution by DeKay is still considered a monumental work pioneering the knowledge of a fauna for which very little had been published up to that time. Yet, it did not lack a number of contemporary critics. For example, some complained that the *Zoology of New-York* contained mostly non-New York species (including the Florida manatee). Yet, had he not included those "extralimital" species, some like the blind cavefish would not have described at that time (see Smallwood 1941 for some insights on this). Also, some were shocked for the alleged cost of the publication ($130,000[362]), an astronomical sum for that time (Welch 1998, p. 99). For many, quality was not necessarily at the level of the expenses and some pounded both the contents and the illustrations, including his emphasis he put in using local or vernacular and Indian names (Dictionary of American Biography Vol. 3:203-204).

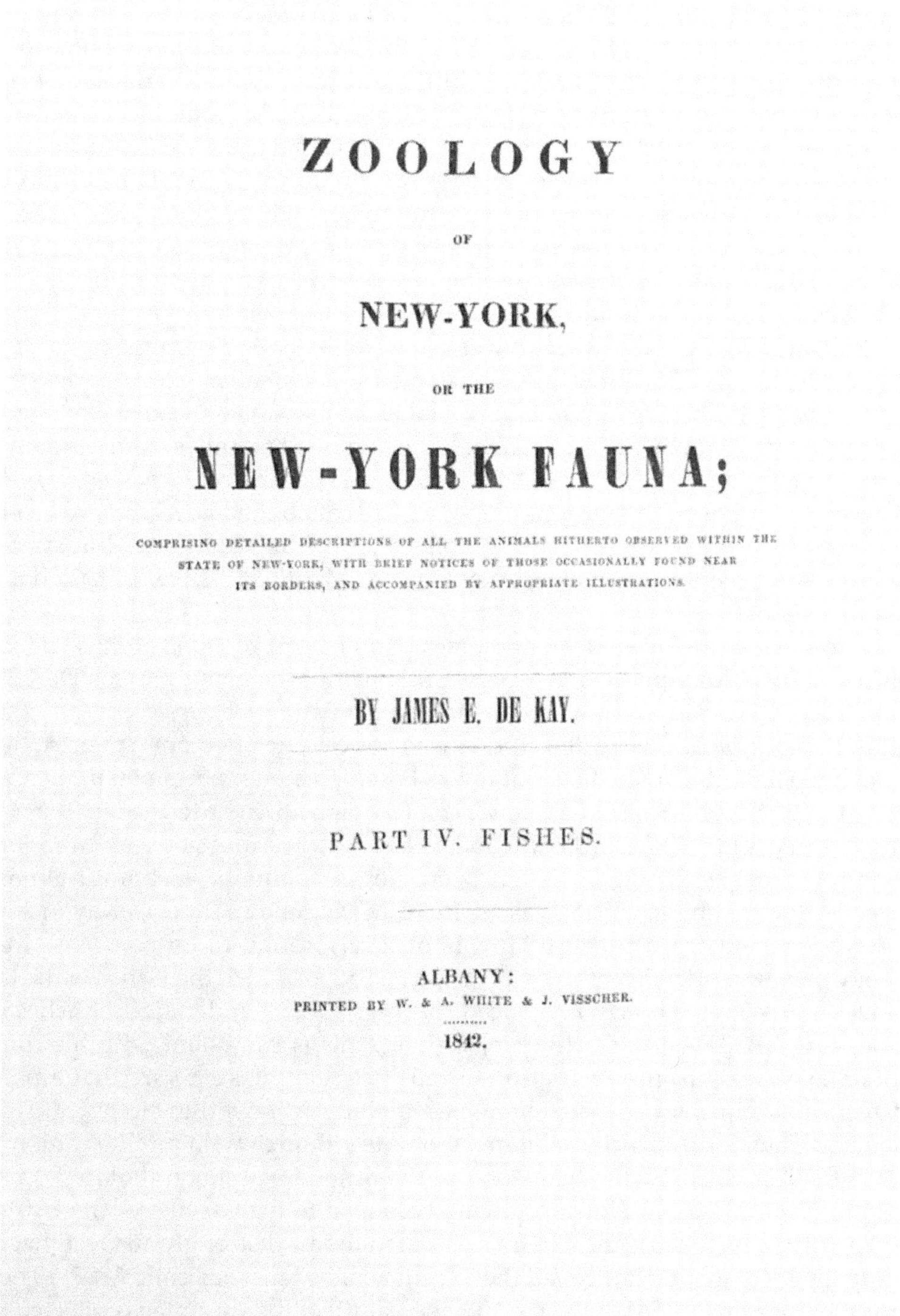

Figure 17.2. Cover of DeKay's *"Zoology of New York..."* (public domain illustration).

---

[362] Almost $2,800,000 in 2020 dollars.

The Blind Cavefish

A number of cavefish tales had been published for China and Europe from the sixteenth throughout the eighteenth centuries (Romero 2001). The first published record of a confirmed troglomorphic fish in the Western Hemisphere was probably that of James Flint (Flint 1822), a Scotchman who lived for several months in Jeffersonville, Indiana, in 1820 and recorded that "a Colonel C – [sic] of Indiana told me that a settler in his neighbourhood [sic] digging a well, penetrated into a stream of water, and found blind fishes in it." He added as a footnote that "Since the above was written, a notice of blind fishes has appeared (if I mistake not) [sic] in the memoirs of the Wernerian Society of Edinburgh"). Yet, such account was never published in that journal. Another early account of a cavefish for North America was by Robert Davidson (1808-1876), who visited Mammoth Cave in Kentucky in October 1836 accompanied by Stephen Bishop (1780-1850). Davidson reported that "*white fish* were found here without eyes" whose existence was already known by some of the locals (Davidson 1840, Romero & Woodward 2005).

The first time that an American troglomorphic fish was mentioned in the scientific literature was in a short note in the *Proceedings of the Academy of Natural Sciences of Philadelphia* (Anonymous 1842). There it was reported that a W. T. Craigie donated to the Academy at the 24 May 1842 meeting a specimen of "a small white fish, also eyeless (presumed to belong to a subgenus of *Silurus*), taken from a small stream called the 'River Styx' in the Mammoth Cave, Kentucky, about two and one-half miles from the entrance." Today, at the collection of the Academy there are three specimens of *Amblyopsis spelaea* in alcohol, that appear linked to this donation. Two are catalogued as ANSP 7964 collected by W.T. Craige, and the other, ANSP 7964, collected by 'Mrs. C.H. Graff, Messrs. Craige & Lambert'. All three specimens were captured in Mammoth Cave, but no dates are given (Romero 2001).

Yet, following the Rules of the Zoological Nomenclature, none of these references count as a scientific description since no scientific name was given. It was DeKay who did so in his *Zoology of New York* where he named the fish "*Amblyopsis spelaeus*" (known today as *Amblyopsis spelaea*). The description was not very detailed nor of a great quality. This could have been due to the fact that it was based on a poor specimen in the Cabinet of the Lyceum of Natural History of New York (Putnam 1872) or to the fact that DeKay was not a trained ichthyologist (Smallwood 1941). Yet, we must be careful in judging scientific procedures with standards that were not in common place until almost 100 years later. Although this cavefish, was captured in Mammoth Cave, DeKay included it in his New York faunal list because "It cannot therefore fail to be perceived that the Ichthyology of New-York will embrace a very large proportion of the Fishes of the United States" (DeKay 1842:iv). He actually placed this new species under a list of fishes under the subheading '(EXTRA-LIMITAL)' [sic]. Again, this is consistent with his romantic views of an expanding frontier but also with his desire of making sure that a potential species whose specimens had been circulated already in scientific circles, did not go unnamed and, therefore, he included it in a footnote although without illustration.

What is less clear is what happened to the original specimen (holotype) used to describe the species. The specimen orginally belonged to the Cabinet of the Lyceum of Natural History of New York and cannot be located today. I strongly suspect that it was lost during the 1866 fire that destroyed the Lyceum collections (Fairchild 1887). The New York Survey Museum (NYSM), which is the depository of the specimens collected by the NY Geological Survey, has two specimens of *A. spelaea*; one NYSM11464, was collected at River Styx in Mammoth Cave on May 1844 by J.A. Granger of Canandaigua, NY. The transferal letter is to T. Romeyn Beck, a physician from Albany, who was head of the Albany Medical College. His brother, Lewis Caleb, was a mineralogist with NYSM. A second specimen at the same collection lacks information. Neither seems to be the one used by DeKay in his description of the first North American cavefish. DeKay would never write again about *Amblyopsis* (or any other fish); however, this fish caught the attention of a number of anatomists who immediately began studying it (Romero 2001).

Dekay's Last Days

From the time of his retirement from the New York Geological Survey in 1844 until his death, DeKay lived at his house in Oyster Bay and did not publish anything else. Some biographical notes seem to indicate that he spent his last years trying to recover from the physical demands of his work on the New York Fauna ("The vast labors, demanded of him in the preparation of his State Reports on Zoology, impaired his health, which he never afterward fully regained," Anonymous 1852). I have not been able to ascertain what was his medical condition nor the causes of his death. He died at Oyster Bay, on 21 November 1851 at the age of 57, a rather above-average age for people at that time. He was buried in St. Georges Churchyard in Hempstead, New York (Anonymous 1851, Welch 1996). According to his testament and last will, DeKay left all his state to his wife.

Although James and Janet had four sons and four daughters (for their names and biographies see Fisher 1973), only four of them survived him. He was described as a man of "uprightness, amiability and cheerful temperament." (Anonymous 1852)

This unlikely pioneer of biospeleology left us with the first scientific description of a cavefish for the Western Hemisphere, a voluminous zoological work, and a sense of science as a romantic endeavor. All three legacies are worth of a man's life dedication to the pursue of knowledge.

Acknowledgements

The following individuals provided useful information: Amy Crumpton (American Association for the Advancement of Science), William R. Massa, Jr. and Christine Connolly (Yale University), Leslie K. Overstreet (Smithsonian Institution Libraries), Tracy Elizabeth Robinson (Smithsonian Institution Archives), Arlene Shaner (New York Academy of Medicine), and Arnott T. Wilson (Edinburgh University). Andrea Romero read an early version of this MS and made valuable suggestions. Robert A. Daniels (New York State Museum) not only provided useful information but also read critically a more advanced version of the MS and made very valuable suggestions.

Literature Cited

Anonymous. 1837. Communication (No. 161) From the Governor. Assembly, February 11, 1837. Relative to the Geological Survey of the State. Albany: State of New-York.
Anonymous. 1842. [Mammoth Cave Blind Crayfish and Fish]. *Proceedings of the Academy of Natural Sciences of Philadelphia* **1**:175.
Anonymous. 1851. Died. *New York Herald* 23 Nov. 1851.
Anonymous. 1852. Obituary (James E. Dekay). *American Journal of Science* **13**:300-301.
Coe, W. R. 1918. A Century of Zoology in America. pp.:391-438. *In*: E. S. Dana (ed.). A *Century of Science in America*. New Haven: Yale University Press.
Davidson, R. 1840. *An Excursion to the Mammoth Cave, and the Barrens of Kentucky. With some Notices of the Early Settlement of the State*. Lexington: A.T. Skillman & Son.
DeKay, J.E. 1838. *Report of Dr. James E. De Kay on the Zoological Department of the Survey*. New York.
DeKay, J. E. 1842. *Zoology of New York or the New-York Fauna. Part IV Fishes*. Albany: W. & A. White & J. Visscher.

DeKay, J.E. 1853. *Catalogue of the Cabinet of the State of New-York and of the Historical Antiquarian Collection.* Albany: C. Van Benthuysen, 1853.

Dictionary of American Biography. 1928-1936. Volumes 1-20. New York: Charles Scribner's Sons.

Dix, J.A. 1836. *Report of the Secretary of State in Relation to a Geological Survey of the State on New-York. Made to the Legislature, January 6, 1836.* Albany: Croswell, Van Benthuysen and Burt.

Eckford, I. & D. Huxley. 1988. *The Eckford Family in Australia - The Early Years.* Eleebana, S.S.W.: Eckford.

Fairchild, H.L. 1887. *A History of the New York Academy of Sciences, formerly the Lyceum of Natural History.* New York: Fairchild.

Fakundiny, R.H. & J.R. Albanese. 1988. New York State Geological Survey, pp. 320-330, In: A.A. Socolow (Ed.). *The State Geological Survey: A History.* Association of American State Geologists.

Fisher, L.M.D. 1973. *The DeKay Family in America.* Downy: Published by the author.

Flint, J. 1822. *Letters from America, Containing Observations on the Climate and Agriculture of the Western States, the Manners of the People, the Prospects of Emigrants, &c., &c.* [sic] W. & C. Tait, Edinburgh.

Koeppel, G.T. 2000. *Water for Gotham: A History.* Princeton: Princeton University Press.

Kohlstedt, S.G. 1976. *The Formation of the American Scientific Community: The American Association for the Advancement of Science, 1848-1860.* Urbana: University of Illinois Press.

Putnam, F.W. 1872. The Blind Fishes of the Mammoth Cave and their Allies. *American Naturalist* **6**:6-30.

Romero, A. 2001. Scientists prefer them blind: The history of hypogean fish research. *Environmental Biology of Fishes* **62**:43-71.

Romero, A. & J. S. Woodward. 2005. On white fish and black men: did Stephen Bishop really discover the blind cave fish of Mammoth Cave? *Journal of Spelean History* **39**(1):23-32.

Smallwood, M.S.C. 1941. *Natural History and the American Mind.* New York: Columbia University Press.

Sterling, K.B. 1999. DeKay, James Ellsworth. *In:* J.A. Garraty & M.C. Carnes (eds.), *American National Biography*, Vol. 6, pp. 356-357 New York: Oxford University Press.

Terrie, P. G. 1997. *Contested Terrain. A New History of Nature and People in the Adirondacks.* Syracuse: Syracuse University Press.

Welch, M. 1998. *The Book of Nature: Natural History in the United States, 1825-1875.* Boston: Northeastern University Press.

Welch, R.F. 1996. James E. DeKay: The Man Behind the Snake. *Reptile and Amphibian Magazine* (Jan-Feb):58-63.

Wilson, J.G. & J. Fiske (Ed.). 1968. *Appleton's Cyclopaedia of American Biography.* 6 volumes. New York: D. Appleton & Co. (in Volume 2:125-126).

Chapter 15. The Life and Work of a Little Known Biospeleologist: Theodor Tellkampf [363]

Introduction

One of the first names associated with the research of cave fauna from Mammoth Cave was that of Theodor Tellkampf. He not only described several of species, some of which are still valid, but also conducted a number of morphological studies. Yet very little is known about this scientist. Below I present what I have been able to gather about him. Although more needs to be known, this paper represents the first attempt to produce a narrative on his life and scientific career.

His Life and Times

Theodor Tellkampf was born on 27 April 1812 in Heinde, Germany. Although Juettner (1909: 98) wrote that he had been born in Bückeburg, his birth certificate shows he was a native of Heinde, where he was christened on 19 May 1812. The family had lived in Bückeburg until sometime between 1808 or 1809 but moved to Heinde where Theodor's father had leased an estate. These two towns are about 70 km apart, a considerable distance at that time (today the village of Heinde is part of the town of Bad Salzdetfurth).

Theodor had five brothers and three sisters. The eldest of the brothers was the son of his father, Johann Georg Diedrich[364] and his first wife Johanna Friederike Catharina Margaretha Werner, whom he married on 28 August 1797. The rest of Theodor's brothers and sisters were the progeny of Johann's second wife Charlotta Rosina Christina Baum[365]. Theodor was the sixth child of this union.

The Tellkampf family name has been changing through time. Spellings of Theodor's earlier ancestors include Tellkamp and Tellkampff. There is also some confusion about the way Theodor spelled his first name. In the parish register of Heinde his name is written as August Otto Theodor Tellkampf. While in America, his first name was sometimes spelled "Theodore." More confusing is his "middle" name. First of all, Germans do not use middle names (with the exception of people living in East Frisia until about 1900). Sometimes the name they use as a first name is the last one of the series of names given when baptized. That explains why he always used "Theodor" or "Theodore" as his name in America. For his middle name he sometimes used "A." which would be an abbreviation of his first christened name August (see, for example, Juettner 1909: 98, White 1884: 240.). In the only letter written by Theodor that I have been able to locate, deposited in the Archives of the Museum of Comparative Zoology at Harvard, he spells his name "Theo A. Tellkampf." Yet the "G ." as a middle name appears in a number of other sources.

Why did he feel compelled to create a "middle" initial for himself? It was

---

[363] Originally published as Romero, A. 2002. The life and work of a little known biospeleologist: Theodor Tellkampf. *Journal of Spelean History* **36**(2):68-76.
[364] *b*. Hannover, Germany, 2 May 1771; *d*. Hannover-Linden, Germany, 25 May 1846.
[365] *b*. Mollenfelde, 1778; *d*. Hannover-Linden, 10 March 1857.

not very unusual for Germans coming to the U.S. to add a middle initial in order to "Americanize" their name. An example was Carl H. Eigenmann, another German by birth and one of the most prolific authors on cave fishes who added an "H" as his middle initial (Romero 1986).

Theodor attended the gymnasium (high school) at Hannover and studied Medicine at Gottingen until the summer of 1838. A 26 August 1838 letter from Adolph Tellkampf (the eldest brother) to Johann Ludwig (another of Theodor's brothers) states that Theodor had studied Medicine in Göttingen until the summer of 1838 and intended to go to the University of Jena in the Fall of 1838 to pursue his doctorate. The University of Jena has no information about him. He attended the University of Göttingen from 28 October 1831 until February 1833. He was a student in the Faculty of Philosophy (where the natural sciences were taught) and studied physics. According to White (1884) he obtained his M.D. at the University of Wurzburg, Bavaria, in 1838. His doctoral dissertation was titled "Beitraege zur Lehre der Hautkrankheiten" (On skin diseases) but it was dated as published in Vienna in 1839. During my initial investigations I thought that he might have graduated from the University of Vienna not only because of the place of the publication of his doctoral ("inaugural") dissertation but also because of two other factors: one, the University of Vienna had one of the most prestigious medical schools in the world at that time, with very strong morphological leanings (the kind of things Theodor emphasized in his papers on cave fauna); the other is that he had a brother, George Hermann Daniel Tellkampf, a merchant, who also had lived in Vienna since at least 1831.

According to the University of Wurzburg, Theodor's dissertation was published in Vienna in 1839. However, the registrar of the University of Vienna did not find any record of Theodor attending that institution nor can a copy of his dissertation can be found in its library.

The question, then, is why was his dissertation published in Vienna if he did his M.D. studies at Wurzburg? It is possible that he may have got his degree from Wurzburg in 1838, but because of the time when it was printed the date on his dissertation is 1839.

Among Theodor's siblings, one achieved notoriety -his brother Johann Ludwig (Louis)[366]. Johann was a lawyer of international reputation who arrived in the U.S. in 1838 and taught political economy at Union College, Schenectady, N.Y. until 1843 when he went back to Germany. He returned to the U.S. again in 1844 where he occupied the Frederick Gebhard Chair at Columbia College (today Columbia University) until 1847 when he was replaced temporarily by Theodor until Columbia found a permanent replacement (Anonymous 1843, 1876, Danton 1946). He studied the American institution of prisons. Maybe that is why Theodor published a book on the health of prisoners (Tellkampf 1844d).

Ludwig returned to Germany in 1884 by invitation of the King of Prussia and was appointed Professor at the University of Breslau. In 1855 he became a member of the House of Lords ("Herrenhaus") of Prussia (among other parliaments in Germany) and was regarded as the leader of the liberal party in that aristocratic body. Another of Theodor's brothers, George Hermann Daniel Tellkampf[367], also lived in the U.S. until 1886. He was a stockbroker in New York City.

Theodor traveled to the U.S. for the first time in 1839. He sailed from

---

[366] *b*. Btickeburg, Germany, 28 January 1808; *d*. Berlin, 15 February 1876.
[367] *b*. Heinde, Germany, 29 January 1810; *d*. Hannover, Germany, 23 November 1893

Bremen to New York on board the ship *New York*. Instead of staying in New York with his brothers, Theodor went to Cincinnati where there was a sizeable German colony and where his services as a doctor may have been very welcomed. He lived there until 1843 where he "spent much time traveling and studying" (Juettner 1909:99). Cincinnati provided him with a location much closer to the Mammoth Cave than New York. He returned to Europe in 1843 and in 1844 he was offered (but declined) a Chair at the University of Berlin and that same year he returned to America and lived in New York until 1880. That year he returned to Germany and died on 7 September 1883 in Hannover.

Little is known of Theodor's descendants. He married Marie von Roth in 1858, who died the following year in New York, but nothing else is known about her. He also had one son named Georg Tellkampf, born in New York in 1858, who later became a physician. There is a record of Georg traveling from Hamburg to New York in June 1876. He was a student at that time, 19 years old. Another source mentioned him living as a physician in New York (year unknown) (Kuwert, pers. comm.).

Scientific Work

Theodor published some of the earliest morphological descriptions of the first cavefish reported in the scientific literature, *Amblyopsis spelaea*. He contributed detailed descriptions of this species and concluded that its eyes and those of blind cave crayfishes had become rudimentary as a result of disuse:

"While it is true, in general, that all animals retain their essential form, and that no species passes over into another by transformation, we know that less material changes of form are produced by external influences such as changes in climate or food, lasting though many generations of the same species".

For Tellkampf the original, unmodified species was still a mystery. Therefore, he did not want to settle this issue until "such species, corresponding with them in all essential points, are found" (Tellkampf 1844b: 393).

Theodor also described two members of the Class Arachnida. In 1844 he described the opilion *Phalangodes armata* new genus, new species, from Mammoth Cave. Both taxa are still valid and the genus became the type for the subfamily (Phalangodiinae), family (Phalangodidae) and superfamily (Phalangodoidea). Also, in 1844, Tellkampf described *Anthrobia monmouthia*, new genus and new species of spider. Keyserling in 1862 corrected the spelling of the specific name to *mammouthia*. This genus and species are still considered valid and are placed in the family Linyphiidae.

Other valid species named by Theodor are two beetles: *Petomaphargus hirtus* and *Ptomaphagus tellkampfi*, and the eyeless crayfish *Oreonectes pellucidus* (Fig. 15.1).

Theodor was frequently cited by contemporaries studying the Mammoth Cave fauna such as Jeffries Wyman[368], Alpheus Spring Packard,

---

[368] *b*. Chelmsford, Middlesex, Massachusetts, 11 August 1814; *d*. Bethlehem, New Hampshire, 4 September 1874.

Jr.[369], and Frederic Ward Putnam[370]. He also belonged to the major scientific societies of his time. By 1844 he appears as a member of Lyceum of Natural History of New York (Winsor 1991: 108), the predecessor of the New York Academy of Sciences. In 1848 he was inducted as a fellow of the Academy of Medicine of New York, the year the Academy was founded.

After his brief return to Germany and establishment in New York as a physician, he apparently abandoned the study of cave fauna altogether. He may still have had some interest in natural history. In the Archives of the Museum of Comparative Zoology, at Harvard, there is one letter by "Theo A. Tellkampf" to William Greene Binney[371] a malacologist graduated from Harvard. In this letter dated 14 Dec. 1867, Theodor discusses ascidian (sea squirt) anatomy but says that he is in no position to help Binney with his research. His name and New York address appeared in the Naturalists' Directory, part I, 1865, edited by F.W. Putnam (p. 28.) and described him as an expert in "Ascidians, Histology."

Theodor also achieved certain notoriety as a physician. For example, Heinrich Schliemann, who discovered the ruins of Troy, says that he used a formula based on quinine devised by "Tellkampf, the German doctor from New York" in order to fight fevers (Schliemann 1995).

As far as I can tell, this is the first biography on Theodor Tellkampf. Some biographical notes have been published as short obituaries in medical journals, some of them inaccurate and never with any reference to his speleological work.

Acknowledgments

Uwe Kunert, from Hamburg, Germany, provided me with very valuable, first-hand information about Theodor and his family. James Cokendolpher provided useful information about Tellkampf' s contributions to arachnology. The following people provided me with valuable information: Jocelyn K. Wilk, Assistant Archivist, Columbia University Archives & Columbiana Library; Toby A. Appel, Yale University; Dennis B. Worthen, Executive Director of The Lloyd Library and Museum in Cincinnati; Mary Person, Curatorial Associates, Harvard Law School Library; Melanie M. Halloran, Reference Assistant, Harvard University Archives, Pusey Library; Ed Morman, Associate Librarian, New York Academy of Medicine; Robert Young, Ernst Mayr Library, Harvard University; Lynn Nyhart, University of Wisconsin-Madison; Fred Churchill, Indiana University. U. Kunert, J. Cokendolpher and Kelly M. Paulson read an earlier version of the manuscript and made valuable suggestions.

---

[369] *b*. Brunswick, Maine, 19 February 1839; *d*. Providence, Rhode Island, 14 February 1905.
[370] *b*. Salem, Massachusetts, 16 April 1839; *d*. Cambridge, Massachusetts, 14 August 1915.
[371] *b*. Boston, Massachusetts, 22 October 1833; *d*. Burlington, New Jersey, 3 August 1909.

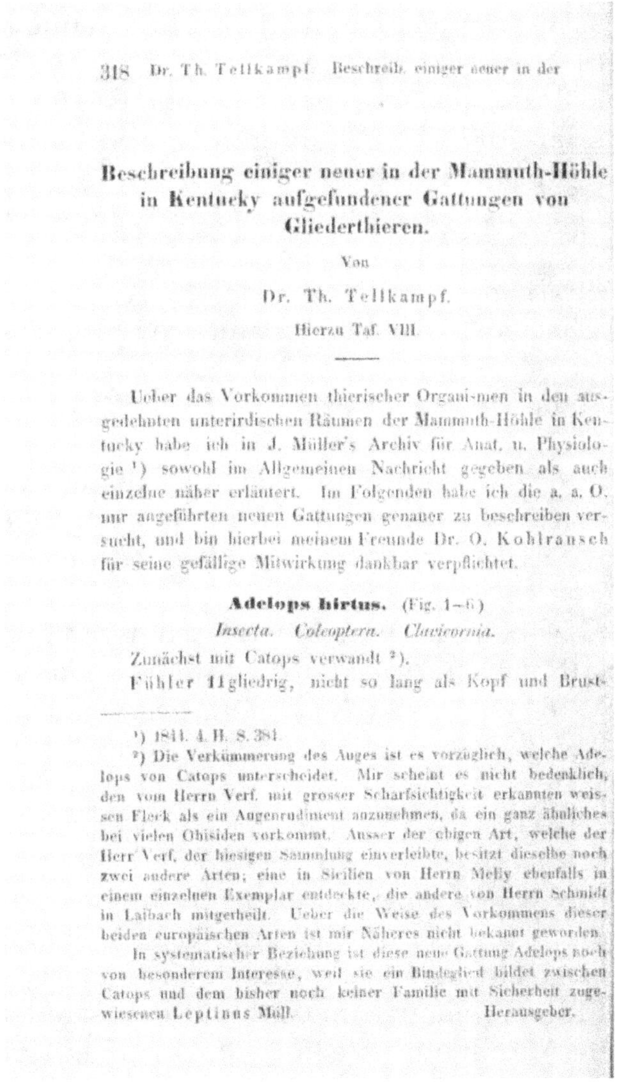

318 Dr. Th. Tellkampf. Beschreib. einiger neuer in der

Beschreibung einiger neuer in der Mammuth-Höhle in Kentucky aufgefundener Gattungen von Gliederthieren.

Von

Dr. Th. Tellkampf.

Hierzu Taf. VIII.

Ueber das Vorkommen thierischer Organismen in den ausgedehnten unterirdischen Räumen der Mammuth-Höhle in Kentucky habe ich in J. Müller's Archiv für Anat. u. Physiologie [1]) sowohl im Allgemeinen Nachricht gegeben als auch einzelne näher erläutert. Im Folgenden habe ich die a. a. O. nur angeführten neuen Gattungen genauer zu beschreiben versucht, und bin hierbei meinem Freunde Dr. O. Kohlrausch für seine gefällige Mitwirkung dankbar verpflichtet.

Adelops hirtus. (Fig. 1–6)

Insecta. Coleoptera. Clavicornia.

Zunächst mit Catops verwandt [2]).

Fühler 11gliedrig, nicht so lang als Kopf und Brust-

[1]) 1841. 4. H. S. 381.

[2]) Die Verkümmerung des Auges ist es vorzüglich, welche Adelops von Catops unterscheidet. Mir scheint es nicht bedenklich, den vom Herrn Verf. mit grosser Scharfsichtigkeit erkannten weissen Fleck als ein Augenrudiment anzunehmen, da ein ganz ähnliches bei vielen Ohisiden vorkommt. Ausser der obigen Art, welche der Herr Verf. der hiesigen Sammlung einverleibte, besitzt dieselbe noch zwei andere Arten; eine in Sicilien von Herrn Melly ebenfalls in einem einzelnen Exemplar entdeckte, die andere von Herrn Schmidt in Laibach mitgetheilt. Ueber die Weise des Vorkommens dieser beiden europäischen Arten ist mir Näheres nicht bekannt geworden.

In systematischer Beziehung ist diese neue Gattung Adelops noch von besonderem Interesse, weil sie ein Bindeglied bildet zwischen Catops und dem bisher noch keiner Familie mit Sicherheit zugewiesenen Leptinus Müll. Herausgeber.

Figure 15.1. First page of Tellkampf's first scientific publication on Mammoth Cave fauna. *Adelopus histus* is a cave beetle.

Literature Cited

Anonymous. 1843. The German Professorship of Columbia College. Typescript at the Columbia University Archives.

Anonymous. 1876. "Information..." Acta Columbiana n/p. at the Columbia University Archives. Photocopy

Danton, G.H. 1946. "A smart flippant little fellow..." Johann Ludwig Tellkampf. *N.Y. History* (October):1-19.

Juettner, 0. 1909. *Daniel Drake and his followers. Historical and biographical sketches.* Cincinnati: Harvey Publishing Co., 496 pp.

Putnam, F. W. (ed.). 1865. *The Naturalists directory*. Salem: Essex Institute, 2 vols.

Romero, A. 1986. He wanted to know them all: Eigenmann and his blind vertebrates. *National Speleological Society News* **44**(11):379-381.

Schliemann, H. 1995. Alla scoperta di Troia. Ed. Grandi Tascabili Newton (http://digilander.iol.it/egidiosiviglia/scoperta.htm).

White, W.T. (Ed.). 1884. *The medical register of New York, New Jersey and Connecticut, for the year commencing June 1, 1884*. New York: G.P. Putnam's Sons.

Winsor, M.P. 1991. *Reading the Shape of Nature. Comparative Zoology at the Agassiz Museum*. Chicago: University of Chicago Press. 324 pp.

Original Publications by T. Tellkampf

Tellkampf, T.A. 1844a. Beschreibung e1mger neuer in der Mammuthöle in Kentucky aufgefundener Gattungen von Gliederthieren. Taf. viii. *Archiv des Vereins der Freunde der Naturgeschichte in Mecklenburg* **10**:318-322.

Tellkampf, T.A. 1844b. Uber den blinden Fisch der Mammuthöle in Kentucky. (Muller's) *Archiv fur Anatomie und Physiologie* **4**:381-395.

Tellkampf, T. A. 1844c. *Ausflug Nach der Mammuthöhle in Kentucky. Das Ausland*. **1844**: 671-672, 675-676, 679-680, 683-684, 687-688, 691-692, 695-696, 699-700.

Tellkampf, T.A. 1844d. *Die Straaflinge in den amerikanischen und englischen Besserungsgefangnissen. Nebst Bemerkungen tiber den Gesundheitszustand der Straflinge in den obigen Anstalten* von Theodor Tellkampf. Appended to Tellkampf, J.L. Uber die Besseru ngsge fa angnisse in Nordamerik und England Nach elgenen Beobachtunge in den Jahren 1838 bis 1843.

Tellkampf, T.A. 1845. Memoires on the blind-fishes and some other animals living in the Mammoth cave in Kentucky. *New York Medical Journal or New York Journal of Medicine*. **1845**:84-93.

Tellkampf, T.A. 1847, German views of English critics. *The American Review* **6**(5):497-503.

Tellkampf, T.A. 1870. Note respecting the eyes of *Amblyopsis spelaeus*. *Annals of the Lyceum of Natural History of New York* **9**:150-152.

Translations by T. Tellkampf

1847. Proofs that the periodic maturation and discharge of ova are, in the mammalia and the human female, independent of coition, as a first condition of their propagation (also titled Tracts on generation) translated with Chandler Robbins. Original title "Beweis der von der Begattung unabhängigen periodischen Reifung und Loslösung der Eier der Saugethiere und des Menschen" by T.L.W. Bischoff. New York: Samuel S. & William Wood. 65 pp.

Taxa Dedicated to Him

Neaphaenops tellkampfi is a troglobitic ground beetle who feeds exclusively on cave cricket eggs, which it sniffs out and digs up.

Chapter 16. Felipe Poey and the Myth of the "Isolated Genius" [372]

Summary

Felipe Poey y Aloy was a Cuban naturalist and intellectual who greatly contributed to the advance of science in Cuba. Despite being an autodidact, he showed a great deal of competence as a scholar, teacher, and organizer of academic institutions in both the public and private spheres. Although being portrayed as an almost "isolated genius," he maintained a very broad and vigorous network of communications with scientists all over the world. He always kept an open mind and adapted to new scientific ideas. Although the narrative of his life and accomplishments has been the subject of much hagiography, we contend that such narratives have obscured, rather than illuminated, his contributions to society and the social, political, and economic contexts under which he developed his work.

Introduction

There are many hagiographies of Spanish or Latin American scientists who lived before the 20th century written by either Spanish or Latin American authors and published as recent as the middle of the 20th century. They are in the literary tradition of religious biographies, particularly prevalent among Catholic authors who mixed both Spanish mysticism as well as nationalism. Parts of those narratives include a subtext in which the scientists are described as "isolated geniuses" who achieved great scientific accomplishments almost exclusively out of some sort of unprecedented inborn virtuosity while being more or less isolated from the scientific networks of their times.

Glick and Quinlan (1975) argued that the myth of the isolated genius in Spanish science was false. Others (Beddall 1983) have stated that "not all Spaniards, or even all Spanish geniuses, are or feel themselves to be isolated…" but that in some cases (e.g., Félix de Azara[373]) they worked "under circumstances of social, scientific, and geographic isolation that by most standards would be considered fairly extreme."

Sometimes the Cuban naturalist Felipe Poey has also been heralded as an example of such isolated geniuses (e.g., Carbonell y Rivero 1928:7-9). The aim of this paper is to discuss the circumstances of the development of Poey's work within both the international and Cuban context from a scientific networking viewpoint. I will argue that despite his very valid and important achievements as a scientist, the way he has been portrayed by some authors until relatively recently (1) did not help to better understand the real significance of his achievements, (2) deemphasized the way he networked his communications with other intellectuals his whole life, and (3) obscured the very fact that he was a very open-minded, autodidact intellectual who did the best he could to keep up to date with new scientific ideas under the circumstances under which he lived.

Poey's Life and Career

For the purpose of the discussion on the impact of hagiographies on the perception of Poey as a scientist in general and as an "isolated genius" in particular, I summarize

---

[372] Romero, A. 2014bg. Felipe Poey, hyperbole, and the myth of the "isolated genius" among Spanish and Latin American scientists. *Polymath* **4**(4):85-95.
[373] *b*. Barbuñales, Huesca, Spain, 18 May 1746; *d*. Barbuñales, 20 October 1821.

below the biographical aspects of Felipe Poey relevant to that discussion. For a more complete biography of Poey that includes numerous primary sources see González López (1999)[374].

Felipe Poey y Aloy (hence Poey, po'-ay)[375] was the son of Juan Andrés Poey y Lacase[376], a Frenchman who while in Cuba was involved with his family in slave trade and María del Rosario Aloy y Rivera[377] of Spanish and Cuban ancestry and a member of the Cuban creole gentry through the ownership of plantations. In 1804, at the age of five, Poey went with his family to Pau, France, a small town in southwestern France not far from the border with Spain. There Poey started his primary education and after the death of his father, less than two years after they arrived to France, his mother moved back to Havana, leaving Felipe in a boarding school in Pau for three more years.

While there he was struck by polio, which partially paralyzed the right side of his body, after which he returned to Cuba. That condition obligated him to write with his left hand and in vertical fashion while his drawings were based on outlines drawn directly from placing the specimens on a piece of paper which resulted in many life-size illustrations. Additionally, this ailment impaired him from fieldwork, making him reliant on others to secure specimens for his work as naturalist (Boss & Jacobson 1975, Romero 2007). Yet, this very disability may have decanted his interest for nature at an early age: as an infant he spent long time on his belly watching the activities of ant colonies (Sanchez Roig 1955:50).

Back in Havana, he attended the *Real Seminario de San Carlos y San Ambrosio*. This seminary was open to people interested on either following a career in the clergy or pursuing secular careers such as law, political economics, mathematics, and the natural sciences. Another characteristic of that seminary was that they rejected scholastic approaches to teaching. There, Poey studied philosophy under Félix Varela y Morales[378] a distinguished priest, politician, philosopher, and educator who introduced new teaching methods in Cuba while establishing the first teaching laboratories for physics and chemistry in that country (Altshuler & Baracca 2005).

Poey received his bachelor's degree in law in 1820. That year he joined the *Real Sociedad Económica de La Habana* (formerly *Sociedad Patriótica*, later *Sociedad Económica de Amigos del País*). This was a learned society of Cubans interested in many subjects but mostly economic, cultural, and social aspects of the island. One of the major activities of this society was to send to other countries commissions that could bring back the latest advances in many different areas of knowledge. Among their major accomplishments were the establishments of the first public library in Cuba, the botanical gardens, the *Escuela Nacional de Bellas Artes "San Alejandro"*, and a number of educational reforms. For a man with the interests of Poey, the *Sociedad* was a place to be, particularly because his former teacher and mentor, Félix Valera, was an active member of that institution. It was then when he began making collections of plants and animals. It was also Valera who encouraged Poey to contribute articles to *El Observador Habanero*, an ephemeral periodical (1820-1822) devoted to politics, the sciences, and literature. There he began his career as a popular science writer, something he would continue doing for the rest of his life.

To please his mother he went to Madrid, Spain, to enhance his studies in law and received his degree with honors as doctor of law at the *Universidad Complutense* in 1822. For a while he taught law in the *Real Academia de Jurisprudencia y Legislación* in Madrid, which at that time had a strong liberal leaning along the lines of his mentor Varela. He

---

[374] Additional general biographical information on Poey can be found in Jordan 1899, Sánchez Roig 1937, Chardon 1947, Vivanco y Díaz 1951, Cruz 1979.

[375] *b*. Havana, Cuba, 26 May 1799; *d*. Havana, 28 January 1891.

[376] *b*. Estos, Pyrénées-Atlantiques, southwestern France; *d*. Pau, Pyrénées-Atlantiques, France, 26 February 1806.

[377] *b*. Havana, Cuba, 6 October 1783; *d*. ?

[378] *b*. Havana, Cuba, 20 November 1788; *d*. St. Augustine, Florida, U.S.A., 27 February 1853.

marginally became involved in politics (he was a liberal sympathizer). At that time the liberal movement of Spain was a bourgeois, nationalist one characterized by an anti-absolutist monarchism, somewhat anticlerical, and welcoming to the idea of the French Enlightenment. In 1823 after the French army invaded Spain and restored Ferdinand III as an absolutist monarch, many liberals fled Spain, mostly for London and France. Varela, who had moved to Spain in 1821 after being chosen to represent Cuba before the Spanish parliament, fled for Gibraltar first and the United States later. Poey returned to Cuba that year and in 1824, married[379], and decided to become a full-time naturalist. That was something he could afford despite his and his wife's somewhat modest wealth. By that time (1822) the Spanish parliament (*Las Cortes*) had commissioned the travel by some Spanish naturalists to explore Cuba, Puerto Rico, and the Philippines to survey their natural resources. They also established the first professorship (*cátedra*) of natural history in Havana and to occupy that position the botanist, politician and writer Ramón de la Sagra[380] was named (Barreiro 1944:120).

Poey never practiced law, a very lucrative profession in Cuba at that time, and supported himself by teaching at the *Colegio of San Cristobal de La Habana o de Carraguao*, well known for its pedagogical innovations. There he taught geography, Latin, and French at the primary school level.

In 1826 Poey traveled to Paris, France, to officially study law at *La Sorbonne*. Yet, he had another agenda since he carried with him specimens of plants, insects, and other animals, including 35 specimens of Cuban fishes in a barrel of brandy plus 85 drawings of fishes from Cuba. Although he might have obtained a law degree, he spent a considerable amount of time working with the famous naturalist Georges Cuvier[381] at the *Muséum national d'Histoire naturelle*. At that time Cuvier and his student Achile Valenciennes[382] were working on the encyclopedic *Histoire naturelle des poisons* (The Natural History of Fishes), which would end up being a 22-volume publication and the most comprehensive reference on fishes for many years. Poey's information about the Cuban fishes was incorporated into Cuvier and Valencinnes's gigantic work. In addition to his work on fishes while in France, he was very active in other scientific circles becoming a founding member of the *Société Entomologique* in 1827. While in France, he published *Centurie des Lépidoptéres de l'ile de Cuba* (a monograph on the butterflies and moths of Cuba, Paris, 1832)[383].

After his return to Havana in 1833 Poey devoted himself entirely to the study of natural history. It is possible that by that time his mother had passed away and he abandoned any pretense of pursuing a career in Law. He did a number of drawings of specimens with his associate, Juan Cristóbal Gundlach[384], while discovering many new species of mollusks that are included Ludwig Karl Georg Pfeiffer's[385] *Monographia Heliceorum Viventium* (1847). In 1836 he published the first of his 18 editions of the *Compendio de la Geografía de la Isla de Cuba*, a textbook on the geography of Cuba.

---

[379] He married Maria de Jesús Aguirre y Hornillos with whom he had six children. His first son, Andrés Poey y Aguirre (*b*. La Habana, Cuba, 15 November 1825; *d*. Paris, France, 4 January 1919) was a Paris-educated meteorologist, naturalist, and archaeologist. He was the director of the *Observatorio Físico-Meteorológico de La Habana* and during the reign of Maximilian he was the founder of an establishment of the same kind of facility in Mexico. He practiced positivism, a quite popular philosophical current in Cuba's second half of the nineteenth century.

[380] *b*. A Coruña, Galicia, Spain, 8 April 1798; *d*. Neuchâtel, Switzerland; 23 May 1871.

[381] *b*. Montbéliard, France, 23 August 1769; *d*. Paris, France, 13 May 1832.

[382] *b*. Paris, France, 9 August 1794; *d*. Paris, 13 April 1865.

[383] The full title was Centurie des lépidoptéres de l'ile de Cuba: contenant la description et les figures coloriées de cent espèces de papillons noveaux et peu connus. représentés d'apres nature, souvent avec la chenille, la chrysalide et plusieurs détails microscopiques.

[384] née Johannes Cristoph Gundlach, *b*. Marburg, Germany, 17 July 1810; *d*. Havana, Cuba, 14 March 1896.

[385] *b*. Kassel, Hesse, Germany, 4 July 1805; *d*. Kassel, 2 October 1877.

In 1842 Poey founded the *Museo de Historia Natural de La Real y Pontificia Universidad de La Habana* (the museum of natural history of the University of Havana) after several years of planning. The University of Havana was run by a Dominican order and was not secularized until 1842 (Romero 2014). It was not until 1863 that the faculty of sciences was created. Because of shortage of funding, they lack labs. On 1876 the university was reinvigorated and the library and a larger natural history museum were organized (Álvarez Conde 1958:204, Pruna 1994)[386]. The very same year the museum was founded (1842), he was appointed professor of comparative anatomy and zoology at the university despite the fact that he lacked formal education in the natural sciences (Sanchez Roig 1955:36-37). By the 1850's he was already recognized as the most erudite naturalist in Cuba. Between 1851 and 1861 he published the two volumes of his *Memorias sobre la Historia Natural de la Isla de Cuba*[387].

In 1861 Poey and his son Andrés appear as founding members of the *Real Academia de Ciencias Médicas, Físicas y Naturales de la Habana*[388] envisioned as a place for experimental learning. The creation of the *Academia* was the result of a number of political and personal circumstances. That institution was not really established to advance science in Cuba but rather to improve relations between the Cuban elites and the metropolis. The Poeys were two of the five members from the natural history section. Other five members were from pharmacy and twenty were physicians.

In 1863 he was commissioned by the *Capitán General* (top representative of the Spanish government) of Cuba to gather specimens for the collections of the *Museo de Historia Natural de Madrid*. That took place during a period (1859-1866) when the *Museo* was making a number of efforts to increase its collections from abroad with uneven success (Barreiro 1944:285-300). In that same year he was appointed Chair of botany, mineralogy, and geology, and later published the *Repertorio Físico-Natural de la Isla de Cuba* (1865-1868). This was a compilation of articles on various fields of natural history by him and other Cuban naturalists. The ichthyological papers were reprinted in the *Anales de la Sociedad de Historia Natural de Madrid*, under the title *Synopsis Piscium Cubensium*, or *Catálogo razonado de los Peces Cubanos* (Annotated Catalog of the Fishes of Cuba, 1868).

By the 1870's he was a very prolific and accomplished author with publications not only on ichthyology, entomology, and malacology, but also in the areas of anthropology, geography, and geology (he also published poems). Those publications had been written in Spanish, French, and English and had been printed in both Cuba and abroad[389].

In 1873 Poey was appointed Dean of the Faculties of Philosophy, Letters, and Sciences, but in 1880, he retained only the post of Dean of Sciences. In his last years at the university, he taught Vertebrate Zoology and the Zoography of the Vertebrates. His greatest work was probably *Ictiología Cubana*. The text, originally written in French, was completed in 1877 and the Atlas in ten large folio volumes plus a supplement, in 1878. The atlas, drawn by him, consisted of 1040 plates with 1300 figures in color and natural size (except sharks) illustrating 758 species of Cuban fishes. The manuscript was presented at the Amsterdam Fair in 1883 and gained the author a gold medal and

---

[386] That museum was moved to its current location at the *Universidad de La Habana* in 1930 and was renamed *Museo de Historia Natural Felipe Poey*.

[387] This work is complicated to cite, particularly when it comes to dates. This is a collection of papers in two volumes. The first set of papers was published between 1851 and 1854 and the second set between 1858 and 1861.

[388] Replaced in 1962 by the *Academia de Ciencias de Cuba* created in 1962. Today it publishes the scientific periodical *Poeyana*.

[389] Despite cataracts at old age, together with his partial paralysis, he never stopped writing until his death. He also published on archaeology, philosophy, linguistics, and jurisprudence while modernizing the medieval methods of education in 19th century Cuba (Dihigo 1915).

honorable mention. King William III conferred on him the decoration of the order of the "Lion Neerlandais" (Dutch Lion) and the King of Spain distinguished him with the title *Comendador de la Order de Isabel la Católica* (Commander of the Order of [Queen] Elizabeth the Catholic).

Although the British Museum and the Smithsonian Institution both bid high prices for the work, Poey preferred to sell it at a much lower price to the Spanish government, which acquired it in 1885 and deposited at the *Biblioteca de Ciencias Naturales* at Madrid (Barreiro 1944:348). In the 1930's, the government of President Gerardo Machado in Havana set aside $10,000 to arrange for its publication, and although two prominent naturalists, Dr. Carlos De La Torre y Huerta[390] and Felipe García Cañizares[391], worked to prepare it for publication, this was prevented, as Álvarez Conde (1958: 232) writes, by "crisis económicas sucesivas" (successive economic crises). In 1955, under the editorship of Mario Sanchez Roig and Federico Gomez de la Maza, a first volume in Spanish of 372 pages appeared in Havana, containing an introduction and part of the text (Sanchez Roig 1955). Further efforts were planned to print Poey's *Ictiología* (Duarte Bello 1962) but it was not finally published in its entirety until 2000 (Poey 2000).

Poey's Networking

There is little question that Poey developed an extensive academic network around the world through correspondence, membership, and other means that included, but was not limited to, the following institutions (alphabetically): Academy of Natural Sciences of Philadelphia (Philadelphia, Pennsylvania), American Academy of Arts and Sciences (Cambridge, Massachusetts), British Museum (London, England), Buffalo Society of Natural Sciences (Buffalo, New York), Entomological Society of Philadelphia (Philadelphia, Pennsylvania) , Essex Institute (Salem, Massachusetts), *Königliche Akademie der Wissenschaften* (Berlin, Germany), Museum of Comparative Zoology at Harvard University (Cambridge, Massachusetts), Lyceum of Natural History of New York (New York), *Real Academia de Ciencias Exactas, Físicas y Naturales* (Madrid, Spain), *Real Academia de Ciencias y Artes de Barcelona* (Barcelona, Spain), *Real Museo de Ciencias Naturales* (Madrid, Spain), *Real Sociedad Española de Historia Natural* (Madrid, Spain), Royal Zoological Society of London (London, England), *Société Entomologique de France* (Paris, France), and the Smithsonian Institution (Washington, DC). He even published some of his works in the journals of some of those institutions.

In those centers, as well as with independent naturalists, he maintained an extensive correspondence with many scientists from those countries as well as from Belgium and Switzerland. A list of those scientists can be found in Vivanco y Díaz (1951), De La Torre (1942:314-315), and González Lopez (1999:186-211, 239). Those scientists were mostly ichthyologists, entomologists and malacologists. His letters were very detailed showing a great deal of competence on what he was doing[392]. He also hosted a number of foreign scientists visiting Cuba such as Juan Cristóbal Gundlach[393], David

---

[390] *b*. Matanzas, Cuba, 15 May 1858; *d*. 19 February 1950, Havana, Cuba.

[391] *b*. Sancti Spiritus, Santa Clara, Cuba, 14 July 1872; *d*. 1953.

[392] Thanks to Susan Sanctis of the Museum of Comparative Zoology at Harvard University for providing copies of some of the letters written by Poey to H.A. Hagen

[393] *b*. ? July 1810, Marburg, Hesse, Germany; *d*. 14 March 1896, Havana, Cuba.

Starr Jordan[394], Carl Friedrich Eduard Otto[395], Ludwig Karl Georg Pfeiffer[396], and Charles Wright[397]. Some of them wrote later on about Poey as both as a scientist and as a person.

As mentioned earlier, within Cuba he was a founder/president/member of numerous institutions in addition to his university posts. Furthermore, he was the public face of science in his country through his numerous articles printed in the popular press of the island[398] (see Poey y Aloy 1888).

Posthumous Recognition

Another measure of whether or not his work was appreciated during his lifetime is in the recognition he received shortly after his death. On 5 June 1907 his remains were exhumed from the Colón cemetery in Havana and reburied in the vestibule of the building of the Faculty of Sciences, which was named in his honor. On 15 January 1909, a mausoleum was raised there with a long Latin inscription and a marble bust of the scholar Fig. (16.1). On 26 May 1913, the *Sociedad Cubana de Historia Natural "Felipe Poey"* was founded and lasted until 1962. It counted as members some of Poey's students and admirers and dedicated itself to the study of zoology, botany, anthropology, mineralogy, geology, paleontology and agronomy, roughly the same areas of interest for Poey (Mestre 1918). It published between 1915 and 1961 the *Memorias de la Sociedad Cubana de Historia Natural Felipe Poey*, which contained some of the most significant studies of Cuban natural history (Montané 1918). Today the *Academia Nacional de Ciencias de Cuba* publishes a periodical titled *Poeyana* in his honor.

Adoption to New Ideas

It is not uncommon that some scientists lose standing during their lifetime among their peers by refusing to accept new ideas that later became mainstream. The most revolutionary and controversial biological idea of the 19th century was Charler Darwin's theory of evolution by means of natural selection.

Although ambivalent at the beginning (Pruna 1994), Poey went on to become an evolutionist based on his studies of blind cave fishes and his belief that cave colonization was not the result of "accidents" but rather an active process (Romero 2007). When the works of Charles Darwin[399], Thomas Henry Huxley[400], and Herbert Spencer[401] were published, he embraced them enthusiastically. He seemed to be particularly impressed with Spencer (Jordan 1899) probably because of Spencer's progressionist ideas. We must remember that Poey was a disciple of Cuvier, a staunch creationist, and an admirer of another creationist and Darwin's rival, Jean Louis Rodolphe Agassiz[402]. Yet, the Cuban naturalist read the arguments after Darwin's publication of *The Origin of Species by means of Natural Selection* (1859) and started to shift his opinion about the notion of evolution.

Thus, by 1868 he started to express doubts toward creationism and by 1886 he had fully embraced Darwin's ideas, a transformation of thought that was parallel to changes in his own religious ideas from being a Catholic to become agnostic and later an atheist

---

[394] *b*. Gainesville, New York, U.S.A., 19 January 1851; *d*. Stanford, California, U.S.A., 19 September 1931. He was one of the towering figures of ichthyology and university administration of his time.

[395] A German botanist and malacologist, *b*. 1812; *d*. 1885.

[396] *b*. Kassel, Hessel, 4 July 1805, Kassel, Hesse, Germany; *d*. 2 October 1877, Kassel, Hesse-Nassau, Germany.

[397] American botanist and collector, *b*. Wethersfield, Connecticut, U.S.A., 29 October 1811; *d*. Wethersfield, 11 August 1985.

[398] see De La Torre (1942), González Lopez (1999), and Vivanco y Díaz (1951), for the titles of those publications.

[399] *b*. Shrewsbury, England, 12 February 1809; *d*. Downe, Kent, England, 19 April 1882.

[400] *b*. Ealing, west London, England, 4 May 1825; *d*. Eastbourne, England, 29 June 1895.

[401] *b*. Derby, England, 27 April 1820; *d*. Brighton, England, 8 December 1903.

[402] *b*. Motier-en-Vuly, Switzerland, 28 May 1807; *d*. Cambridge, Massachusetts, 14 December 1873.

(Pruna and González 1989:35, 63-64, 181). For Poey this intellectual development took place rather smoothly. After all, the introduction of Darwinian ideas in Latin America was progressive and without much resistance at least among academic circles (for an overview of this issue see Pruna & García González 1989).

As Jordan (1884) stated "There is no characteristic of Professor Poey's work more striking than his entire lack of prejudice, or, in other words, his teachableness. (…) Among all the eminent zoölogists of our time, I know of none so ready to learn, whatever the source from which information may come. He has no theories which he is not ready to set aside when a better suggestion appears."

Figure 16.1. The author next to Poey's statue at his mausoleum in the patio of the University of Havana. Picture by Ana Romero.

Discussion and Conclusions

Autodidactism

Despite the fact that Poey's education in law were of first order in three different institutions of higher education in three different countries, he never had a formal education on what was his real profession: naturalist, and that is not surprising. The very term biology was not used for the first time until 1802 (McLaughlin 2002). In fact, for the first half of the nineteenth century there were very few people around the world making a living out of being biologists, and there were usually associated with either universities or museums of natural history, and that happened only in European countries and the United States. Therefore, during Poey's lifetime to acquire a formal education in life sciences was far from a guarantee for a steady income. The profession of biologist was yet to be institutionalized globally and more so in Cuba. Although he spent time with some of the most distinguished zoologists of his time, he seemed to learn more by intellectual osmosis and reading than by any other method.

Even in Cuba he lacked some role models to follow. For example, his only real predecessors in the study of Cuban fishes were Antonio Parra y Callado[403] who published in 1787 *Descripción de Diferentes Piezas de Historia Natural* (Description of Different Items of Natural History) describing 71 species of fishes and Ramón de la Sagra who published *Historia física, política y natural de la Isla de Cuba* (Physical Political, and Natural History of the Island of Cuba) (1838 for the French edition and 1857 for the Spanish edition)[404]. Parra had been commissioned to collect plants and animals to be sent to the *Reales Gabinetes de Historia Natural y Jardín Botánico* in Madrid for which he was paid (Chardon 1949:231, García González 1995) and his book, although interesting for its time, was a far cry from the type of methodology for scientific descriptions that Poey had to use during his time. Parra's book was rather an inventory of species with some narrative about their natural history while Poey had to use the rigor imposed after Linnaeus's 10th Edition of his *Systema Naturae* (1758). As per de la Sagra's book, as mentioned earlier, the section on fishes was written by someone who had never been to Cuba.

When it comes to colleagues in Cuba during his lifetime, things were not much better even when he was a professor at the University of Havana. According to Jordan (1884) by that time there were only two students pursuing botany and probable the same number in zoology. The library resources Poey used in Cuba were mostly his and he had the only zoological laboratory in the island.

Isolation

Poey would have been the first to recognize that he had a wide network inside and outside Cuba and that his intellectual background was very wide and shaped by whatever he could learn by means of either reading published literature or through personal correspondence. Because of the dates and places of birth and death he was technically a Spaniard, not matter how Cuban he felt. That was reinforced by the fact that during his 91 years of life he spent only a few years outside Cuba (in France and Spain). Yet, he considered himself "cosmopolitan." Jordan (1922: vol. 1: 285) who met him in 1884 mentioned that he used to say: *Comme naturaliste je ne suis pas espagnol – je suis cosmopolite*[405].

In terms of historical periods, Poey fits into the second of the tree periods described in Basalla (1967) when it comes to the history of science in developing countries. For Basalla there were three periods: (1) exploration by Europeans, (2) period of dependence from European Science, (3) independence from it. In many ways Poey played all three roles but, mostly, he was part of the classical for period (2). Despite his physical disabilities he was able to secure a great deal of specimens and observations that were new to science, just as other European explorers would do for period (1). For that he had to overcome lots of the challenges faced by earlier explorers such as the lack of local institutions and infrastructure to support him at the beginning. He also initially lacked local colleagues with whom he would share his interests although later on he did have a few students such as De La Torre with who he could work. We must remember that when he died, he was still a subject of the Spanish crown living in a Spanish colony, and nationalism did not play a part in a transition to a true intellectual independence since he

---

[403] *b*. Tavira, Portugal, 27 June 1739; *d*. ? (García González 1989, pp. 79-82).

[404] The section of fishes is on volume 4, pages 145-255, and was actually written by Antoine Alphone Guichenot (*b*. Paris, France, 31 July 1809; *d*. Cluny, France, 17 February 1876). He was a zoologist who collected specimens for the *Muséum national d'histoire naturelle* in Paris as an assistant to the Chair of reptiles and fishes. He never visited Cuba.

[405] "As a naturalist I have never been a Spaniard, I have been cosmopolitan."

was still subject of whatever intellectual currents were prevalent in both the U.S. and Europe. Also, because Cuba, as a Spanish colony, had had so little tradition in integrating science into society, the national context in which he worked was almost void. We are not sure whether or not he found resistance to science as an additional obstacle. What we do know is that he was respected because of his personality and expertise on fishes but was probably seen as a respectable curiosity by the Cuban society at large. Although he was given university posts and the Spanish government did ask him to collect specimens to be sent to Madrid, Poey was most likely chosen for those roles by default given that there were no other naturalists of his prestige in Cuba at that time. So, his social role was not that important except for his position as educator and founder of some institutions in civil society. It is not clear that his work received much of state financial aid or encouragement.

These contentions are supported by a comment made by Jordan after visiting him in Cuba in 1884: "Although Professor Poey is evidently held in very high respect in the university, in which he has long been dean of the faculty of science, I cannot imagine that he ever received much help or sympathy in his scientific work from that quarter, or indeed from any other in Cuba. His friends and countrymen are doubtless glad to be of assistance to so amiable a gentleman as the Señor Don Felipe, but for the claims of science the people of Cuba, as a class, care very little" (Jordan 1884).

Poey did introduced and/or expanded the teaching of science at all levels as an author of textbooks, university administrator, and as a teacher himself. He founded and/or participated in all major scientific organizations of his time that were aimed at promoting the sciences. He clearly maintained channels of formal and informal national and international scientific communication and the use of the languages most employed at that time in science (French, English) were not a problem for him.

Given all of the above it is hard to argue that Poey was an "isolated genius." On the contrary, he was as well connected and as well-known as any other scientist of his generation among his peers. Although he had a few students who were competent at what they did, none of them –and no other contemporary biologist in Cuba for that matter- came close to him when it came to knowledge and insight on both the natural history of the island or insights on biological ideas as a whole. Thus, he may have felt internally isolated from an intellectual viewpoint but externally connected through his network of international contacts.

Hagiography

Hagiography of Spanish and Latin American scientists (and Poey was both since Cuba did not gain independence from Spain until 1898, seven years after his death) predating the 20th century is quite common. For example, to describe the Venezuelan mathematician and astronomer Juan Manuel Cagigal y Odoardo[406] the following rhetoric has been used: "On November 22, 1844, arrives back to Caracas. He is already a sleepwalker. And that colossus who had in the brain a lighthouse, which flashed by itself as many glares, which is always recreated with the geometric poem of the stars and toured the brilliant course of the constellations, died in the darkness of dementia"[407] (Grisanti 1956:43).

Descriptions of Poey along similar lines are not much different including claims that he preceded Darwin in his evolutionary ideas and others in the discoveries of fossil

---

[406] *b*. Barcelona, Venezuela, 10 August 1802; *d*. Yaguaraparo, Venezuela, 10 February 1856.
[407] The original is Spanish reads as follows "El 22 de noviembre de 1844 arriba de nuevo a Caracas. Es ya un sonámbulo. Y aquel coloso que tenía en el cerebro un faro, que despidió de sí tantos fulgores, que se recreó siempre con el poema geométrico de las estrellas rutilantes y recorrió el fulgurante derrotero de las constelaciones, murió en las tinieblas de la demencia."

humans, claims for which there are not any basis: "He had the glory to anticipate other scholars claiming the existence of fossil man before prehistory was recognized by science; and before Darwin made known his theory of the 'Origin of the species,' he established the doctrine of 'continuous is the transformism', principle on which rests the evolution in the formation of the organic species"[408] (Carbonell y Rivero 1928:8).

This trend regarding Poey had been noticed by others. Boss & Jacobson (1975), for example, wrote "In an effulgent fashion, the Cuban writers (Dihigo, 1915; Jaume, 1955; Mestre, 1891; Montane, 1918; Torre, 1942) shower reams of almost hysterical praise upon Poey. At times the reader has to struggle through many pages of dithyrambic prose before he can garner a single sober fact. Nevertheless, it must be admitted that this talented and industrious man added much to the intellectual life of his country. He contributed significantly to the knowledge of Cuba's bewilderingly rich fauna and he deserves to be better known."

As a Scientist and Intellectual

Poey's contributions to science were important, widely published, and well known by many both in Cuba and abroad. He belonged, founded and/or presided over the most important scientific and intellectual societies of his time in both Cuba and abroad while receiving a number of international recognitions. The list of people with whom he corresponded in scientific matters, reads like the "who is who" of researchers of his time in the areas of ichthyology, ornithology, entomology and malacology. Foreign naturalists visiting Cuba not only met with him but also wrote articles about him as a scientist and as a person. His writings reveal that he was open to (and later adopted) the scientific ideas such as the Darwinian theory of evolution by means of natural selection. Despite not having a formal education in the sciences his knowledge and pedagogical abilities were recognized to the point that he ended his career as Dean of the School of Sciences at the Universidad de La Habana.

Thus, how can we best define Poey as a scientist? We could say that he was a man of intelligence and commitment, which helped him to overcome not having a formal education in the sciences and his own physical disability that prevented him from doing fieldwork. The many areas on which he worked and published are evidence of his intellectual curiosity and breath. He had an open mind that allowed him to learn new facts and adapt his thinking to new ideas for which he earned the respect of his peers internationally and from many of his compatriots during his lifetime. In other words, he did the best anyone could have done giving the geographic, political, and social limitations of his time; therefore, there is no need for more hagiography on him. We could even argue that hagiography is the result of lack of understanding of the actual contributions of the person being biographed by substituting comparative analyses with hyperbole.

Acknowledgements

Thanks to Susan Sanctis of the Museum of Comparative Zoology at Harvard University for providing copies of some of the letters written by Poey to H.A. Hagen dated on 1 May 1885. An anonymous reviewer provided useful comments.

---

[408] The original Spanish reads as follows: "Tuvo la gloria de anticiparse a otros sabios afirmando la existencia del hombre fósil antes de que la prehistoria fuera reconocida por la ciencia; y antes de que Darwin diera a conocer su teoría del 'Origen de la especies' él estableció la doctrina de 'lo continuo es el transformismo', principio en que descansa la evolución en la formación de las especies orgánicas."

Literature Cited

Altshuler, J. and A. Baracca. 2005. The development of University Physics in Cuba, 1816-1962. Pp. 5-15, *In: History of the Development of Physics in Cuba.* The development of an advanced scientific system in an underdeveloped country (A. Baracca, Ed.). Berlin: Max Planck Institute für Wissenschaftsgeschichte.

Álvarez Conde, J. 1958. Felipe Poey y Aloy (1799-1891), pp. 213-248, *In: Historia de la zoología en Cuba.* La Habana: Publicaciones de la Junta Nacional de Arqueología y Etnología.

Barreiro, A.J. 1944. *El Museo Nacional de Ciencias Naturales.* Madrid: Consejo Superior de Investigaciones Científicas.

Basalla, G. 1967. The spread of western Science. *Science* **156**:611-622.

Beddall, B.G. 1983. The isolated Spanish genius – myth or reality? Féliz de Azara and the Birds of Paraguay. *Journal of the History of Biology* **16**:225-258.

Boss, K.J. and M.K. Jacobson. 1975. Felipe Poey with a catalogue of the Mollusca described by him. *Occasional Papers on Mollusks* **4**:105-132

Carbonell y Rivero, J.M. 1928. *La ciencia en Cuba.* La Habana, Cuba: Montalvo y Cárdenas.

Chardon, C.E. 1947. *Los naturalistas en la América Latina.* Ciudad Trujillo (Santo Domingo), República Dominicana.

Cruz, M. 1979. *El ingenioso naturalista Don Felipe de La Habana.* La Habana, Cuba: Editorial Gente Nueva.

De La Torre, C. 1942. Don Felipe Poey, pp. 313-345, *In: Figuras cubanas de la investigación científica.* La Habana: Publicaciones del Ateneo de La Habana.

Dihigo, J.M. 1915. Poey en su aspecto literario y lingüístico. *Memoria de la Sociedad Cubana de Historia Natural* **1**:111-131.

García González, A. 1989. *Antonio Parra en la ciencia hispanoamericana del siglo XVIII.* La Habana: Editorial Academia.

García González, A. 1995. La obra botánica de Antonio Parra. *Asclepio* **47**:143-157.

Glick, T.F. and D.M. Quinlan. 1975. Félix de Azara: The Myth of the Isolated Genius in Spanish Science. *Journal of the History of Biology* **8**:67-83.

González López, R.M. 1999. *Felipe Poey y Aloy. Obras.* La Habana: Imágen Contemporánea.

Grisanti, A. 1956. *El sabio Cagigal y su familia.* Caracas: Imprenta Nacional.

Jaume, M. L. 1955. Poey, padre espiritual de los naturalistas cubanos. *Memorias de la Sociedad Cubana de Historia Natural* **22**:93-96.

Jordan, D.S. 1884. Sketch of Professor Felipe Poey. *Popular Science Monthly* 25:2-7.

Jordan, D.S. 1899. *Science Sketches.* Chicago: A.C. McClug and Company.

Jordan, D.S. 1922. *The Days of a Man. Being Memories of a Naturalist, Teacher, and a Minor Prophet of Democracy.* 2 Vols. New York: Yonkers-on-Hudson.

McLaughlin, P. 2002. Naming biology. *Journal of the History of Biology* **35**:1-4.

Mestre, A. 1891. Elogio de F. Poey. *Revista Cubana* **13**:1-16.

Mestre, A. 1918. La vida de la "Sociedad Poey" de 1916 a 1917. *Memorias de la Sociedad Cubana de Historia Natural "Felipe Poey"* **3**(1):3-18.

Montané, L. 1918. Alrededor de la psicología de Poey. *Memorias de la Sociedad Cubana de Historia Natural* **3**:21-29.

Poey, F. 1858. Memorias sobre la Historia Natural de la isla de Cuba, acompañadas de sumarios latinos y extractos en francés, etc. La Habana, Cuba: Barcina.

Poey, F. 1865. Repertorio Físico-Natural de la Isla de Cuba. La Habana, Cuba: Gobierno y Capitanía.

Poey, F. 1868. *Synopsis Piscium Cubensim,* pp. 279-484, Vol. 2, *In*: Reporte Físico-Natural de la Isla de Cuba. La Habana, Cuba: Gobierno y Capitanía.

Poey y Aloy, F. 2000. *Ictiología Cubana* (3 vols.). La Habana: Imagen Contemporánea.
Poey y Aloy, F. 1888. *Obras literarias de Felipe Poey*. La Habana, Cuba: La Propaganda Literaria.
Pruna, P.M. 1994. National Science in a colonial context. The Royal Academy of Sciences of Havana, 1861-1898. *Isis* **85**:412-426.
Pruna, P.M. and A. García González. 1989. *Darwinismo y sociedad en Cuba. Siglo XIX*. Madrid: Consejo Superior de Investigaciones Científicas.
Romero, A. 2007. The discovery of the first Cuban blind cave fish: the untold story. *Journal of Spelean History* **41**(131):16-22.
Romero, A. 2014. Constituciones De La Real Y Pontificia Universidad De San Gerónimo: Fundada En El Convento De San Juan De Letran, Orden De Predicadores, De La Ciudad De San Cristobal De La Habana, En La Isla De Cuba – Primary Source Edition. (Facsimile Reproduction) Charleston, NC: Nabu Press 2013. 136 pp. $16.66 paperback (ISBN 978-1289778668) *Polymath* (In Press in this issue).
Sánchez Roig, M. 1937. *Felipe Poey, el máximo naturalista de Hispanoamérica*. La Habana, Cuba: Imprenta Molina y Cía.
Sanchez Roig, M. 1955. *La monumental obra de don Felipe Poey y Aloy, Ictiología Cubana*, vol. 1, Habana, Cuba.
Vivanco Y Díaz, J. 1951. *Don Felipe. Su vida y su obra*. La Habana, Cuba: Imprenta "El Siglo XX".

Chapter 17. Cope, Caves, and Skeletons in the Closet [409]

Introduction

The next time someone tells you that the lives of scientists are boring, tell the story of Professor Cope. Edward Drinker Cope was born in Philadelphia on July 28, 1840, to a wealthy Quaker family. His interest in natural history was spared when he was just six years old, when his father took him to see an exhibition of skeletons and fossils mounted by Dr. Albert Koch. In 1845, Koch had excavated in Alabama the remains of at least five different individuals of two different genera of fossil whales. Instead of separating the pieces by individuals, he put all the bones together and "reconstructed" a single specimen 114 feet long, about 30 feet longer than the blue whale, the largest mammal believed to have ever lived on Earth. He claimed that the skeleton belonged to a sea serpent (Fig. 17.1). This bizarre reconstruction was exhibited in several American cities, including Philadelphia. A few days later, Cope wrote a letter to his grandmother telling her how impressed he was with this great skeleton of a serpent. He had seen the light. He was to become a naturalist.

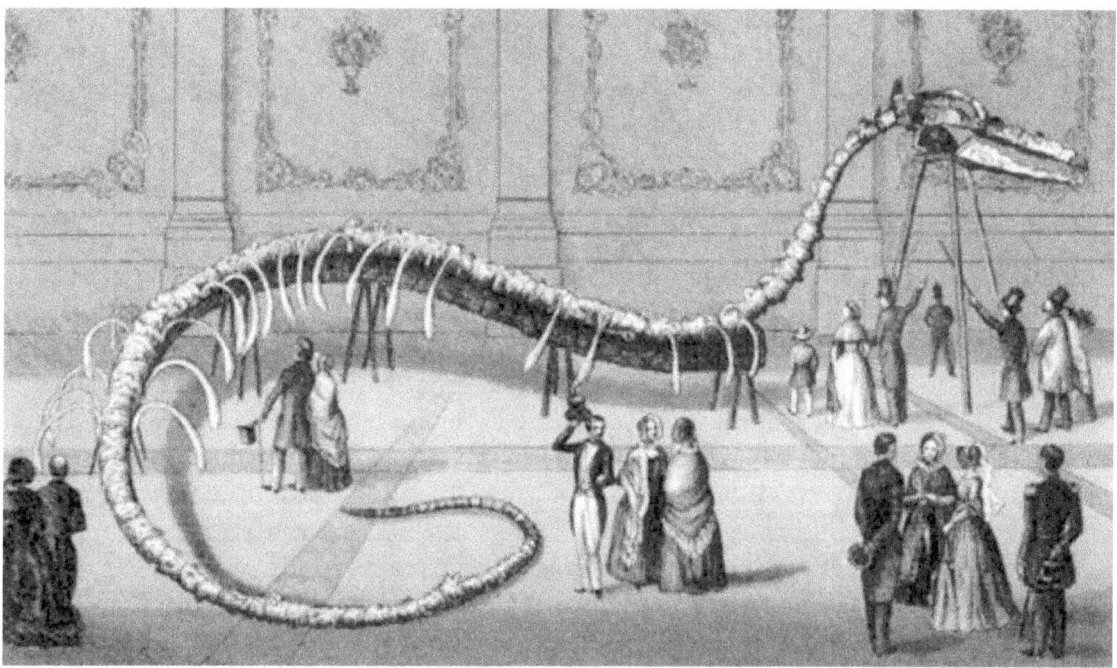

Figure 17.1. Alleged "sea serpent" mounted by Albert Koch. Artist unknown. Public domain image.

Cope did not start going to school until the age of nine, and he continued intermittently until he was nineteen, being taught many times by private tutors. His

---

[409] Based on *Romero, A. & Andrea Romero. 1999. Cope, Caves, and Skeletons in the Closet. National Speleological Society News* **57**(11):341-343.

father wanted him to become a farmer. He was sent to the farms of relatives during the summer months, experiences that furthered his interest in natural history to the point that in 1858 he began working for free at the Academy of Natural Sciences of Philadelphia. He published his first paper when he was eighteen.

His only formal university education was in 1860 and 1861, when he took just one class at the University of Pennsylvania, comparative anatomy, taught by Dr. Joseph Leidy, the founder of American paleontology. When he entered the university, he had already published thirty papers.

To keep him out of the Civil War, his father, a sincere Quaker with pacifist and abolitionist preferences, sent him to study in Europe in 1863. There he studied the natural-history collections of the most important museums in Berlin, Leyden, Munich, Vienna, Paris, and London. On his return to the United States in 1864, Cope became Professor of Comparative Zoology and Botany at Haverford College, a position he kept until 1867. In July 1865, he married his distant cousin Annie Pim. In the summer of 1867, he took his family, including one-year-old daughter Julia, to Virginia, where he explored the local caves.

His great interest in fossils led him to begin field work in the western United States in 1871. His wealth enabled him to support his own expeditions until 1880. Cope was one of the last zoologists to explore and collect specimens under the threat of hostile Native-Americans (and sometimes whites). He often traveled with soldier escort, but he even took his wife and daughter on one of these trips. In 1876 he camped in Sioux territory within a day's ride of the major encampment of Sitting Bull, shortly after the battle of Little Big Horn.

He lost most of his inheritance of a quarter of a million dollars in a mining fraud in Colorado and New Mexico. To pay for his expeditions, he became a popular scientific lecturer. He also had to reduce his field work in the western fossil fields, and in 1894 he sold his valuable scientific collection to the American Museum of Natural History. Fortunately, the University of Pennsylvania appointed him to a teaching position in geology and mineralogy in 1889, and in 1896 he was promoted to Chairman of Zoology and Comparative Anatomy, a post he retained until his death.

Cope also had an interesting personal life. Wealthy and handsome (Fig. 17.2), he was the center of attention wherever he went. He spoke German, French, Spanish, and, to a limited extent, some native American languages. He was noted for his wit and his sparkling conversational style. Because he was a workaholic and regarded himself as a perfectionist, he expected perfection in others, and many people thought him arrogant. He had an oddly fiery temperament, being aggressive and abrasive. He tended to interrupt colleagues while they were presenting papers. Contemporary observers called him churlish, pugnacious, and quarrelsome. Even in the circles he frequented the most, like the Academy of Natural Sciences and the American Philosophical Society, he made many enemies. Once he got a black eye (and gave one, too) in a fight with Persifor Frazer (a member of one of the most prominent families in Philadelphia) in the halls of the Philosophical Society.

But Cope was one of the most prolific American naturalists. He published about fourteen hundred papers and books and named more than twelve hundred vertebrate species. Because he was always in a hurry to publish his research, he wrote many things that were not quite right from a scientific point of view. He also loved speculation.

Cope was not a simple character, and he did not have an ordinary life. He not only made important contributions to speleology, but was embroiled in one of the most bitter and public scientific controversies in the nineteenth century. And his private life was also the subject of gossip and innuendo.

Cave Explorations

Cope published twenty-seven papers pertaining to speleology, which made him one of the pioneers of the science in the United States. His first contribution to speleology, however, was no contribution at all. In 1864 he published a paper in which he claimed to have discovered a new species of blind cave fish. In 1863 he had received two specimens of catfish from Conestoga Creek, a tributary to the Susquehannah River in Pennsylvania. He decided, mostly on the basis of what he thought were rudimentary eyes, to create a new genus and species, *Gronias nigrilabris* (Fig. 17.3). To Cope, this was an example of "animals deprived of the sense of sight," and he wrote that his new species "is supposed to issue from a subterranean stream." However, ichthyologist after ichthyologist who have since examined these specimens have come to a different conclusion. The specimens of Cope's alleged new species are nothing more than representatives of a very widespread North American freshwater fish, the brown bullhead *Ameiurus nebulosus*. Contrary to Cape's report, the eyes are present, but asymmetrically developed due to some error during the fishes' growth (Romero 1999).

When it came to bones and fossils, however, Cope had a better sense of what is important and what is not. He knew that caves are always a good place to scout for them, and most of his contributions to speleology came from his efforts to identify and analyze fossil remains from caves. Cope's first search for bones in caves was in June 1864, when he journeyed to the Catskill Mountains in southeastern New York and visited Howe's Cave, where he found a number of fossilized bones. For two months in the summer of 1867, he explored caves in southwestern Virginia, where he found a bear's lair in one cave and bones in another, as well as remains of a blind beetle and big cats and other mammals. Among the caves visited in that area were Erhart's Cave in Montgomery Country and Spruce Run and Big Stony Creek caves in Giles County.

But one of his most interesting discoveries came from a cave he never visited. In 1868 he examined bones and teeth found in cave soil brought from the Caribbean island of Anguilla, probably from Cavanaugh Cave. Cope found those remains extremely remarkable. He thought that these bones belonged to a chinchilla-like animal. Actually, it probably resembled a large capybara more than anything else, although the two animals are not closely related. What was really impressive about the remains was the size of the animal, comparable to that of an American black bear. Recent studies by Donald McFarlane at Claremont College have shown that the individuals represented by the bones weighed from fifty to two hundred kilograms. Cope named this new species *Amblyrhiza inundata* (Fig. 17.4). The genus name roughly translates into "strange root," for to Cope it was an aberrant beast. In order to explain how such a large animal could be found on such a small island, he came up with the hypothesis that all those islands had been connected to South America by land bridges that had disappeared when the sea level rose about 125,000 years ago. That is why the species name is *inundata*, flooded.

Figure 17.2. Edward Drinker Cope. Picture courtesy of the Smithsonian Institution Archives.

Figure 17. 3. "Gronias nigrilabris" Picture courtesy of the Academy of Natural Sciences of Philadelphia.

Figure 17.4. Ameirus nebulosus. Picture from Fishes from Texas by Joseph Tomerelli[410]

Subsequently, it has been learned that there was an ice age at the time the rodent lived, and that the lower sea level had made Anguilla, St. Martin, and St. Barts into a single large island capable of supporting such a large animal.

Figure 17.4. Reconstruction of *Amblyrhiza inundata* by Dan Bruce. Courtesy of R.D.E. MacPhee.

---

[410] At http://www.fishesoftexas.org/taxa/ameiurus-nebulosus Retrieved 5 November 2020.

In 1869, Cope visited Jefferson County in eastern Tennessee. There he found cave insects and centipedes and bones of a cave rat. In 1871, he went to Wyandotte Cave, Crawford County, Indiana, where he claimed to have gone 9.5 miles into that cave in what was to be his last cave exploration.

But he was not through with speleology. In 1871, Cope studied bone remains from a cave discovered by limestone miners, Port Kennedy Cave in what today is Valley Forge National Historical Park. He described thirty- four species of mammals from that cave, thirteen new to science. Additionally, other species have been described since, but the precise location of the cave is unknown. Its entrance has apparently been filled, and the National Park Service is trying to relocate it.

Another speleological contribution by Cope came with the description of a fossil bear from California. This species is now known as the North American short-faced bear, *Arctodus simus* (Fig. 17.5). Since he had found it in a cave, he thought he had discovered a species of cave bear. These bears were the largest land carnivores in North America during the last two million years, 1.5 meters tall when walking and 3.4 meters tall when standing on their hind legs. They would have been able to reach 1.2 meters above a basketball hoop. They ranged from Alaska and the Yukon to Mexico and from coast to coast. A recent estimation of the autumn weight, with full complement of fat for the winter, of a large short-faced bear is 700 kilograms. (The largest polar bear ever weighed was 660 kilograms.) The common name of the bear derives from its lack of a well-defined forehead and its short, broad muzzle, more resembling that of a lion than living North American bears. The muscles from the broad cheek bones to the lower jaws were extremely well developed, as would be expected in a carnivore adapted to crushing bones for marrow.

Figure 17.5. Artist's reconstruction of *Arctodus simus*.[411]

Rather than being "pigeon-toed" and waddling like a modern bear, *Arctodus simus* had toes extending straight forward, presumably allowing it to move more easily in bursts of speed or long-distance travel in search of prey or carcasses. It was probably a rather solitary predator or scavenger of large

---

[411] From https://paleo-media.fandom.com/wiki/Arctodus?file=Arctodus.jpg Retrieved 5 November 2020.

herbivores such as bison, musk-oxen, caribou, deer, horses, and ground sloths. The wide face allows great width between the canine teeth for a good grip on prey, and it might also have allowed a wide snout for a keenly developed sense of smell. In this respect it resembled the polar bear, the most carnivorous living bear.

Additional remains of these bears have been found in caves. Bones of eight individuals from Potter Creek Cave, California, were all females, suggesting that they denned there for reproductive purposes. However, bones of both sexes have been found in other caves. They occasionally suffered from diseases such an osteomyelitis, tuberculosis-like condition, and possibly syphilis- like infections. *Arctodus simus* died out toward the close of the last glaciation, perhaps partly because of the earlier extinction of some of its large prey and partly because of increased competition with brown bears, which apparently appeared in North America about 150,000 years ago. Of course, Cope did not know all these details, but his pioneering work on the fossils of this species opened the door to these and many other discoveries.

One of Cape's last letters, dated March 10, 1897, was on cave fauna. He wrote to his daughter, "I am finishing the Fort Kennedy investigation and will soon turn the papers in for publication. I have now identified 56 species of Vertebrata of which 50 were of Mammalia. Of the latter all are extinct except six species, which are still living, and are conspicuous species of the present inhabitants of the wild of this country."

Cope was also the leading neo-Lamarckian in the United States. Together with two other American naturalists interested in cave faunas, Alpheus Hyatt and Alpheus Spring Packard, he thought that cave organisms lost their eyes and pigmentation simply as a result of disuse and then passed that loss on to their offspring in an example of the now- discredited notion of inheritance of acquired characteristics.

Despite my searches at the American Museum of Natural History in New York and the Academy of Natural Sciences in Philadelphia, where many of Cope's papers and memorabilia can be found, I could not locate a single picture of him doing speleological work. This is not surprising. Not only was field photography uncommon at that time, but much of his cave research took place with specimens from caves he never visited.

The Bone Wars

No biography of this American scientist can be complete without mentioning two aspects of his life for which he is better known than for his speleology. One is the bone wars, and the other is the story of his own bones.

The American paleontologist Othniel Charles Marsh[412] (Fig. 17.6) seemed to have a lot in common with Cope. He too went to Europe to escape from the Civil War. He had an interest in fossils, in his case sparked by diggings in the Erie Canal near his home. Marsh had a millionaire uncle, George Peabody, who financed his education and bequeathed one hundred thousand dollars to Yale, which offered Marsh a professorship there in the new Peabody Museum. The deal was good; Marsh would receive his salary directly from his uncle, so he did not have to teach and could spend more time studying fossils. He became the first official professor of paleontology in the country.

---

[412] *b*. Lockport, New York, U.S.A. 21 October 1831; *d*. New Haven, Connecticut, U.S.A., 18 March 1899.

Marsh met Cope at the University of Berlin in 1863. They started out as friends and later hunted fossils together in the eastern United States. Trouble began when Marsh visited one of Cape's digs in New Jersey and covertly paid Cope's crew to send future finds to him. To make things worse, Marsh was a major critic of Cope' s most famous scientific blunder. Cope had reconstructed the skeleton of a marine reptile in a way that looked very odd, with a very long tail and a short, stubby neck. This animal looked so strange that Cope thought he had discovered a species quite different from any other fossil reptile, and he rushed to publish his findings. Unfortunately, he had mounted a plesiosaur skull on the tip of its tail, not on the end of its neck. Cope was so mortified at his mistake that he tried to buy up all the printed copies of the publication, but Marsh was unforgiving. In 1870, he pointed out Cape's mistake and made Cope look like a sloppy scientist who was so untrained he could not get his vertebrae straight. Cope would never forgive nor forget, and what followed in the next twenty-five years or so was to be known as the Great Bone Wars.

Figure 17.6. Othniel Charles Marsh (public domain image).

Partially to rebuild his reputation, Cope embarked on a race to show who was the better scientist, a race that would consume him for the rest of his life. Obviously his having published fourteen hundred papers by the time of his death at the age of fifty-six meant that he not only had to write quickly, but also that he became sloppy and superficial, a feature of Cope's character that even his most loyal friends recognized. For example, many times he sent papers by telegram, leading to misspellings. Cope had difficult handwriting, and a telegrapher once mistakenly sent the name *Lefalophodon*, a name previously claimed by Marsh for another fossil.

Cope and Marsh became paranoid of each other: their crews open quarries, strip them in months, and destroy what they could not remove, so that the other could not get anything from those sites. They both coded messages containing news of finds or deals. They used false names and even laid dummy trails to confuse spies. Crates of specimens awaiting shipment were put under heavy guard. Both had double agents working for their competitor.

Marsh was not a pleasant character either. He was pugnacious and quarrelsome and had few, if any, friends, while the number of his enemies was quite large. This earned him the nickname The Great Dismal Swamp. He was autocratic and a tightwad, someone who never paid his employees on time. Trying to further damage Marsh's reputation, Cope passed on to a reporter from the *New York Herald* damaging information on Marsh. They publicly exchanged accusations and insults. Cope wrote of Marsh, "I suspect that a Hospital will yet receive him." Marsh responded by saying, "... I had some doubts about his sanity." Cope suggested that Marsh, who had never married, was a homosexual by saying that he was not "normally or properly constituted." While Marsh accused Cope of incompetence, Cope accused Marsh of plagiarism, the ultimate scientific crime. This accusation was based on the fact that, unlike his rival, Marsh rarely ventured into the field. He instead directed field excursions from his home in New England. Many discoveries credited to Marsh were actually made by his field associates, but he rarely mentioned their names in his papers. Marsh, in return, tried to have Cope fired from the University of Pennsylvania. He told the university, William Perry, that Cope had libeled him in the press to the president.

Despite the confrontations, the Bone Wars generated one of the most productive eras in the history of paleontology. Cope and Marsh described 136 species of dinosaurs. Before them, only nine had been named. Among their discoveries were dinosaurs very familiar to us today, such as triceratops, allosaurus, diplodocus, stegosaurus, and what used to be called brontosaurus.

Skeletons in the Closet

Because of Cape's celebrity status in his time and the large number of enemies he had, some rumors about him have persisted until today. Cope separated from his wife sometime in 1894 or 1895. Although the letters between them during the time of separation do not show any hostility, rumor had it that she left him because he was a philanderer. Those rumors have not been substantiated, which does not necessarily mean they were not true. Another rumor is that he died of syphilis. Fortunately, this rumor can be investigated, because Cope donated his brain and skeleton to the American Anthropometric Society, hoping he could become a type-specimen for human beings. The skeleton was lost for many years, but it reappeared in 1966 in a collection of primate material in Pennsylvania's Museum of Archaeology and Anthropology. Dr. Jane Pierce

Davidson, a professor of history at the University of Nevada in Reno, who recently wrote a biography of Cope, arranged for a forensic pathologist to examine the skeleton for evidence of syphilis. None was found. Cope had died on April 12, 1897, probably from prostatitis, a condition he carried for many years.

Acknowledgments

Dr. Donald A. Mcfarlane provided information and useful advice on the section about *Amblyrhiza*. Dr. Ross D. E. MacPhee of the American Museum of Natural History graciously gave us permission to reproduce the illustration of *Amblyrhiza* painted by Dan Bruce. Dr. William Turnbull of the Field Museum of Natural History at Chicago provided also very useful information on the *Arctodus*, the "cave" bear discovered by Cope. Mr. William G. Saul of the Academy of Natural Sciences of Philadelphia provided the authors with useful information about the alleged cave fish *"Gronias nigrilabris"* deposited at that institution We also thank Prof. James Stewart, at Macalester College, with· whom we discussed some aspects of American history during Cope's times.

Literature Consulted

Biknevicius, A.R., R.D.E. MacPhee & D.A. McFarlane. 1993. Body Size in *Amblyrhiza inundata* (Rodentia: Caviomorpha) an Extinct Megafaunal Rodent of the Anguilla Bank, West Indies: Estimates and Implications. *American Museum Novitates* (3079): 1-25.

Bowler, P.J. 1977. Edward Drinker Cope and the Changing Structure of Evolutionary Theory. *Isis* **68**(242):249 -265.

Davidson, J.P. 1997. *The Bone Sharp: The Life of Edward Drinker Cope*. Philadelphia: The Academy of Natural Sciences of Philadelphia. 237 pp.

Grady, F.V. 1987. Edward Drinker Cope's Contributions to Speleology. *Journal of Spelean History* **21**(3 -4):35-37.

McFarlane, D.A., R.D.E. MacPhee & D.C. Ford. 1998. Body Size Variability and a Sangamonian Extinction Model for Amblyrhiza, a West Indian Megafaunal Rodent. *Quaternary Research* **50**:80-89.

McNierney, M. 1982. The Great Bone Wars. *The American West* **19**(3):52-61.

Osborn, H.F. 1931. *Cope: Master Naturalist*. Princeton, New Jersey: Princeton University Press. 740 pp.

Romero, A. 1999f. Myth and reality of the alleged blind cave fish from Pennsylvania. *Journal of Spelean History* **33**(4):67-75.

Werkley, C.E. 1975. Professor Cope, not Alive but Well. *Smithsonian* **6**(5):72-75.

Chapter 18. He Wanted to Know Them All. Eigenmann and His Blind Vertebrates

Summary

When *Cave Vertebrates of North America, A Study in Degenerative Evolution*, was published in 1909, it instantly became a sort of bible for all who had any interest in cave animals and those interested in the loss of vision. That book reflects the author's qualities in his lifelong research efforts: it is comprehensive, detailed, and authoritative. This is the story of the man, Carl H. Eigenmann, and the work that led him to produce one of the "classics" of speleological literature.

Early Years

Little is known about Carl Eigenmann's childhood. He was born March 9, 1863, in Flehengen, a small village near Karlsruhe, Baden, Germany. His mother died when he was a child, and his father remarried soon afterward. When he was in his early teens, Carl came to the U.S. with his uncle and settled in Rockport, southern Indiana.

Working and studying energetically, Carl Eigenmann finished high school in only two years and was admitted to the University of Indiana, intending to study law. However, during his sophomore year, the famous American naturalist David Star Jordan, a natural history professor, overturned the traditional system of courses in the classics and instead offered his students a choice of studying either Latin or biology. The new system was a tremendous success.

With his characteristic enthusiasm, Eigenmann took biology. His accomplishments must have been quite impressive since Jordan, writing about the first group of students taking the biological alternative, said: " ... the leader of these, Carl H. Eigenmann, found Zoology the passion of his life" and that his work was "of the highest order" (Jordan 1922). A year later, Eigenmann was posted as Instructor of Zoology. During 1885, his junior year, Eigenmann published his first scientific paper (co-authored with Jordan) and submitted many more.

First Cavefish

In his "*Cave Vertebrates,*" Eigenmann stated that "my first experience with blind vertebrates was in 1886 when Superintendent Punk sent to Indiana University a living blind fish which had been taken from a well at Corydon, Indiana, and which proved to be a new species, *Typhlichthys wyandotte,* the only representative of the genus so far taken north of the Ohio River." Eigenmann didn't describe this "new" species until 1905. Later it became evident that this fish and his "*Typhlichthys osborni*" were misidentifications of the southern cavefish, *Typhlichthys subterraneus,* a species previously described by Girard in 1859.

---

[413] Based on Romero, A. 1986. He wanted to know them all: Eigenmann and his blind vertebrates. National Speleological Society News 44(11):379-381 1986.

Figure 12.1. Carl Eigenmann. Photo courtesy of the National Museum of Natural History, Smithsonian Institution.

Miss Smith

In 1886, with a bachelor's degree under his arm, Eigenmann took a long trip to California. He arrived too late to apply for a job as a school principal, as he had intended, so he moved on to San Diego to visit Miss Rosa Smith[414] (Fig. 12.2), an ichthyologist with whom he had maintained a professional correspondence. Miss Smith must have been a remarkable woman. Born in Illinois on October 7, 1858, she was one of the first American women to become a scientist, studying fish and spiders. She was the first woman president of Sigma-Xi, the national scientific research society. From 1880 to 1882, she attended Indiana University, where she may have met Eigenmann. Whatever the circumstances, Eigenmann, five years her junior, fell in love with her, and they were married in San Diego on August 20, 1887.

Figure 12.2. The Eigenmanns in 1889.

---

[414] *b.* 7 October 1858, Monmouth, Illinois; *d.* 12 January 1947, San Diego, California.

New Horizons

The newlyweds were offered a chance to study the immense Brazilian fish collections made by Louis Agassiz in 1865-1866 and Agassiz and Franz Steidackner during the Hassler Expedition. This was an expedition that encircled South America in 1871-l872. The Eigenmanns moved to Cambridge in 1888 to undertake this tremendous task at the Museum of Comparative Zoology of Harvard University.

With his brand-new master's degree from Indiana, Eigenmann and his wife spent long hours working on the taxonomy and development of those South American fishes. In 1888 they began publication of a series of papers on the subject. During the summer, they went to the Marine Biological Laboratory at Woods Hole, Massachusetts, where they spent the rest of the year researching.

The next year, 1889, Eigenmann received his doctoral degree from Indiana and returned to California as Curator of the San Diego Natural History Society while establishing the San Diego Biological Station.

The California Blind Fish

The fish of the California coast became a new focus of interest for the Eigenmanns. Mrs. Eigenmann took her husband to a place named Point Lorna, where a blind fish (*Typhlogobius californiensis*) could be found among the rocks. As Eigenmann put it: "when a stay in southern California came in prospect, a study of the blind fish, *Typhlogobius*, living under rocks along the base of Point Loma, was one of the first definite plans (for the study of blind animals) formed." (Fig. 12.3).

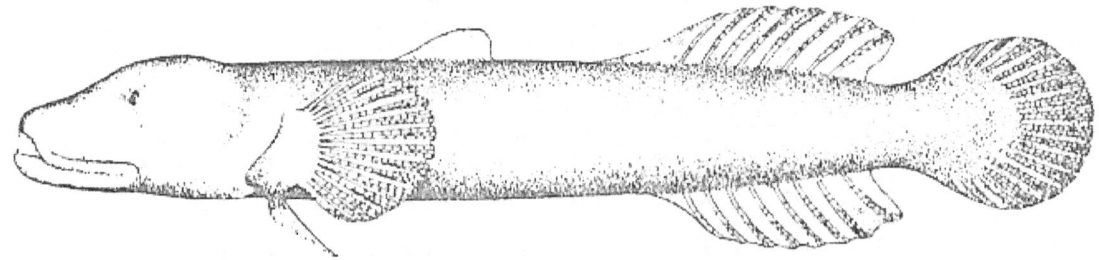

Figure 12.3. Eigenmann's drawing of the California blind gobid (*Typhlogobius californiansis*). (Public domain illustration).

This new subject impressed him so much that the question of how some vertebrates became blind was one of the centerpieces of his research for the next few years. He had started working on that fish when a new event allowed him to broaden his study area.

In 1891 Jordan became Stanford University's first president. Before leaving Indiana University, he appointed Eigenmann as professor of zoology at Bloomington, Indiana. The German-American professor was within the North American fish family's geographical range, Amblyopsidae, and many other cave vertebrates. By then, many of the world's blind vertebrate fauna had already been reported from the Mississippi plateau area.

From 1897 to 1909, much of Eigennann's effort was devoted to comprehending the process of the loss of visual structures in cave vertebrates. In May 1896, he visited Dalton's Spring (actually a cave-stream), where he secured 20 specimens of the northern cavefish *Amblyopsis spelaea*. This became his favorite collecting locality. By 1903 the state legislature of Indiana placed the land, where the entrances and exits to this and other caves are located under the care of the trustees of Indiana University.

In 1898, Eigenmann published the description of a new species of cavefish he named after his wife, *Typhlichthys rosae*, the Ozark cavefish. Today it is recognized as *Amblyopsis rosae*. Its rudimentary eyes characterize this species from southwestern Missouri.

Eigenmann extensively visited the caves of Indiana, Kentucky, Texas, and Missouri, searching for specimens for his work. At Mammoth Cave, he became " impressed with the value of the scientific problem the cave presented" and (after walking for one hour or two from the entrance) he noted, "... it begins to impress one very forcibly."

Cuba

There are two species of cavefish (of marine origin) from Cuba, which are quite distinct from those in North America. Felipe Poey described these Cuban blind cavefish in 1856 (Romero 2007). The two species are known today as *Lucifuga subterraneus* and *Stygicola dentatus*. Eigenmann wanted to know how much convergence occurred among cave animals and that the Cuban species offered an excellent opportunity. He wanted to know all cave species (Fig. 12.4).

In March 1902, Eigenmann visited Cuba for the first time lo secure specimens for his comparative studies. He worked on fish reproduction in the past and quickly recognized that these two fish species were viviparous (have live births instead of laying eggs). Therefore, he planned a second visit during October- November when, as he calculated, pregnant, blind fish could be collected with the young embryos still in their early development stages. This lime he planned to collect individual fish alive to be placed in cages located in a well-lighted cave entrance. His idea was to determine what happens when light penetrates the rough adult body walls and reaches the embryos. Eigenmann and his assistants were not very lucky securing specimens with embryos, so he embarked again for Cuba, arriving on December 18, 1903, but no fish with young embryos were then found.

With his typical persistence, Eigenmann went back to Cuba on August 15, 1904, and obtained two females with young, but by this time, the cages that had been built and placed two years earlier were in terrible shape. So, in September, he took the fish to Indiana, where low temperatures caused high mortality. In 1905 three of his co-workers went to Cuba again trying to secure more live females with young; success was also not complete.

Contrary to Mammoth Cave, Eigenmann found the localities for the Cuban blind fish "monotonous in the extreme." From 1906 to 1907, he conducted studies in Europe, mostly in Germany at the laboratory of Prof. Robert Wiedersheim -who specialized in vestigial organs- with the Cuban specimens he collected.

During this period, he made plans to visit the Yucatan Peninsula. There had been persistent rumors of varied cave fauna in that part of the world. He never made that trip. However, in the '30s, many scientific expeditions unearthed an impressive number of cave organisms from the Yucatan Peninsula, many of them fish studied by Carl Hubbs.

From 1898 to 1905, Eigenmann published 39 papers on cave vertebrates, dealing mostly with developmental and anatomical aspects of loss of vision in blind fish, salamanders, lizards, and the blind rat in an attempt to understand the underlying process of blindness among these animals. All this research was summed up in his "*Cave Vertebrates of North America*" (1909), containing 341 pages and 30 plates.

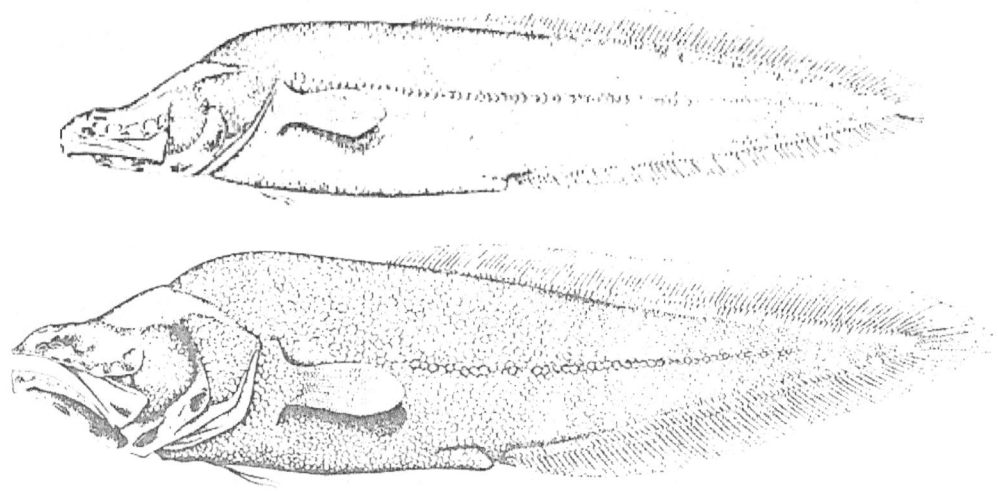

Figure 12.4. Drawings of Cuban blind cavefish, *Lucifuga subterraneus*, and *Lucifuga* (formerly *Stygicola dentatus*) collected by Eigenmann in 1902 (public domain illustrations).

The Meaning

Even though he was a taxonomist by training, Eigenmann never stressed classification problems of cave vertebrates -after all, there were not that many that had been discovered at that time-. Instead, he was concerned with the question of the origin and evolution of the cave faunas.

To understand Eigenmann's contributions to biospeleology, we have to take a historical perspective. Although Darwin had put forward his ideas about organic evolution by means of natural selection several decades before Eigenmann's interest in cave fauna, evolution by means of natural selection was far from being widely accepted because among other things, Mendel's crucial papers explaining the basis of genetics were not rediscovered until the turn of this 20th. Century. To make things even more complicated, Darwin himself had argued for the Lamarckian mechanism (disuse combined with the inheritance of acquired characteristics) to explain blindness in cave animals (Chapter 3 in Romero 2009).

There is little question that Eigenmann's evolutionary explanations for the reduction and/or disappearance of organs were a mixture of neo-Lamarckism and Darwinism. On the one hand, he stated that "the bleached condition of animals living in the dark, an individual environmental adaptation, is transmissible and finally becomes hereditarily fixed." On the other hand, he was quick to affirm that " ... ornamental secondary sexual characters not being found in blind fish are, when present, probably due to visual selection," apparently implying natural selection based on sexual selection.

Such an apparent contradiction was not unusual at his time (indeed, Darwin himself held similar concepts). The dismissal or the idea of inheritance of acquired characteristics would not occur until the late 1920s when the new ideas of population genetics were put forward.

Eigenmann was an early supporter of Herbert Spencer's ideas that cave faunas were not the result of "accidents" but rather the product of an active process of colonization.

He also argued that the reduction or disappearance of organs among cave animals was a convergent evolution case. In other words, the underground environment's well-defined conditions facilitate evolutionary changes leading to blindness and depigmentation in a variety of different vertebrate and invertebrate organisms. This seemingly self-evident fact is still overlooked or discounted by many researchers.

Eigenmann was quick to point out that the lack of pigmentation had to be understood as combining genetically fixed and epigenetic (environmental) characteristics. In other words, although a lack of pigmentation is a characteristic genetically determined, its degree may vary under certain light conditions.

Finally, for his time, he was closer than anyone else in singling out standard development than developmental abbreviation as the descriptive mechanism for understanding the evolution of blindness among cave animals.

South America and More Cavefish

Although Eigenmann maintained a profound interest in cave faunas, he kept working on other fish. During his most active speleological period (1898-1905), he published 26 papers on different topics. Later he spent almost all of his efforts on discovering and describing South American freshwater fishes. The contributions here were as significant as those on cave faunas. Again, he wanted to know and study them all.

Even so, he found more time and opportunities to contribute to our current knowledge of cave animals. He kept publishing articles on the subject until 1919 when he described a new species of blind fish, *Trogloglanis pattersoni,* from the artesian waters of San Antonio, Texas. This is a remarkable herbivorous toothless catfish, which shares its habitat with another blind catfish, *Satan eurystomus,* a carnivore.

Epilogue

Back in the 1910s, South America was not ideal, even for a mature man of apparently endless energies. During his 1912 trip to Colombia, he was affected by fevers of unknown origin, and while climbing the Chilean

Andes in 1919, he had to quit. His exploration days were over, at the relatively young age of 59. In 1926 he went to sunny California, hoping to recover his health. However, on April 24, 1927, he died at a private hospital at Chula Vista, San Diego County. His wife died on January 12, 1947, at the age of 89. Four daughters and a son survived them.

During his lifetime Eigenmann received many honors. He was most proud to be a member of the National Academy of Sciences. He published 229 papers (some of them extensive monographs) and described nearly 400 species of fish. But few of those monographs rival "*Cave Vertebrates of North America*" because in a few of them did he try so hard to show that he had intended to know them all.

Figure 12.5. Picture of Carl H. Eigenmann taken on 15 September 1914 with photographic effect. Courtesy of the Indiana University Archives.

Acknowledgments

Thanks to Dr. Stanley H. Weitzman of the National Museum of Natural History, who helped me locate some of the illustrations that accompany this article and read the M.S., making valuable suggestions.

Suggested Readings

Eigenmann, C.H. 1890. The Point Loma Blind Fish and its Relatives. *Zoe* **1**:65-72.

Eigenmann, C.H. 1899. Explorations in the caves of Missouri and Kentucky. *Proceedings of the Indiana Academy of Sciences* (1898):58-61.

Eigenmann, C.H. 1909. *Cave Vertebrates of North America: A Study in Degenerative Evolution.* Publications of the Carnegie Institute (104):341.

Henn, A.W., 1927. Professor Carl H. Eigenmann. *Annals of the Carnegie Museum* **17**(3-4): 409-414.

Hubbs, C.L., 1971. Rosa Smith Eigenmann. *Notable American Women 1607-1950: A Biographical Directory,* E.T. Jonas, ed. Vol. I. Cambridge: The Belknap Press of Harvard University Press.

Jordan, D.S., 1922. *The Days of a Man.* 2 vols. New York: World Book Company.

Myers, G.S. 1928 Carl H. Eigenmann - Ichthyologist. *Natural History* **28**(1):98-101.

Romero, A. 2007. The discovery of the first Cuban blind cave fish: the untold story. *Journal of Spelean History* **41**(131):16-22.

Romero, A. 2009. *Cave Biology: Life in Darkness.* Cambridge: Cambridge University Press.

Stejneger, L., 1938. Carl H. Eigenmann. *National Academy of Sciences of the United States, Biographical Memories.* Pp. 306-324.

Chapter 19. The Unsung Heroes of Speleology [415]

At the beginning of the Twentieth Century, the Congo was like no other European colony in the African continent. Instead of being the property of the state, the 2,345,410 square kilometers of tropical forest full of mystery and intrigue was the personal property of King Leopold II of Belgium. Known as the world's largest private ranch, it would epitomize the African jungle and inspire literary masterpieces such as Joseph Conrad's *Heart of Darkness*.

The Belgian king saw the importance of scientific research in this area. He instructed voyagers and explorers to collect information about the Congo, its people, flora, and fauna. Not surprisingly, many Belgian officials there collected and took to Europe zoological specimens that had never been studied. The Congo became such a famous place that a great number of specimens of the fauna and culture of that part of the world were put on display at an international exposition in Brussels in 1897. This was such a success that Leopold created the Congo Museum in Tervuren, Belgium. Leopold gathered the brightest scientists in the country to advise him about how to manage the collections. Among those was one of the most famous naturalists that Belgium has ever produced, George Albert Boulenger.

Boulenger was born on October 19, 1858 in Brussels. He was the only child of Gustave Boulenger, a Belgian public notary, and Juliette Piérart de Valenciennes. After graduating in 1876 from the Free University in Brussels with a degree in natural science, he started working at the Museum of Natural History of Brussels as an assistant naturalist studying amphibians, reptiles, and fishes. The reason why at that time this heterogenous group of vertebrates were given to the person in many natural history museums around the world, was because these animals were preserved using similar techniques.

 Because of the lack of appropriate collections and well-stocked libraries in Belgium, Boulenger made frequent visits to the *Muéee National d'Histoire Naturelle* in Paris and the British Museum in London. This young man must have greatly impressed his colleagues, because at the age of 22, in 1880, Boulenger was invited to work at the London museum by one of the foremost zoologists of the time, Albert Gúnther[416]. Boulenger was to prepare a new catalog of amphibians in the national collection there. A position at the museum meant he had to be a British civil servant, so he became a naturalized British subject. In 1882, he was appointed first-class assistant in the Department of Zoology, a position he held until his retirement in 1920.

---

[415] Based on Romero, A. & K. Benz. 2000. The Unsung Heroes of Speleology. *National Speleological Society News* **58**(4):106, 126.
[416] *b*. 3 October 1830, Esslingen, Germany; *d*. 1 February 1914, Kew, England.

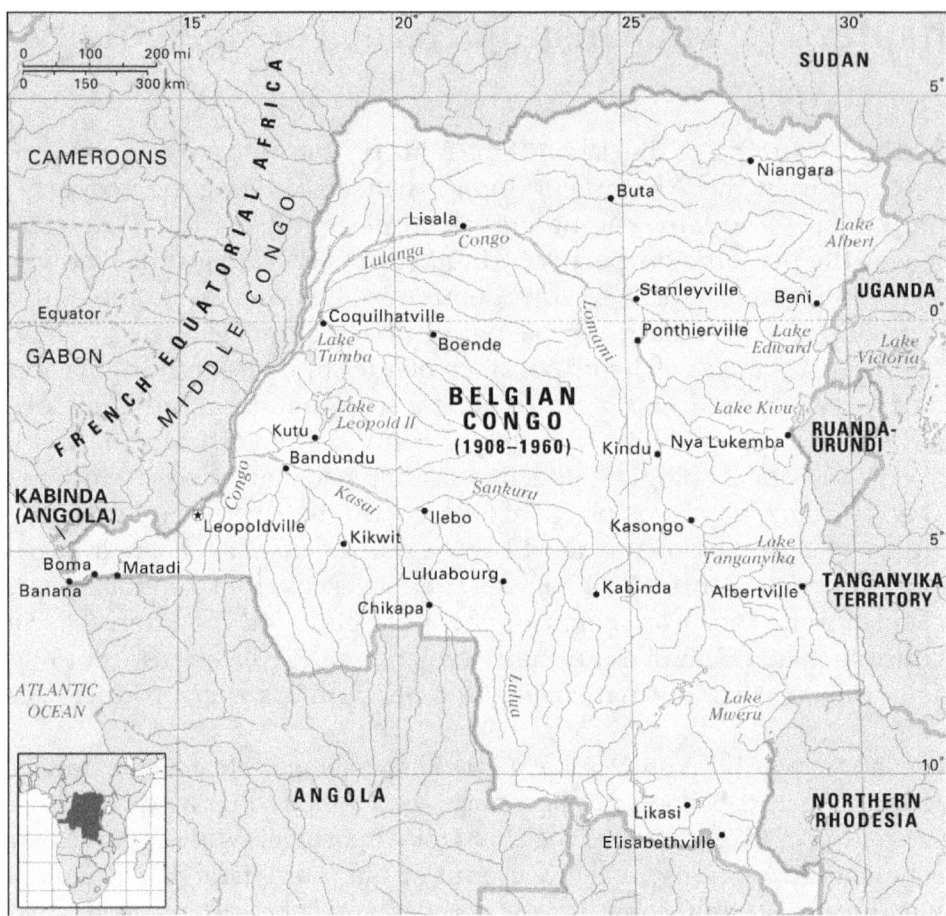

Figure 19.1. Colonial map of Belgian Congo (public domain map).

Figure 19.2. George Albert Boulenger at his lab. Photo courtesy of the Natural History Museum, London Library and Archives.

After he retired, he switched specialties and immersed himself in the study of roses. By the time of his death in 1937, he had published thirty-four papers on botanical subjects and two volumes on the roses of Europe. But besides having this unusual career, Boulenger was an unusual character. He was incredibly methodical, but what impressed people was an amazing memory that enabled him to remember every specimen and scientific name he ever saw. He also had extraordinary powers of writing effortlessly. He seldom made a second draft, and his manuscripts showed but few corrections. They went straight to the publisher. They were never typed, as he never employed a typist.

Boulenger was austere, played the violin, and loved the theater, particularly the operetta. French and English were his mother tongues, but he could speak German and was able to read Spanish and Italian and a bit of Russian. He had a working knowledge of both Greek and Latin. He was known as a kind man, especially toward children, who were fascinated by the fact that a chimp shared his family home in London with his wife and three sons.

Despite his varied interests, Boulenger was first and foremost a scientist. By the time of his retirement in 1920, Boulenger had published 877 papers totaling more than five thousand pages, as well as 19 monographs on fishes, amphibians, and reptiles. He described 1,096 species of fish, 556 species of amphibians, and 872 species of reptiles. He was famous for his monographs on amphibians, lizards and other reptiles, and fishes, as well as for his monographs on the fishes specifically from Africa. His scientific contributions were recognized all over the world. He was a member and distinguished officer of scientific societies in Europe and the United States. In 1935 the American Society of Ichthyologists and Herpetologists elected him as its first honorary member, and in 193 7 Belgium conferred on him the Order of Leopold, its highest honor for a civilian.

From the Congo Caves with Love

In 1897, when King Leopold II was recruiting naturalists for a commission in charge of the Congo Museum. Boulenger, despite the fact that he was living in London, was named chairman. This is when he put his extraordinary energy and productivity into the study of the freshwater fishes of Africa. He encouraged travelers and residents in various parts of Africa to send home specimens. In 1921, he received a strange specimen from the Congo. It was eyeless and lacked pigmentation. He was quick to recognize that this was something nobody had seen before. It was not closely related to any existing surface species in Africa. Despite the fact that he was retired and no longer working in zoology, he decided to write a brief paper describing this new species of cavefish, the first ever described from Africa. He named it *Caecobarbus* geertsi, from *caeco* (blind) and *barbus* (barbed) and for the mysterious person, M. Geerls, who provided him with the specimen. Today it is popularly known as the Congo or African blind barb. Boulenger's only previous publication on cave animals had been a short note on eye development in the European cave salamander *Proteus* published in the prestigious scientific journal *Nature*.

While Boulenger is recognized as the scientific discoverer of this species, the fact is that others were the true discoverers of the tiny cave fish. For one thing. Boulenger had never been to Africa. Who was the true discoverer of this biological relic? Here is where the story becomes as obscure. Details are sketchy and holes in the story are big, but here is what we know

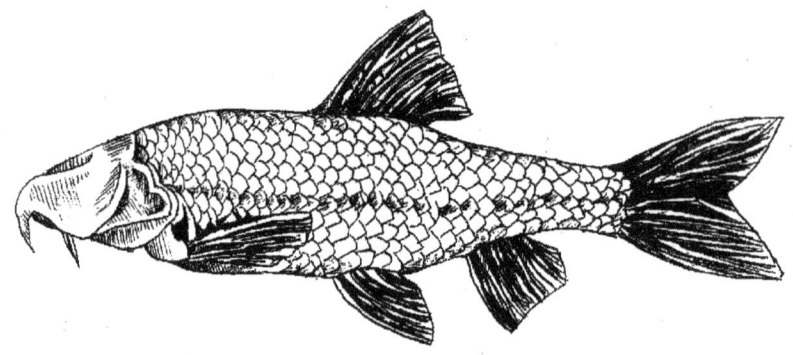

Figure 19.3. Caecobarbus geertsi from Proudlove and Romero 2001.

In 1917, during the dry season, a party of amateur cave explorers penetrated 500 meters into a limestone cave situated about 700 meters above sea level in the Lower Belgian Congo near Thysville (today Kankan, near Mbanza-Ngungu, 5°18' S, 14°50' E). There they found the blind and depigmented cave fish. One of those amateur explorers, M. Geerts, waited until after World War I, when it was safe for him to go back to Belgium, and carried with him a few specimens of this bizarre fish. The choice for receiving these specimens was obvious, for Boulenger was not only the foremost specialist in African fishes, but was also a Belgian. In his 1921 paper, Boulenger, grateful and following a common practice among naturalists, named the species after the person who had brought it to him. Unfortunately, we know nothing about this mysterious M. Geerrs except for the fact that he was not a professional scientist.

Even more mysteriously, it seems that in 1915, two years before the fish had been collected by Geerts, another explorer, M. Delporte, may have actually been the first European to see the fish. Nothing is known about Delporte. To add to the list of unknown pioneers, we also learned that one E. Randour found the same fish in other caves in the same area during the 1920s. All we know about him is that he wrote a little-known booklet on the Congo that is difficult to find in any American library.

Soon this extraordinary fish became a celebrity in the world of cavefishes. It was easy to transport alive, and many European scientists started to work on this species. It became so famous that in May 1951 it was exhibited at the New York Aquarium. Today it is one of the best-studied species of cavefish in biospeleology. Thanks to these studies, we have a better understanding of the adaptations of cave animals to the subterranean environment. But without the efforts of cavers Geerts, Delporte, and Randour, these studies would not have been possible. While they are not as famous as Boulenger, without them and others like them, our knowledge of cave biology would be far poorer than it is today. Professional scientists owe a great deal of gratitude to such people. They are the unsung heroes of speleology.

Acknowledgments

Dr. G. Teugels, of the *Musée Royal de l'Afrique Centrale* in Tervuren, Belgium, provided useful information and specimens of *Caecobarbus geertsi*.

Literature Cited

Atz, J.W. 1951. Sightless Cave Fishes from the Bas Congo. *The Aquarist* **15**(12):236-237.

Boulenger, G.A. 1921. Description d'un Poisson Aveugle decouvert par M.G. Geerts dans la Grotte de Thysville (Bas-Congo). *Revue de Zoologie Africaine* **9**:252-253.

Heuts, M.J. & N.Leleup. 1954. La Géographie et l'Ecologie des Grottes du Bas-Congo. Les Habitats de *Caecobarbus* geertsi Blgr. *Annales du Musée Royal du Congo Beige Teruuren.* Ser.8. Sciences Zoologiques **35**:1-71.

Jankowska, M. & G. Thinès. 1982. A comparative study of group density in cave and epigean fishes. *Behavioral Proceedings* **7**:281-294.

Norman. J.R. 1938. Dr. G. A. Boulenger, FRS. *Nature* **141**:16-17.

Poll, M. 1968. George-Albert Boulenger 1858-1937. *Academie Royale de Belgique* pp.: 906-927. Academie Royale de Belgique, Classe des Sciences.

Proudlove, G.S. & A. Romero. 2001. Threatened fishes of the world: *Caecobarbus geertsi.* Boulenger, 1921 (Cyprinidae). *Environmental Biology of Fishes* **62**(1-3): 238-238.

Smith, M.1938. George Albert Boulenger, 1858 -1937. *Copeia* **1938**(1):1-3.

Chapter 20. Charles Marcus Breder, Jr. 1897-1983 [417]

Dr. Charles M. Breder, Jr., the ichthyologist who conducted a series of field and laboratory studies on the blind cave form of the fish *Astyanax fasciatus mexicanus*, died October 28, 1983, in his home at Englewood, Florida.

Born June 25, 1897, in Jersey City, Breder's formal education was interrupted by World War I. He obtained a University degree in 1938 when he received an honorary doctorate from Newark College, today part of Rutgers University. By that time, Breder was already a prestigious biologist within the scientific community. He not only pushed a considerable number of papers but also because of the impact of his publications in ichthyology.

Figure 20.1. Charles Breder at his lab. *Picture courtesy of the American Museum of Natural History.*

---

[417] Based on Romero, A. 1984. Charles Marcus Breder, Jr. 1897-1983. *National Speleological Society News* **42**:270-271.

Between 1919 and 1921, Breder worked as a science assistant and fisheries expert at the U.S. Bureau of Fishes in Washington. From 1921 until 1940, he worked at the New York Aquarium, where he started as an aquarist. He left as Assistant Director after being Acting Director from 1937 to 1939.

While at N.Y. Aquarium he planned an expedition to Mexico, under the auspices of the New York Zoological Society, to study a new genus and species of cavefish (*Anoptichthys jordani*) recently described by Carl L. Hubbs and Williams T. Innes. This species was later recognized as a blind cave population of the eyed surface fish *Astyanax fasciatus mexicanus*. As a result of this expedition and the live fish brought back to New York, Breder published about twenty papers on this cavefish, most of them between 1942 and 1947. This made this species one the best-known cavefishes of the world, still a subject of intense study in universities in the U.S. and Europe today.

During this intense research period, Breder took a job at the American Museum of Natural History in New York, where he remained as Chairman and Curator of Fishes and Aquatic Biology until his retirement. After that, he continued an active research program in marine fishes, mostly at the Mote Marine Laboratory in Sarasota, Florida.

In addition to his significant contributions to the study of cavefishes, Breder was regarded as an international authority on fish schools. His book "Modes of Reproduction in Fishes," co-authored with Donn E. Rosen and published in 1966, is still one of the most comprehensive reference on the subject.

Breder left two sons, nine grandchildren, and thirteen great-grandchildren. He is survived by his widow, Priscilla Breder, formerly P. Rasquin, who collaborated with him in the study of *A. fasciatus* and made essential contributions to this species' biology herself. Despite his contributions to biospeleology, Breder is little known among the speleological circles in the United States partially because he was not a speleologist *per se* but rather a biologist who found cave fishes an exciting subject of research, and partly because his research topic was outside of the United States.

Chapter 21. Herbert L. Needleman 1927-2017 [418]

On July 18, 2017, Dr. Herbert L. Needleman, of Squirrel Hill, Pittsburgh, passed away. He was a pediatrician and a child psychiatrist who greatly contributed to improve environmental health worldwide by demonstrating in the late 1970s that children exposed to even small amounts of lead could suffer intellectual and behavioral disorders. He was living at Weinberg Village, an assisted living home in Glen Hazel, neighbor in Pittsburgh, Pa., where he had spent the last two years while suffering from Alzheimer's disease. Dr. Needleman's work prompted regulations that limited or banned the metal in a range of common products, like gasoline and paint, and set a standard for the modern study of environmental toxins. He was an elected member of the Institute of Medicine and the founder of the Alliance to End Childhood Lead Poisoning, later known as the Alliance for Healthy Homes which has since merged with the National Center for Healthy Housing. The cause of his death was pulmonary edema. He was 89.

Biographical Timeline

Herbert Leroy Needleman was born on December 13, 1927, in Philadelphia, one of two sons of Joseph and the former Sonia Shupak who had both immigrated to the U.S as children. His father was a furniture salesman. His mother, was an immigrant from Russia whose father had built the Shupak Pickle factory that still stands in Philadelphia, building a pushcart operation that later merged with Vlasic in 1968, one of the most popular pickle brands in the U.S. Dr. Needleman worked there when he was young. His mother was the one who ran the household. Both of them were Jewish. Dr. Needleman was not an observant Jew but on certain occasions he would use some Yiddish expressions.

Dr. Needleman graduated from Muhlenberg College in Allentown, Pa., in 1948, making him the first college graduate in the family. In 1952, he received a MD from the University of Pennsylvania. He trained in Pediatrics at the Children's Hospital of Philadelphia and served as Chief Resident. He completed a fellowship in Pediatric Cardiology and Rheumatic Fever through the National Institutes of Health. After practicing family Pediatrics in Philadelphia and Neonatology at Pennsylvania Hospital, he completed a residency in psychiatry at Temple University Health Sciences Center in Philadelphia, Pa. From 1970 to 1972, he was an assistant professor of psychiatry at Temple University. From 1972 to 1981 he occupied the same position at Harvard University Medical School, in Boston, MA. Since 1981, he had been a professor of child psychiatry and pediatrics at the University of Pittsburgh School of Medicine.

He served in the Army, at Fort Meade, Md., and in the Army Reserve, attaining the rank of captain. He had to delay his entrance in the Army due to a knee injury he suffered while playing football.

Dr. Needleman's first marriage, to Shirley Weinstein, ended in divorce. He is survived by his wife of 54 years, the former Roberta Pizor in Pittsburgh; a son from his first marriage, Samuel Needleman, MD, of Pittsboro, North Carolina; two children from his second marriage, Joshua Needleman, MD, of the Bronx, New York City, and Sara Needleman Kline of Arlington, Va.; as well as seven grandchildren and three great-grandchildren.

---

[418] Originally published as Romero, A. 2018. In Memoriam: Herbert L. Needleman (1927-2017). *Environmental Research* **165**:507-509. (DOI: https://doi.org/10.1016/j.envres.2017.11.048).

Scientific Research

Dr. Needleman's interest in studying the effect of lead in children began while he was working at a community psychiatric clinic in North Philadelphia after medical school. There he met a youngster who approached him and explained his ambitions, which were large, even as the boy struggled with words. The child was bright and open; nonetheless he had deficits that struck Dr. Needleman as similar to those found in children with lead poisoning. "I thought, how many of these kids who are coming to the clinic are in fact a missed case of lead poisoning?" he said in a later interview.

His clinic office overlooked a school playground which sparked him with an idea. At that time, it was well known that exposure to high doses of lead caused mental slips, even permanent brain damage, and death. But what about the low-level exposure that many children, perhaps the ones playing in the yard, absorbed every day just by living in older urban neighborhoods thick with lead paint and industrial contamination?

Given that no one could study the effects carefully, there was no clear answer to that question; after all, the available tests for lead exposure were on hair, blood, or fingernails, which were not entirely reliable. Bone is the most accurate long-term repository since once absorbed into the body, lead circulates in the blood and accumulates in the skeleton. However, sampling bone tissue among living individuals is not only painful but also too intrusive and that makes the procedure unjustifiable, particularly in the case of children.

Inspired by an earlier study on lead poisoning based on measuring lead exposure in teeth from a small sample of individuals, Dr. Needleman got the idea of using these to as biomarkers of lead exposure. After all, teeth are a part of the human skeleton and young children shed them. Dr. Needleman then carried out a series of studies in both Philadelphia and, later, in a much larger project in the Boston area. What he did was to offer children aged 6 and 7 small rewards for their loose teeth, once they had fallen out. "That was the insight that changed everything," said Dr. Bernard Goldstein, former dean of the University of Pittsburgh's graduate school of public health. "Herb became the Tooth Fairy." The results of those studies showed that children living in poor urban neighborhoods had lead levels five times higher, on average, than those of their peers in the suburbs.

While still at Harvard, Dr. Needleman and his collaborators published in 1979 a paper in *The New England Journal of Medicine* that showed that children whose accumulated exposure to lead was highest in the group scored four points lower on an I.Q. test than children whose exposure was at the lowest end. The study was based on samples from more than 2,000 children.

Teachers rated the high-exposure children as having a host of classroom issues, including attention deficits and behavior problems. In subsequent studies Dr. Needleman demonstrated that there was a correlation between high lead levels and reading delays.

"It's not like you can look at one kid and spot a four-point difference in I.Q., and say, 'O.K., we know lead caused this,'" said Linda Birnbaum, director of the National Institute of Environmental Health Sciences, in Durham, N.C. "It's a population effect; you have to have a population of the right kids and ask the right questions. That's what Dr. Needleman did, and it has become a model" for subsequent research.

Attacks on his Scientific Integrity

Thanks to the studies by Dr. Needleman, federal regulators imposed stiffer regulations of lead in gas, tin cans, paint, household pipes and many other products, including the complete elimination of lead in gasoline. Today, federal health authorities consider lead at any level unsafe for children.

Needless to say, Dr. Needleman's work generated a strong adverse reaction by the lead industry which resorted in attacking both his scientific integrity and his personal character using all kind of tactics. For example, in the 1980's a pair of psychologists approached him for his data from the 1979 study, as part of a court case in which they were testifying on behalf of a lead smelting company. They then accused him of scientific misconduct before the newly formed federal Office for Scientific Integrity (OSI), then part of the National Institute of Health.

Dr. Needleman testified under oath that although he may have made some math mistakes in his analysis, those errors were minor and did not alter the conclusions of his research. Investigators from the OSI eventually agreed and dismissed all charges. Yet, the University of Pittsburgh, where he was then on the faculty, conducted its own investigation and locked him out of his own files, putting bars on his file cabinets. Although he was also cleared in that investigation, that was a personal burden for him and his family. "Even some of his university's colleagues that supported him in private, avoided to be seen with him in public" says his son Josh.

Dr. Philip Landrigan, the dean for global health at the Icahn School of Medicine at Mount Sinai in New York, said, "You have no idea what he went through. He swung in the wind for those years, but never backed down. I don't use this word often, but hero is appropriate in Herb's case." The whole thing took a toll on him and his wife.

In a 2005 interview, Dr. Needleman was asked whether the attack on his credibility was meant to scare off other researchers looking into environmental toxins. "If this is what happens to me, what is going to happen to someone who doesn't have tenure?" he replied. "I'm worried that people who are trying to get a niche and don't have tenure are asked to do things they question the ethics of," he continued, "will be intimidated. It's a real force."

Figure 21.1. Dr. Herbert Needleman (left) with the author in 1992. Picture by Ana Romero.

Other Humanitarian Work

Dr. Needleman's contributions to science and public health need to be understood into a larger context. In the 1960s, while teaching at Temple University, he became horrified by the impact of the Vietnam War on the civilian population, particularly its children. He became an active opponent of the war, which at least once led to his detention when he was arrested and spent a night in jail with Benjamin Spock, a renowned pediatrician, for their participation in an antiwar protest. In 1996, while at Temple University, he helped to found the Committee of Responsibility (COR) to Save War-Burned and War-Injured Vietnamese Children. COR enlisted a list of Nobel prize winners and doctors from around the country to support its goal of bringing Vietnamese children to the U.S. for medical treatment of war injuries. Dr. Needleman was its chairman from its beginning until the end of the war in 1975. "In that decade, COR brought about 200 children to the U.S. for treatment, and helped another 300 get treatment in Vietnam," said John Balaban, whom Dr. Needleman hired in 1968 to work with him at COR. "The goal was both to help the children and help end the war," said Mr. Balaban, now an English professor at North Carolina State University: "The basic ethos of COR was to bring the children into our communities and let people know about them. He believed there was a basic decency in Americans that could not support the destruction of innocent people," added Balaban. "One of those children lived with us, at our house," said his son, Joshua. "I was only 4 years old, but I remember."

He had a strong sense of right and wrong and was fearless when it came to make those sentiments public. He never stepped away from helping others in case of need. For full disclosure, that was something this author learned firsthand when his own scientific integrity was attacked by the gasoline and lead industries and government authorities in his native country, Venezuela, in the 1990's (Romero 2008, Romero 2010).

Legacy

Dr. Needleman played a key role in securing some of the most significant environmental health protections achieved during the 20th century. He has been credited with having played the key role in triggering environmental safety measures that have reduced average blood lead levels by an estimated 78 percent between 1976 and 1991. All that, despite generating stiff resistance from related industries, who targeted him with frequent attacks. Yet, he persisted in campaigning to educate interested parties, including parents and government panels, about the dangers of lead poisoning.

Figure 21.2. Dr. Needleman testifying before the U.S. Congress in 1991 (public domain image).

Acknowledgements

I thank Dr. Joshua Needleman and Mr. John Balaban who provided firsthand accounts about the life and work of Dr. Needleman. An anonymous reviewer provided useful comments. Other biographical information was summarized from the following sources:

Literature Cited and Consulted

Balaban, J. 1991. *Remembering Heaven's Face.* Athens: University of Georgia Press
Carey, B. 2017. Dr. Herbert Needleman, Who Saw Lead's Wider Harm to Children, Dies at 89. *The New York Times* 27 July 2017.
https://www.nytimes.com/2017/07/27/science/herbert-needleman-dead-lead-poisoning-in-children.html?_r=0 (Retrieved 6 August 2107).
Committee on Measuring Lead in Critical Populations, Board on Environmental Studies and Toxicology, Commission on Life Sciences, National Research Council. 1993. *Measuring Lead Exposures in Infants, Children, and other Sensitive Populations.* Washington, DC: National Academy Press.
Denworth, L. 2008. *Toxic truth. A Scientist, a doctor, and the battle over lead.* Boston: Beacon Press.
Ernhart, C. & H. Needleman. 1987. Lead levels and child development. *Journal of Learning Disabilities* **20**:262-265.
Glymour, C. 2010. *Galileo in Pittsburgh.* Cambridge: Harvard University Press.
Kennedy, D. 1997. *Academic duty.* Cambridge: Harvard University Press.
Langer, E. 2017. Herbert L. Needleman, pediatrician who exposed dangers of lead poisoning, dies at 89. *The Washington Post* 20 July 2017.
https://www.washingtonpost.com/local/obituaries/herbert-l-needleman-pediatrician-who-exposed-dangers-of-lead-poisoning-dies-at-89/2017/07/20/3bc644ba-6d53-11e7-b9e2-2056e768a7e5_story.html?utm_term=.ca698adbf783 (Retrieved 6 August 2017).

Markowitz, G. & D. Rosner. 2002. *Deceit and denial. The deadly politics of industrial pollution.* Berkeley: University of California Press.

Markowitz, G. & D. Rosner. 2013. *Lead wars. The politics of science and the fate of America's children.* Berkeley: University of California Press.

Michaels, D. 2008. *Doubt is their product. How industry's assault on science threatens your health.* Oxford: Oxford University Press.

Needleman, H.L. 1984. Testimony before the Committee on Environment and Public Works, U.S. Senate (98th Congress, 3nd Session) on S.2609: Airborne Lead Reduction Act of 1984. Senate hearing: 98-978, June 22, 1984.

Needleman, H.L. 2000. The removal of lad from gasoline: Historical and personal reflections. *Environmental Research* **84**:20-35.

Needleman, H.L.; C. Gunnoe, A. Leviton, R. Reed, H. Peresie, C. Maher & P. Barrett, 1979. Deficits in Psychologic and Classroom Performance of Children with Elevated Dentine Lead Levels. *New England Journal of Medicine* 300:689-695.

Needleman, H.L.; O.C. Tuncay & I.M. Shapiro. 1972. Lead levels in deciduous teeth of urban and suburban American children. *Nature* **235**:111-112.

Romero, A. 1996. The environmental impact of leaded gasoline in Venezuela. *Journal of Environment and Development* **5**(4):434-438.

Romero, A. 2008. Nobody's dolphins, pp. 11-19, *In*: Trauth, J. & A. Romero (Eds.). *Adventures of the wild: experiences from biologists from the Natural State.* Fayetteville: University of Arkansas Press.

Romero, A. 2010. The invisible enemy, pp. 7-12, *In*: LaFond, L., C. Berger & A. Romero (Eds.). 2010. *Adventures in the Academy: Professors in the Land of Lincoln and Beyond.* Edwardsville: College of Arts and Sciences, SIUE.

Romero, A. 2017. Tenure carries both privileges and responsibilities. *The Edwardsville Intelligencer* 31 July 2017, p. 3.

Chapter 22. Columbus, Christopher [419]

The place and date of birth of Columbus is still a matter of conjecture but he might have been born in Genoa around 1451 and died at Valladolid, Spain, on 20 May 1506. Most of Columbus's life prior acquiring fame is shed in mystery, many times by Columbus himself. We do know that he started sailing probably in his teens and the places included from Iceland to the western coasts of Africa. He was probably mostly self-educated reading as much as he could on cosmography and travels, including those of Marco Polo.

He probably developed early on the idea of sailing to the Far East going westward. He tried to gain support for that endeavor first in Portugal at no avail. Then moved to Spain and after several years seeking support, he secured it from the Spanish monarchs. He made four voyages to the American continent. In the first one (1492-1493) he visited what is known today as the Bahamas, Cuba, and Hispaniola. In his second voyage (1493-1496) he visited Dominica, Marie Galante, Jamaica, Cuba, and Hispaniola. In his third voyage (1498-1500) he visited Trinidad, the northern coasts of South America, and Hispaniola.

Columbus was unquestionably a man of genius. He was a bold, skillful navigator, better acquainted with the principles of cosmography and astronomy than the average skipper of his time, a man of original ideas, fertile in his plans, and persistent in carrying them into execution. The impression he made on those with whom he came in contact even in the days of his poverty, such as FGray Juan Pérez, the treasurer Luis de Santangel, the Duke of Medina Sidonia, and Queen Isabella herself, shows that he had great powers of persuasion and was possessed of personal magnetism. His success in overcoming the obstacles to his expeditions and surmounting the difficulties of his voyages exhibit him as a man of unusual resources and of unflinching determination.

His image today is that of a man that caused the death of millions of Native Americans through disease brought to the Americas as well as through wars and enslavement.

---

[419] Based on Romero, A. & S.D. Kannada. 2008. Columbus, Christopher. Pp. 163-164, In: *Encyclopedia of Tourism and Recreation in Marine Environments* (M. Lück, Ed.). Oxfordshire, UK: CAB International.

Christopher Columbus (public domain image).

Chapter 23. Vespucci, Amerigo [420]

Amerigo Vespucci was born in Florence, Italy on 9 March of 1451 and died in Seville, Spain on 22 February 1512. From an early age he showed a great deal of interest in astronomy, mathematics and cartography. In 1492 Vespucci left Florence, where he worked for the Medici family, and arrived in Seville, Spain, where he became the director of a company that supplied ships for long voyages. With this connection, he made five voyages to the American continent between 1497 and 1505, sometimes in the service of Spain and sometimes in the service of Portugal. On his voyages, Vespucci visited many new lands in the Caribbean and the Gulf of Mexico. He also made voyages to coastal areas from southern Argentina to the Labrador Peninsula as well as to Cape Verde and the Falkland Islands.

Although Columbus is better known than Vespucci today, that was not the case in the early sixteenth century, when the voyage narratives of Vespucci were more widely disseminated, by far, than were those of the voyages of Columbus. Furthermore, Florence, his native city, was the center for diffusion of news of discovery of the New World. Thus, it is not surprising that when the crew of the ship he commanded in his fourth trip of 1504 returned back to Spain, they told the story of their discoveries to cartographers and in 1507 Martin Waldseemuller, a German mapmaker, suggested to call the new lands America, a name that rapidly spread among other mapmakers. Thus, America was named after Amerigo Vespucci.

---

[420] Romero, A. & S.D. Kannada. 2008. Vespucci, Amerigo. P.170, *In: Encyclopedia of Tourism and Recreation in Marine Environments* (M. Lück, Ed.). Oxfordshire, UK: CAB International.

Amerigo Vespucci (public domain image)

Section IV. Book Reviews

This is section contains a collection of book reviews I have published on the history and philosophy of science—all of them (except for the last) published in peer-reviewed journals. Sometimes the book's original language was Spanish, but, in most cases, they were in English even when the topic dealt with Spanish or Latin American issues.

Most were about books recently published, but two were about books published centuries ago but whose analysis resulted useful, especially under a science historian's perspective. In most cases, I wrote very positive reviews with the only exception. That was the case in Chapter 30: "The Gene. An Intimate History." Although I always try to give constructive criticism, I found so many flaws in that book that I was very critical. I found all the others informative and useful for providing interesting views on the topics they dealt with. The last one of these reviews ("Chapter 33: Salvador Gilij's narrative about the Orinoco fauna") is different for many reasons. First, this was a coauthored review (a rarity). Second, my coauthor did most of the legwork. Third it is a series of annotated comments on different passages of the book that might be useful to both science historians and biologists. More on that in the comment I added at the beginning of the chapter.

Chapter 24. Las plantas del mundo en la historia: llustraciones botánicas de cinco siglos [421]

Despite its title, which gives the impression that it concerns world history as a whole, this book is basically a collection of historical essays on the works by Spanish botanists. Plant illustrations are loosely used as the unifying theme.

The first essay, by the distinguished scholar Jose Maria López Piñero, provides a reasonable historical account of the better-known Spanish botanists. References are made to major events in the history of botany in countries other than Spain. However, they are sparse and superficial. López Piñero ignores even some foreign botanists who visited Spain and its colonies. He mentions Linnaeus's disciple Pehr Löfling, for one, only in passing, even though Löfling conducted botanical studies not only in Spain but also in Venezuela. Thus, the essay is essentially a good summary of the botanical achievements by Spaniards up to the first half of the nineteenth century. Readers seeking an introduction to this topic will find this chapter an excellent place to start. Some specialists, however, will not always agree with López Piñero's assessment of the impact of the contributions he describes.

Manuel Costa Taléns is the author of the second essay, a mixture of history and concepts of flora and biogeography found in botanical text-books. Like López Piñero's essay, it is good for the non-specialist but contains some assessments that not all experts will agree upon. For example, Costa Taléns, like many of his compatriots before him, feels that foreigners such as Linnaeus and Philip Barker Webb failed to appreciate the botanical studies produced in Spain. There is no question that many of the scientific expeditions the Spanish Crown sent around the world included talented naturalists, while other good botanists remained in the Spanish peninsula and carried out their own research programs. But the failure to publish promptly and in full the results of those efforts, as well as the late acceptance of the Linnaean system in Spain (where for a while botanists preferred that of Joseph Tournefort), contributed to the image of Spain as a country where botanical research was not at the cutting edge of the times. Lack of scientific communication and the Inquisition, which took a hard stance against books written by non-Catholics, did not help to improve that image.

The third essay is a very interesting and illuminating contribution by Felipe Jerez Moliner on the evolution of illustration techniques up to the nineteenth century; anyone curious about this topic should read it. The next essay, by Jesus Ignacio Catalá Borges and Cristina Sendra Mocholi, provides a general introduction to systems of classification. And in the last essay, María Jose López Terrada examines the impact of the world of plants on Spanish art. I was happily surprised to find that the editors included this topic, as it is always fascinating to see the ways in which artists perceive and depict nature and scientific themes.

The rest of the book (more than half) superbly reproduces illustrations of many of the sources mentioned in the text. They include the typical one-species drawings and morphological illustrations (even microscopic ones) of plants and fungi, some in color. The explanatory text that accompanies these illustrations is highly informative.

In general, this book represents a diversity of approaches. The essays are furnished with the appropriate bibliographic sources, but unfortunately the volume lacks a

---

[421] Originally published as Romero, A. 1997. Las Plantas del Mundo en la Historia. *Isis* **88**(4):699-700 (Book Review in English).

subject/name index. This omission and the inaccurate title ("en la historia" should be "en la historia de España") are the book's only two major weaknesses. It is a very useful resource, particularly for those interested in exploring the work of Spanish botanists.

Chapter 25. Del diluvio al megaterio: Los orígenes de la paleontología en España

It is uncommon for a pocket book to be so rich in information and so useful for understanding the development of a particular branch of science in a particular country. Understanding the development of paleontology in Spain is what Francisco Pelayo's book is all about.

To be sure, Pelayo is not working in a vacuum. He carefully researched the history of paleontology, particularly in reference to considerations of the origin and nature of fossils since the ancient Greeks and explanations of why marine fossils were found on dry land, especially at great altitudes. Ideas from the universal flood to catastrophism are reviewed. More importantly, Pelayo carefully describes the influence of non-Spanish authors on Spanish scientists by

---

[422] Originally published as Romero, A. 2000. Francisco Pelayo. Del diluvio al megaterio: Los orígenes de la paleontología en España (Book review in English). *Isis* **91**(1):134-135.

analyzing their correspondence. He puts special emphasis on the influence of the French among Spanish intellectuals, an influence that began in 1700 with the attempt of Phillip V to imitate the French absolutist model and reached its pinnacle with the translation into Spanish of Buffon's *Histoire naturelle* in 1785.

In chronological order, Pelayo describes the works of Spanish researchers who wrote on fossils, including Benito Jeronimo Feijóo, Antonio José Rodriguez, and Fernando López de Cárdenas. But it is Jose Torrubia (1698- 1761) who receives the most attention, and rightfully so. Torrubia was a Franciscan monk who collected fossils not only in Spain but also in the Philippines, Cuba, and continental America. At a time when most authors either uncritically copied what others wrote or indulged in unsubstantiated speculation, Torrubia insisted on the importance of fieldwork, the use of modern (and at the time little used) instruments such as the microscope, and the comparison of one's own specimens with those from other collections, something he did thoroughly with collections from Italy and France. Yet all his care did not keep him from believing in fossils of inorganic origin or from suggesting, on the basis of fossil bones from giant ground sloths, that America, particularly Patagonia, had been populated by "giant" humans ("*gigantes*").

For Pelayo, Georges Cuvier's 1796 scientific description of the fossil ground sloth, the Megatherium, marks the end of major myths in the development of paleontology, not only in Spain but in the Western world more generally. Curiously, as Pelayo points out, Thomas Jefferson -who some call the first American paleontologist and who later became president- had the opportunity to describe this fossil from Argentina but missed the chance.

This is an extremely well-researched book that represents a key contribution to the early history of paleontology not only in Spain and its colonies but also in Europe. If there is a criticism to be made it is that it lacks a subject and author index. This is unfortunate in a volume in which more than a hundred authors are mentioned.

Chapter 26. From Popular Medicine to Medical Populism: Doctors, Healers, and Public Power in Costa Rica, 1800–1940 [423]

This is an excellent book that covers the evolution of medical practice in Costa Rica from colonial times to the point at which conventional (licensed) medicine became available to most Costa Ricans. It illustrates how changes in health practices worldwide influenced medicine in this small and (in terms of medicine) largely peripheral country. Although focused on Costa Rica, *From Popular Medicine to Medical Populism* contains numerous references to medical practices elsewhere in Latin America, so readers interested in the general topic of medicine in this part of the world will find it very valuable. The volume is well organized, and Steven Palmer writes with enormous clarity while exploring the subject thoroughly. He has produced a highly readable book that achieves an excellent combination of depth and breath.

One of the major strengths of this book, and what makes it quite different from other medical historiographies of the region, is that the author fully understands the role of unconventional medicine in Latin American countries. Thus, he explores the influences of Native American, Afro-Antillean, and Chinese traditions and the extent to which the role of the *curanderos* (healers) was embedded in the psyche of the local population.

Between the time the country achieved independence in 1821 and 1870, almost all licensed medical doctors in Costa Rica were European immigrants. Many of these "doctors" had "lost" their diplomas, and their "credentials" were accepted at face value in a country that was desperate for physicians. Slowly but steadily these physicians were replaced by Costa Rican doctors, who were mostly educated in the United States and in Europe; at the same time, the practice of medicine was increasingly institutionalized within hospitals.

Unlike other Latin American countries, where conventional medicine developed mainly in the large urban centers, Costa Rica remained largely rural. This socioeconomic condition set the stage for what the book describes so admirably: the *mestizaje* (race mixing) of medical practice— that is, the way the whole range of medical practitioners, from academically educated doctors to *curanderos* and healers, pharmacists of all kinds, and nurses and midwives, commanded the attention of all citizens in need of medical care.

The hybridization of these practices was not easy, however, because of the constant rivalries among different factions within each of those professions as well as among them. This resulted in what the author calls medical populism, where less academically oriented practitioners absorbed some of the techniques of the more academically trained ones and vice versa. Although this intellectual hybridization parallels the ethnic *mestizaje* of Costa Rica and other Latin American countries, the application of the word "populism" is quite adequate to describe the phenomenon, since populism is itself such a pervasive cultural practice in Latin America, in domains ranging from politics to religion.

For example, it has not escaped Palmer's attention that many academically trained doctors in Costa Rica occupied positions of great political influence (including the presidency) in the country, since populism is the easiest ticket into

---

[423] Originally published as Romero, A. 2006. Steven Palmer. *From Popular Medicine to Medical Populism: Doctors, Healers, and Public Power in Costa Rica, 1800-1940. Isis* **97**(2):369-370.

Latin American politics, it is not surprising that influential doctors accepted a form of populism in the practice of their own profession.

The author's excellent introduction provides the basic material for understanding the content of the book. There is mention of a subject that still awaits further studies: the reception and development of eugenic ideas in Latin America. The reader may be surprised to see how immigration laws and practices in Costa Rica (developed to keep "undesirables" out) mirrored those of the United States at the same time.

I was disappointed that there is no treatment of a medical practice generally illegal but common in Latin America—abortion—particularly because its frequency and the way it is practiced are closely linked to social class. Despite this shortcoming, however, Palmer achieves what he set out to do: a study of the evolution of medical practice and politics for an entire Latin American country.

Chapter 27. Floating Gold: A Natural (and Unnatural) History of Ambergris [424]

Ambergris is one of the most fascinating and sought-after natural goods, being its origin a most unexpected one: it is a byproduct of the digestion by sperm whales (*Physeter macrocephalus*). These marine mammals main source of food are squids and when their beaks (as well as possibly other indigestible material) block their duodenum (the first section of the small intestine), the intestinal wall absorbs water from the feces-impregnated mass causing solidification. As time goes by, the mass grows by accretion (Clarke 2006). The resulting material ends up having an irregular shape, a waxy nature, and can weight up to 420kg. The name ambergris is of French origin from the combination of two words: *ambre* (amber) and *gris* (gray), as a way to describe its appearance. Although they look solid, ambergris pieces float once they are expelled because of their specific gravity being lighter than water.

Though they have been described many times as "whale vomit" the ambergris pieces are most likely defecated or just released in the ocean when they become too big and kill the sperm whale because of intestinal obstruction. Once the whale dies, the ambergris pieces just float on the ocean. While some ambergris has been found on beaches and even collected floating on the ocean, most pieces of ambergris have been obtained directly from sperm whale intestines that have been caught in whaling operations. Since sperm whale whaling is prohibited by the International Whaling Commission (a prohibition that is violated by a few countries), findings of ambergris pieces in natural conditions are about the only way we can increase the not too large collection of these pieces in museums and private collections around the world. Its odor changes according with the time it has been exposed to air: from almost fecal smell to a one of a fragrance of rubbing alcohol, and it is this odor that makes it so attractive.

Ambergris has been known since antiquity with Arabs and Chinese first using it for perfumery or to burn it as incense. Because of its unique odor it was used in Europe during the plague to "sanitize" the air, which was believed (incorrectly) to be at the root of the epidemic. Some also used it as an aphrodisiac and for other medicinal purposes (none of them with confirmed positive effects).

Because of its rarity (only about 1% of sperm whales seems to produce ambergris) its commercial value has always been considered very high. Its monetary worth used to reach several tens of thousands of dollars, but not anymore in this United States. Since it is a "whale product" it is protected under the Marine Mammal Protection Act, therefore, cannot be commercialized.

Floating Gold is about ambergris, but those expecting to read a scientific book, even a standard popular science book when going through its pages, will be disappointed. In many ways, that is not very surprising since we know very little about the natural history of this substance. Besides technical aspects about its chemistry, what was summarized above is pretty much all what we know (or we think we do since it is largely based on hypotheses) about the biology of ambergris. After all, nobody has observed the process of formation or even expelling of ambergris.

Kemp's book can even be interpreted as a collection of just-so stories without a subject index (which is too bad) or a bibliography. That does not mean that there is lack

---

[424] Originally published as Romero, A. 2013ad. Floating Gold: A Natural (and Unnatural) History of Ambergris. Christopher Kemp. Chicago and London: The University of Chicago Press, 2012. 187 pp. $22.50 cloth (ISBN – 13: 978-0-226-43036-2). *Polymath* **3**(2):56. (Book review).

of scientific rigor in it. Kemp took a route of a more casual narrative that will please many readers not really interested in scientific aspects of ambergris as a first motivation for reading it. Yet, they will be surprised because Kemp, a molecular biologist, interjects scientific facts with ease and in very informative way. Kemp is also a popular science writer and does a good job as such. He does not sacrifice attention to the facts when telling human stories about ambergris, and that is where the book excels: describing human nature and behaviors which sometimes brings the best and worst of people when they encounter things that are very valuable and not necessarily well understood.

The reader may ask: is ambergris that indispensable today to the perfumery industry? The answer is no; this substance –or rather the one with its fixative properties for perfumery- is produced today synthetically and at a low price. Yet, if you happen to find one on a beach, keep it or donate it to a science museum: it is always a rare natural product that not many people can find while walking on a beach.

Literature Cited

Clarke, R. 2006. The origin of ambergris. *Latin American Journal of Aquatic Mamm*als **5**:7-21.

Chapter 28. Constituciones De La Real Y Pontificia Universidad De San Gerónimo: Fundada en El Covento De San Juan De Letrán, Orden De Predicadores, De La Ciudad De San Cristobal De La Habana, en La Isla De Cuba) [425]

The University of Havana is not only the oldest institution of higher education in Cuba, but also one of the oldest in the Americas. It was founded in 1728 as "Real y Pontificia Universidad de San Gerónimo de la Habana" (Royal and Pontifical University of Saint Jerome of Havana). As is the case with many universities in Europe, it had to have the backing of both the State (represented in this case by the King of Spain, who at that time was Felipe V (Philip V), and the Church (in this case Pope Innocent XIII). This was a reflection of both the ties that European universities had with the Church since medieval times and the need feudal states had to educate the people who ran state affairs.[426] By 1842 the university became secular changing its name to "Real y Literaria Universidad de La Habana" (Royal and Literary University of Havana) and during the twentieth century the university was at the center of many of the political movements in Cuba, always on the progressive side.[427]

Now Nabu Press, a publisher that specializes in printing books that because of their date of publication are in the public domain, has produced a facsimile version of the 1833 policies and rules of that university approved by Fernando VII [Ferdinand VII] then King of Spain. This is a reproduction of an original that was deposited at the Harvard University Library in 1925. As a result of the U.S. embargo against Cuba, it is not easy for researchers outside that Caribbean island to access this kind of material and, therefore, reproductions like this are always a welcome addition to primary sources that can be accessed easily.

This document reveals a number of historical details that are useful to better understand the history of institutions of higher education in Spanish America. One is that the new Cuban university was modeled after the one in Santo Domingo on the island of Hispaniola (today known as the "*Univerisidad Autónoma de Santo Domingo*") and founded in 1538, which makes it the oldest universities in the Americas). Yet, when it came to writing their policies, they could not find those for the university of Santo Domingo, so they were compelled to create original policies, which were developed with the assistance of the faculty and alumni indicating a very early manifestation of an advanced practice of shared governance.

Still, in the preamble of the book it is recognized that some of the policies were derived from other Spanish institutions of that time like the Universidad de Alcalá (in Madrid) or the Universidad de Salamanca, probably because some of the faculty in Cuba had studied in those institutions.

---

[425] Romero, A. 2014. Constituciones De La Real Y Pontificia Universidad De San Geronimo: Fundada En El Convento De San Juan De Letran, Orden De Predicadores, De La Ciudad De San Cristobal De La Habana, En La Isla De Cuba – Primary Source Edition. (Facsimile Reproduction) Charleston, NC: Nabu Press 2013. 136 pp. $16.66 paperback (ISBN 978-1289778668). *Polymath* **4**(4):96 (book review).

[426] See pp. 206-213 in David C. Lindberg (1992) *The Beginnings of Western Science*, Chicago, University of Chicago Press, 455 pp.

[427] See http://www.uh.cu/universidad/historia-de-la-uh (accessed 14 February 2014) for a more detailed history of the institution.

The policies made clear that the university was closely tied to the members of the convent of San Juan de Letrán who would occupy all the positions in the upper administration of the institution. The document goes on to establish all procedures to a fine level of detail including for non-academic functions such as how funerals of faculty should be conducted, salaries determined, and other minutiae.

Except for page 71 of the facsimile reproduction (which is barely legible), all the pages are well reproduced. Almost half of this book also contains *"Reales Disposiciones"* (Royal Decrees) that go as far as detailing attendance policies. It seems that the Spanish monarchy felt compelled to micromanage the university from about 7,500 km (about 4,660 miles) away. This reflects the strict sense of control that the Spanish crown had regarding what was going on in its territories. So much for shared governance.

In summary, this facsimile reproduction represents a valuable source of information for those interested in the history of universities in Spanish America and of Cuba in general, particularly for those residing outside Cuba.

Chapter 29. The New Celebrity Scientists. Out of the Lab and into the Limelight [428]

In the last couple of decades, we have seen the widespread ascendancy of the phenomenon of celebrity in society. Celebrities as a cultural manifestation are not necessarily something new. We saw that notion in the twentieth century being exploited by Hollywood through their "star system" as well as by sports teams hungry to increase their revenues. Now that phenomenon has expanded into areas that we would not have imagined decades ago, and one of them is in the field of science. With the advent of social media and the relaxation of social views regarding stereotypes, we have seen the rise of the figure of the celebrity scientist.

In a very timely and well written book, Declan Fahy analyzes this phenomenon in depth and provides us with an understanding into this trend. His book is divided into ten chapters. The first one, "A Brief History of Scientific Celebrity," gives as a historical background that shows that the concept of scientific celebrity is not a new one. His first example is Charles Darwin. Before publishing his book *On the Origin of the Species by Means of Natural Selection* (1859) Darwin was unknown beyond some scientific circles. But by publishing a book that defied the conventional wisdom generated by Christian beliefs and written in a language that anybody could understand Darwin produced a firestorm to the point that the first edition composed of 1250 copies sold within the first few days. Darwin's ideas were not only hotly debated but his image through caricatures (often representing his head in a monkey's body) became commonplace.

However, I do not think Darwin was the first scientific celebrity. We could mention that Galileo Galilei also received a great deal of notoriety in Europe because his ideas defied the Ptolemaic view of the universe. Galileo presented a new model in which the earth was not at the center of the solar system. That was a sharp contrast with the model supported by the church in which the sun was orbiting our planet. Further, Galileo presented his ideas with a provocative stance against authority in a very accessible way through the publication of his 1632 book *Dialogo sopra i due massimi sistemi del mondo* (Dialogue Concerning the Two Chief World Systems).

Other examples of early scientific celebrities include people like Benjamin Franklin whose experiments with electricity made him the first internationally famous American scientist, although many forget his scientific contributions because of his great role in politics. In fact, he was known during his time as the U. S. ambassador in France as *L' ambassador electrique* because of his scientific experiments with electricity. Other countries developed their own scientific celebrities like the Spaniard Santiago Ramón y Cajal who by winning a Nobel Prize in Physiology and Medicine in 1906 launched him into national fame.

In any case Fahy does include other earlier scientific celebrities such as Albert Einstein, the British astronomer Fred Hoyle who reached great deal of fame because of his radio broadcasts on scientific topics, and Carl Sagan whose TV show *Cosmos* made him a star. Yet Fahy reminds us that celebrity has its costs for those who are scientists and probably Sagan was the clearest example of that. When Sagan became famous in the 1980s his colleagues started to criticize him for his notoriety talking about the "Sagan Effect" defined as "fame is inversely proportional to the quality of their research." That was highly unfair because he had had a solid scientific career with hundreds of

---

[428] Originally published as Romero, A. 2016ad. *The New Celebrity Scientists. Out of the Lab and into the Limelight*. Fahy, Declan. Lanham: Rowman & Littlefield, 2015. 287 pp. $38.00 hardcover (ISBN 978-1-4422-3342-3). *Polymath* **6**(1):11-15. (Book Review).

publications in the scientific literature and his undeniable defense of science against the threats of pseudoscience and other beliefs. Not only that but his membership to the National Academy of Sciences was denied by his own peers because they were jealous of his public profile.

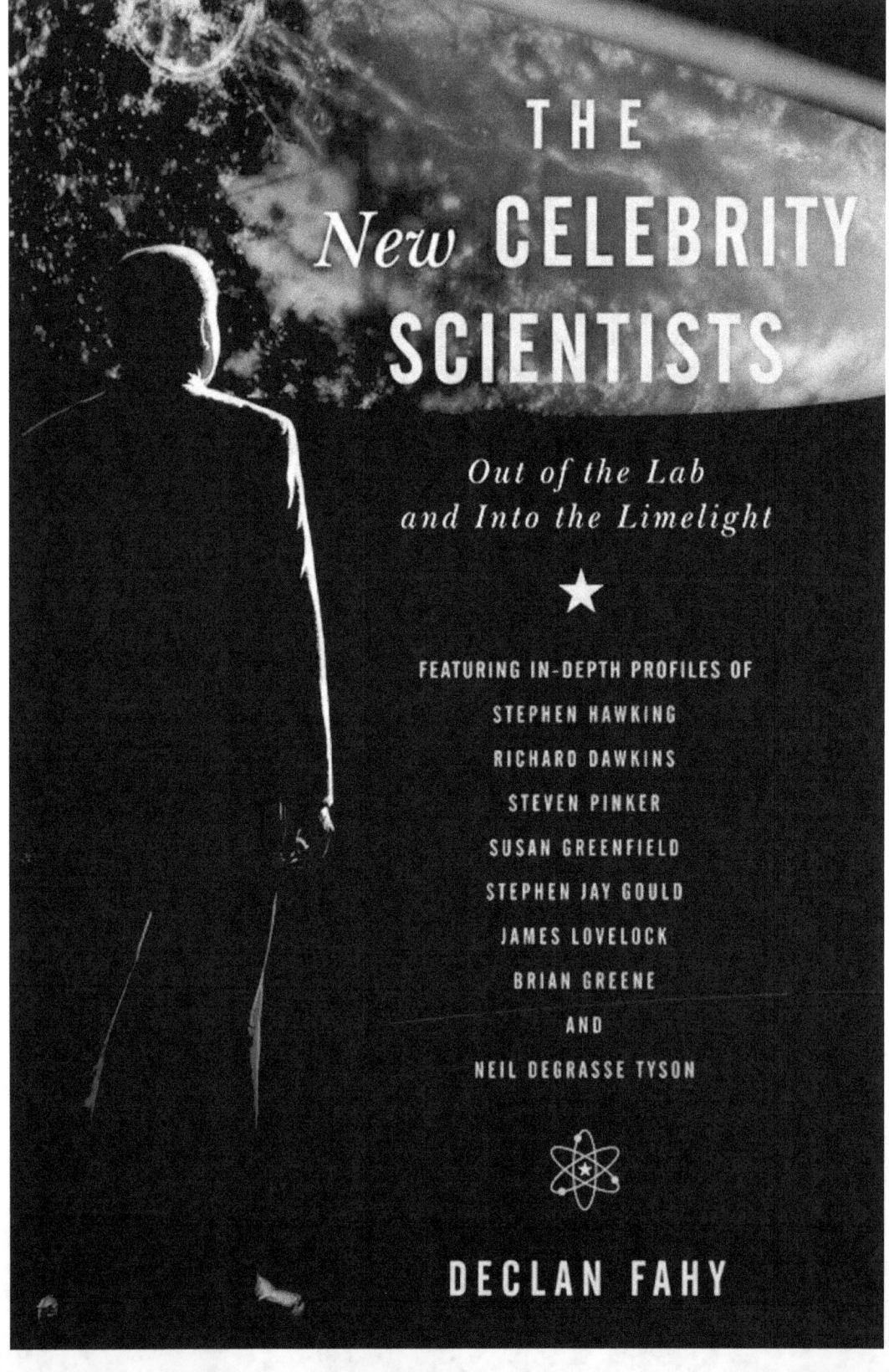

Although other scientists like the anthropologist Margaret Mead and the ecologist Paul Ehrlich are mentioned in this introductory chapter, this section could have benefited further by discussing the cases of environmentalist Rachel Carson whose book *Silent Spring* had a great impact on environmental policy worldwide or the chemist Linus Pauling whose scientific and political stances made him a celebrity in his own right.

Fahy also discusses in this chapter what he calls the bases of "celebrification" which he describes as the result of three processes: (1) an individual whose physical appearance is easily recognizable (e.g. Einstein, Sagan), (2) becoming a cultural commodity (e.g. Einstein seen as a genius, Sagan because his numerous TV appearances, particularly on Johnny Carson's *Tonight Show*, as well as Darwin through the caricatures of him), and (3) becoming the face of science before the general public. In any case almost all of them seem to have used the media and/or images to express their views.

This is a good introduction to the next eight chapters in which Fahy analyzes one living scientific celebrity per chapter. First is the cosmologist Stephen Hawking for whom timing and physical appearance combined to make him famous. His research on cosmology coincided with an increasing public interest in the 1970s not only for that topic in general but also for black holes and the origin of the universe in particular. Because of that, he became a media favorite. Part of the public interest in scientific developments in cosmology was because there were obvious religious implications of his work. Hawking was diagnosed as a twenty-one-year-old undergraduate with of amyotrophic lateral sclerosis or Lou Gehrig's disease. Between 1978 and 1984 there was an explosion of articles about him in the media emphasizing contrasts between his mind and body. The media portrayed Hawking as an icon of a brain independent from his physique. But what really launched him was the publication of his 1988 book *A Brief History of Time*. Thanks to the wise advice of the editor, his book was written in a way that was palatable to the general public, making it a best seller for many months and selling more than nine million copies. Hawking was from early on in a wheelchair and incapable of speaking except through a computer. Despite his physical disabilities or even because of them, Hawking not only welcomed his celebrity status but also has always been very image conscious as documented by the professional photographers that have had picture sessions with him.

The next chapter is dedicated to the British evolutionary biologist Richard Dawkins. He was known early on for his contributions in the area of behavioral ecology but gained prominence by emphasizing the centrality of genes in evolutionary processes, something he explained with both rigor and fluidity in his 1977 best seller *The Selfish Gene*. There he proposed that genes tend to increase their own chances of survival in competition with other genes. In his case timing also helped for two years earlier the Harvard evolutionary biology Edward O. Wilson had created a firestorm by proposing that genes had a great deal of influence on animal behavior. The premise was hotly criticized by many of his colleagues (particularly at Harvard) on the belief that such a proposal justified the idea that behavior (particularly among humans) was predetermined and that nothing could be done to change it. Dawkins also achieved notoriety for the development of the concept of a "meme," that is, that ideas, behaviors, or styles spread through society as part of the culture with memes behaving just as genes would do in the biological arena. He later became a strong defender of evolutionism against creationism as well as atheism though his book *The God Delusion* (2006).

Other cases of scientific celebrities detailed in the book are those of the Canadian-American psychologist and linguist Steven Pinker, the paleontologist and evolutionary biologist Stephen Jay Gould, the British neuroscientist Susan Greenfield, the British environmentalist and futurist James Lovelock, the theoretical physicist Brian Greene, and the astrophysicist Neil deGrasse Tyson. Although Fahy makes an effort to present these examples as a diverse pool, one wonders why he did not include other scientists who are better known to the general public such as the British primatologist Jane Goodall. In any case, the author makes his points with the examples he selected. For instance, he says that the process of "celebrification" occurs in two steps: developing into a public intellectual and the course of being transformed into a celebrity. He contends that scientists need to be available to the media because they work in areas of concern and/or interest to the general public. Yet it is clear through these examples that these celebrity scientists many times face conflicts with their private lives and public revelations, since some press like to focus on more or less "scandalous" disclosures.

In chapter ten titled "A New Scientific Elite" Fahy concludes that now many scientists are common features in the media, including social media, and recognizable beyond their own scientific circles. "When scientists are celebrities, they give science a face, force, and an impact in public life" he writes. Fahy also says that we sometimes forget that unless a scientist has been born with a silver spoon in his or her mouth, they need public support to do their work, whether from government agencies or private sources. Given the decreasing financial support from the public sector, private ones have become more and more important; hence, the need for celebrities to attract that kind of attention.

Other points made in the book are well substantiated such as that of scientists carrying ideas to communities beyond science, that the celebrity scientists show how science works and the joy of doing science despite personal fights among scientists exhibiting that they are humans after all. Also, that scientific celebrities develop the power to influence citizen understanding, public culture, and scientific life. These scientific celebrities tend to be brilliant synthesizers of science and celebrities draw a crowd. Fahy finally concludes that "The Sagan Effect" has almost vanished and that now a legion of scientists seeks to have a strong media presence and that can only be positive.

Given that a number of studies show that scientific literacy in the United States is below of that of many other developed countries in the world, we need more scientists engaged with the public so they can attract more interest in science instead of pseudoscience and the case studies presented in this book are good examples of that.

In summary, this is a book that is judicious, well researched and written, and balanced in its appreciation and understanding of the people portrayed in the case studies. I particularly liked the graphs that show how many times his subjects were the object of citation in the media related to events in their lives. I recommend this book to anyone interested not only in science and its protagonists or communication studies but also in a field that has been little considered but can provide us with many insights into today's culture: the sociology of science.

Chapter 30. The Gene. An Intimate History [429]

Heredity has always been, in one form or another, at the center of biological research. There is little doubt that the first scientific experimentation took place about 15,000 years ago, when humans started breeding plants and animals in order to domesticate them. Today the science of genetics seems to advance at a such pace that even the experts have trouble keeping up with all the developments. Now we are witnessing the expansion of our understanding of this realm of science to levels unimaginably just a few decades ago.

Among the questions related to heredity that researchers have been trying to understand from the beginning of times are: (1) how does fertilization take place? (2) what is exactly transmitted during copulation that leads to conception? (3) is spontaneous generation possible? (4) is sexual reproduction the only way to produce new individuals? (5) what are the respective contributions to the characteristics of a child made by the father and the mother? (6) does the mother make a "genetic" contribution in addition to nursing the developing embryo? (7) are the gametes (sex cells) formed throughout the body or in specific organs? (8) how is the sex of the offspring determined? and (9) how heritable characters are influenced by external factor such as the environment or even use and disuse?

Although today we have clear answers to these questions, those answers came through centuries of trials and the employment of the latest technologies available by both amateurs and professional scientists from all over the world. Therefore, it is not surprising that heredity has always been a major puzzle to both scientists and those who try to make science understandable to the general public.

The scientific bases of modern genetics as a predictive science were not established until the 1860's with the work of the Austrian Augustinian monk Gregor Mendel; yet, his seminal ideas about heredity were not understood until 1900 when the Dutch botanist Hugo de Vries, the German botanist Carl Correns and -to certain extent- the Austrian agronomist Erich von Tschermak, rediscovered Mendel's work and made them well known within the scientific community. By all accounts Charles Darwin did not know of Mendel's work. After all, he always acknowledged that his theory of evolution by means of natural selection lacked the understanding of the phenomenon of heredity. By the same token it is widely accepted that Mendel did know Darwin's ideas. Mendel marked the copy of "The Origin of Species" that he kept at the library of his monastery with an exclamation point next to the passage "There are many laws regulating variation, some few of which can be dimly seen."

In fact, the whole story of why Darwin never knew about Mendel's work is quite fascinating and has been the subject of a great deal of research[430]. The general consensus of these and other researchers is that having Darwin read Mendel's article, he would have found a detailed analysis of the frequencies observed for different inherited traits from generation to generation of the edible pea. Yet, these results were presented in a mathematical form and that might have been unpleasant for Darwin who once said that mathematics in biology is like a scalpel in a carpenter's shop – there is no use for it.

---

[429] Originally published as Romero, A. 2016. Siddhartha Mukherjee. *The Gene. An Intimate History*. New York: Scribner, 2016. 593 pp. $32.00 cloth (ISBN 978-1-4767-3350-0). *Polymath* 6(2):79-83.

[430] See, for example, Bishop, B.E. 1996. Mendel's opposition to evolution and to Darwin. Journal of Heredity 87:205-213; Galton, D. 2009. Did Darwin read Mendel? Q. J. Med. 102: 587–589; and Singh, R. S. 2015. Limits of imagination: the 150th Anniversary of Mendel's Laws, and why Mendel failed to see the importance of his discovery for Darwin's theory of evolution. Genome 58:415-421.

Darwin might have also found Mendel's conclusions unacceptable. Mendel argued that the transference of characteristics amongst cultivated plants occurred by discrete integral steps and could "transform" it into a different species, which ran contrary to Darwin's belief in blending inheritance.

Later, it took the work on the structure and function of the nucleic acids, first by the Swiss physician Friedrich Miescher in 1874 and then by the American James Watson and the British scientists Francis Crick, Maurice Wilkins and Rosalind Franklin as well as scientists working on population genetics, that the full concept of gene was really developed. Despite the importance of these discoveries, many science popularizers, including the famous Isaac Asimov, had trouble transmitting the scientific basis of heredity to the general public in a way that was easy to understand in its complexity.

Now comes Mukherjee's book on the history of the gene. With nearly 600 pages of text, I read the book with great anticipation. First the whole history of genetics is a fascinating one for a number of reasons. First, contributors to this branch of science came from all over the world and at different stages of the history of science. Second, the development of ideas on heredity were not always linear. Third, the history of genetics is full of personal stories including scientists showing the worse of themselves, from back stabbing to appropriation of others' work. Fourth, the science of heredity has been misused for political or ideological reasons to justify even mass murder. Therefore, the history of genetics and its central element, the gene, always provides us with many captivating stories that would captivate both the specialist and the general public.

There have been many other books and hundreds of scholarly articles dealing with different aspects of the history of heredity; there are just too many to mention here. Thus, the book by Mukherjee, an assistant professor of medicine at Columbia University Medical Center and a 2011 Pulitzer Prize winner for his book *The Emperor of All Maladies: A Biography of Cancer*, created all sort of expectations. However, those expectations failed to materialize for a number of reasons.

One of the problems of this book is that by focusing exclusively in the term gene, it misses a lot of the context one needs in order to understand its history. In fact, the term gene was coined by the Danish botanist Wilhelm Johannsen in 1909, that is, nine years after the rediscovery of Mendel's work and about four decades after the work itself. And that does not mean that all the research prior to the twentieth century is irrelevant.

There are even many interesting and revealing developments in the history of heredity prior Mendel. From the domestication of plants and animals as far as 15,000 years ago (not mentioned in the book), to many famous ancient Greek thinkers including -but not exclusively- Pythagoras, Plato, Anaxagoras, or Hippocrates who rightly or wrongly proposed influential ideas about heredity. While there were mostly wrong, they were still believed by many all the way to Mendel's time. Yet, Mukherjee fails to mention their ideas.

Furthermore, the portrait of Aristotle contributions to the notions of heredity in Mukherjee's book are not only incomplete and misleading but also, he ignores many more contributions made by others between the ancient Greeks and Mendel's times such as those by Theophrastus of Eresus, Herophilus of Chalcedon, Galen, Avicenna, to Leonardo da Vinci who debunked many false ideas about heredity.

Members of the movement that created period in the Seventeenth Century known as Modern Science such as William Harvey, Nehemiah Grew or Anton van Leeuwenhoek are also ignored. Eighteenth Century ideas such as epigenesis or parthenogenesis are never mentioned, nor are the contributions of plant breeders of the Seventeenth and Eighteenth centuries. And the same can be said about Charles Naudin, a clear precursor of Mendel's ideas.

But lack of acknowledgement to major historical developments and pioneers are not the only mistakes made in the historical analysis of heredity in this book. Lack of describing the influence of certain ideas is also rampant. Probably one of the most egregious examples is that the author completely obviates the impact that the book "The Selfish Gene" by Richard Dawkins had on our thinking on how genes ultimately work. Forty years ago, Dawkins proposed that genes strive for immortality and that organisms, from bacteria to humans, are just the carriers of such struggle for survival. Not only that but that all things related to life as a phenomenon, such as behavior, serve the ultimate goal of passing information from one generation to the next. And here Dawkins did use appropriately a good metaphor: genes are selfish. Thus, at the end of the day is not a particular organism the one that is trying to survive but the genes within the organisms, and that is an "intimate" notion of genes. Although some of Dawkins ideas are mentioned sporadically, his seminal book and its influence on our understanding of evolution are nowhere to be found in this book.

In addition to the lack of acknowledgement to these and other ideas and precursors, some of the historical characterizations of the individuals mentioned in the text are quite false. On page 28 Mukherjee describes Charles Darwin as a "young clergyman." The problem is that this statement gives the impression that Darwin was some kind of a churchman, cleric, minister, or preacher but that was not the case at all. After dropping out from Edinburgh University school of medicine, which he attended following a long family tradition of producing medical doctors, he dropped out because he could not withstand, among other things, witnessing surgical procedures performed on other human beings without anesthesia. Then, his father pretty much forced him to go to the University of Cambridge in 1828, when he was 18 years old, in order to pursue theological studies so he could become a clergyman, but although he graduated two years later from Christ's College with the Bachelor in Arts degree, a precondition to be ordained, he never took the vows. Actually, he spent most of his time at Cambridge collecting animals, plants, and geological specimens. When Darwin returned to his family home, he had little interest in pursuing a religious career and jumped to the opportunity to go on board of the *Beagle* as an unpaid naturalist. In fact, his father opposed the idea of his son embarking in that voyage telling him "You care for nothing but shooting, dogs, and rat-catching, and you will be a disgrace to yourself and all your family."

Another fundamental problem with this book is the scope given by the title: the history of the gene. Unlike a scientific idea such as evolution, heredity, or ecology, gene is a concept based on a physical entity. As such, we are dealing with a natural object. That is why there are so few books with the title of "history of the atom". It is easier to talk about the atom as an idea later shaped by science than as an object out of context. If you were to encompass genes as part of a history then you should address it as a "History of Heredity" because heredity is an idea, a concept, not an object.

Thus, Mukherjee's book gets off the rail in its historical approach and has plenty of other superficialities and inaccuracies that do not make it to par with many other books that have been published on the subject.

The book's narrative style does not help either. Mukherjee uses metaphors and similes all the time and if you are not a trained biologist you will get lost in trying to understand the real meaning of his attempts of describing facts. For example, on page 150, the author writes "A single strand of DNA consists of a backbone of sugars and phosphates, and four bases –A, T, G, and C- attached to the backbone, like teeth jutting out from a zipper strand." A single illustration of the world's most famous molecule, would have worked better than this simile. And examples of these unintelligible –and sometimes even misleading- comparisons are everywhere in the book.

In conclusion, the history of heredity, genetics and the very concept of gene is a gold mine to be exploited and clearly explained to the general public. Mukherjee's book does not fulfill those goals.

Chapter 31. The Beautiful Brain: The Drawings of Santiago Ramón y Cajal [431]

Santiago Ramón y Cajal is one of the scientific giants of the twentieth century. By all accounts he was the founder of neuroscience. But these two phrases fall short of understanding not only his impact on science but also his unique talents.[432]

Ramón y Cajal (that is his full surname) was born on May 1, 1852 in a very small town called Petilla de Aragón, in northeastern Spain. He was the son of the local surgeon. As a teenager, he rebelled against any form of authority including his father's wishes of becoming either a shoemaker or a barber while his real interest was in becoming an artist for which he showed early signs of talent. Hoping to interest his son in a medical career, Ramón y Cajal's father took him to graveyards to find human remains for anatomical study during the summer of 1868. Sketching bones was a turning point for him and, subsequently, he did pursue studies in medicine at the University of Zaragoza while also showing a great deal of interest in gymnastics and philosophy. After graduating as a medical doctor in 1873, he was drafted as a medical officer and sent to the war in Cuba during 1874 and 1875, in which the Spanish crown was fighting the rebels. There he contracted malaria and later tuberculosis.

Soon after he returned from Cuba, Ramón y Cajal got his first academic position as "auxiliary professor" of anatomy at the University of Zaragoza. In 1877, he received his doctorate in medicine in Madrid. He married Silveria Fañanás García in 1879, an uneducated young woman, who stood at his side for the rest of their lives (she died in 1930). They had four daughters and three sons (two of whom died in their childhood). He was awarded professorship positions at the universities of Valencia (1883), Barcelona (1887) and Madrid (1892). He was also the director of the Zaragoza Museum (1879), director of the *Instituto Nacional de Higiene* – (*National Institute of Hygiene*, 1899), and founder of the *Laboratorio de Investigaciones Biológicas* (*Laboratory of Biological Investigations*, 1922), later renamed in his honor as the *Instituto Cajal* (Cajal Institute).

In 1877, he purchased with his own funds an old-fashioned microscope thus beginning his research career. Initially, he was interested in inflammation and on the structure of muscle fibers. In 1885, during his tenure as Professor at the University of Valencia, the provincial government of Zaragoza, in recognition of his labor during a cholera epidemic, awarded him a modern Zeiss microscope.

Yet, it was not until 1887—at 35 years of age—that he started what was going to be his great contributions to science. That year he learned about the method of cell staining developed by the Italian Camillo Golgi. This method greatly enhanced the ability to study nerve cells and Ramón y Cajal himself modified the Golgi method in order to improve its capabilities. He presented the results of his new observations at the 1889 Congress of the German Anatomical Society, which were greatly appreciated by his fellow scientists.

Based on his observations, he concluded that unlike the prevalent belief that the nervous system was made up a network of continuous elements, it was actually made of basic units represented by individual cellular elements (later named "neurons"). This conclusion is the modern basic principle of the organization of the nervous system.

He also proposed "the law of dynamic polarization," stating that the nerve cells

---

[431] Originally published as Romero, A. 2017ao. Newman, Eric A., Alfonso Araque, Janet M. Dubinsky, Larry W. Swanson, Lyndel Saunders King, and Eric Himmel. The Beautiful Brain: The Drawings of Santiago Ramón y Cajal. New York: Abrams, 2017. 207 pp. $40.00

[432] Romero, A. 1975. Ramón y Cajal: Génesis y Evolución de Un Científico. *Algo* 274 (1975): 8-12.

are polarized, receiving information on their cell bodies and dendrites, and conducting information to distant locations through axons, which turned out to be a basic principle of the functioning of neural connections. Ramón y Cajal and Golgi shared the Nobel Prize in 1906 for their studies on the nervous system. Other contributions from his more than one hundred scientific papers and many books are just too many to summarize for this book review.

Ramón y Cajal (who died in Madrid on October 17, 1934) was also an accomplished photographer and his medical illustrations are legendary even today. Hundreds of his drawings illustrating the delicate arborizations of brain cells are still in use for educational and training purposes today. Even by current standards, they are considered among the most remarkable illustrations in the history of science and that is what this book, *The Beautiful Brain: The Drawings of Santiago Ramón y Cajal*, is all about.

The book contains several very well written chapters by distinguished academics about Ramón y Cajal's life and work. In addition to many of his photographs (a number of which are self-portraits), the book reproduces 75 of his drawings, all of them accompanied by an explanation by the contributors to the book which are very helpful for understanding their scientific significance. Further, the book also contains a few modern photographs of the nervous system that help to contextualize Ramón y Cajal's accuracy of observations as well as his vision that allowed him see things that others were unable to observe.

If there is a book that presents science greatly benefitted by art and insight, this is it.

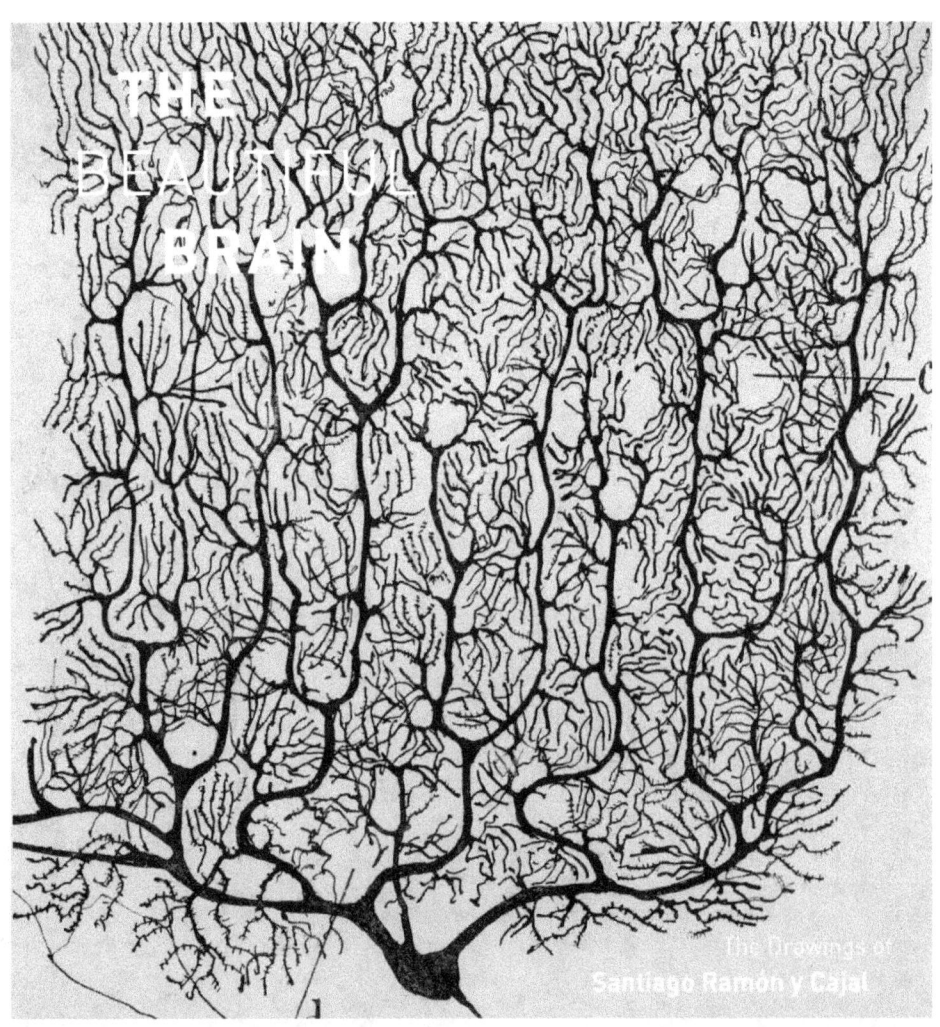

Chapter 32. The Glass Universe: How the Ladies of the Harvard Observatory Took the Measure of the Stars [433]

Women in science have been and are still facing numerous obstacles. According to the American Association of University Professors, despite the fact that 60 percent of all doctoral students (the main pipeline for academia) in this country are women, only 46 percent of assistant professors, 38 percent of associate professors, and 23 percent of full professors are female. On top of that, women faculty in colleges and universities in the United States earn on average 10 percent less than their male counterparts.[434] A number of studies have shown that women in academia suffer from lower expectations for intelligence, so when they coauthor papers with male counterparts the assumption is that the males were the ones who did the actual work.[435] According to a new report recently released by the College and University Professional Association for Human Resources (CUPA-HR) there is a significant gender gap at the top levels of higher education leadership. Women administrators in higher education earn 80 cents on the dollar when compared to men. And despite claims by institutions of higher education that they are egalitarian and politically correct; this disparity has changed little over the last fifteen years.

As we can imagine, discrimination against women in academia is not something new. Hypatia of Alexandria was the first woman scholar for which we have any records. She was a Greek mathematician, astronomer, and philosopher in Egypt, then a part of the Byzantine Empire. She served as the head of the Neoplatonic school at Alexandria. During a civil revolt Hypatia was singled out by Christians for being a pagan and stoned to death. Her body was dismembered and her remains burned.
Now comes a book that although not as dramatic as Hypatia's story, is a bittersweet reminder of what women in academia have to endure. In *The Glass Universe*, Dava Sobel brings to the public eye the overlooked history of the women astronomers at the Harvard College Observatory around the turn of the twentieth century.

The story is about how the director of the Harvard Observatory from 1877 to 1919, Edward Charles Pickering, started to hire women to do the calculations needed in order to run the research at that facility. Obviously, that was well before actual computers came into existence. As Sobel recounts the story in this very well researched book, Pickering hired the Scottish émigré Williamina Fleming, a former teacher, as a maid after her husband abandoned her in a "delicate condition." However, Pickering recognized that Fleming's education and abilities were more suited for her to work at the observatory instead of as a maid. As soon as Fleming began to demonstrate her excellence at mathematical calculations, she was also given the responsibility of hiring dozens of other women to perform astronomical calculations.

The job was difficult not because mathematics has been stereotypically viewed as "not for women" but because it required workers to spend long, cold nights at the observatory and operate heavy telescope equipment, far from what was considered "womanly" at that time. In fact, that team of female mathematicians was called

---

[433] Originally publishes as Romero, A. 2017ap. Sobel, Dava. The Glass Universe: How the Ladies of the Harvard Observatory Took the Measure of the Stars. New York: Viking, 2016. 324 pp. $30.00 hardcover (ISBN 9780670016952). *Polymath* **7**(1):35-38. (Book Review).
[434] NSF. 2015. Women, Minorities, and Persons with Disabilities in Science and Engineering 2015. Washington, DC.
[435] Leslie, A.-J.; A. Cimpian, M. Meyer & E. Freeland. 2015. Expectations of brilliance underlie gender distributions across academic disciplines. *Science* **347**:262-265.

"Pickering's Harem" in a time when political correctness did not exist. Yet, the very fact that such a description was used indicates the view of women as scientists at the turn of the twentieth century. In a time when no electronic equipment existed, those calculations had to be done by hand for months, a really wearisome work.

Although Sobel is a woman, she does not use the narrative in her book to champion any feminist cause. On the contrary, her description of the facts is very sober and fair. This is not new for this author whose previous books on the history of the physical sciences include *Longitude*, *Galileo's Daughter* (for which she was nominated for the 2000 Pulitzer Prize for Biography or Autobiography), *Letters to Father*, *The Planets*, and *A More Perfect Heaven*. All these previous works display the same characteristics we find in this book: study of original documents, avoidance of judgment, descriptions of human behavior, and lack of historical relativism.

All that does not mean that Sobel's narrative is dull or colorless. It is through her well-written prose that the different characters in the story emerge with the impression from the reader that what you see if what you get. For example, Pickering comes across as a judicious and fair individual who respected women and provided them with the opportunities they deserved. That does not mean that he did not carry with him some of the flaws of his times. For example, he paid women less that he paid their male counterparts.

That is evidenced by some of the documents Sobel unearthed. In a journal kept by Williamina Fleming, one can read "He [Pickering] seems to think that no work is too much or too hard for me, no matter what the responsibility or how long the hours. But let me raise the question of salary and I am immediately told that I receive an excellent salary as women's salaries stand (...) Does he ever think that I have a home to keep and a family to take care of as well as the men? But I suppose a woman has no claim to such comforts. And this is considered an enlightened age!" Just to put things into context, Pickering was paying her $1,500 a year while the male assistants were being paid a thousand dollars more.

Yet, Pickering was fully committed to the advancement of women in science. For example, the first PhDs in astronomy at Harvard went to women under Pickering's mentorship.

Thanks to Fleming and other women astronomers such as Annie Jump Cannon, Henrietta Swan Leavitt, Antonia Maury, and Cecilia Payne, hundreds of thousands of stars were detected, classified, and cataloged. Thanks to their tedious work, later astronomers such as Edwin Hubble were able to accurately measure for the first time the size of the universe. It is interesting that this book comes on the heels of another story of women mathematicians—this time African-American—who were crucial as NASA employees who calculated orbital trajectories from the time of its creation to the Apollo 11 mission. The book *Hidden Figures* by Margot Lee Shetterly, was made into a Hollywood movie in 2016. Like their Harvard peers, the story of these women had been neglected but that is no longer the case.

Sobel's book is excellent at providing good historical background and scientific explanations that anyone can understand while being entertaining. It is divided into three parts and fifteen chapters. In addition to 32 plates, the book includes sections for sources, a historical chronology of the Harvard Observatory, glossary, biographical blurbs of Harvard astronomers, assistants, and associates, remarks, bibliography and index.

The Glass Universe (a title that makes a reference to the fact that the photographic plates at that time were made of glass although one cannot avoid thinking of the glass ceiling as a metaphor) is a worthwhile read for being informative, well written, and pleasurable.

Chapter 33. Salvador Gilij's Narrative about the Orinoco fauna

Introduction

In 1989 I was the Executive Director of BIOMA, the Venezuelan Foundation for the Conservation of Biological Diversity. I had hired Alfredo Paolillo as my first director of scientific information services. One day he told me that he had been contacted by the editors of a journal titled *Revista Montalbán*, published by the Universidad Católica Andrés Bello in Caracas. They had asked him to analyze the description of the fauna of the Orinoco River watershed in both Venezuela and Colombia.

Filippo Salvatore Gilii (better known by his Spanish name spelled as Felipe Salvador Gilij) was an Italian Jesuit priest. He was born in Legogne, Umbria, Italy, in 1721. He entered the Jesuits in Rome in 1740 at the *Collegio Romano* (the Vatican University). He continued his studies the following year in Seville at the old Jesuit College of Saint Hermenegildo. In 1743 he was assigned to the missions of the Nuevo Reino de Granada under another Jesuit priest called José Gumilla. Gumilla, who is mentioned by Gilij, had been involved in missionary and exploration activities in northern South America. Both sailed to South America, and Gilij studied theology in the Universidad Javeriana in Bogotá. He was ordained priest, after which he traveled to the Orinoco River. In 1749 he established a mission known San Luis de la Encamarada.

When the Jesuits were expelled from Spanish possessions in 1767, he returned to Rome, where he completed the two books he remembered today. One was a Grammar and Dictionary of the Languages of Various Native Tribes. The other was "Saggio di Storia Americana, o sia Storia Naturale, Civile, e Sacra De regni, e delle provincie Spagnuole di Terra-ferma nell' America meridionale" ("Essay on American History...") first published in four volumes in 1768. Gilij died in Romero on March 19, 1789. In his "Essay," besides of talking about the accomplishments of the Jesuits in what would later be called Venezuela, he also wrote numerous observations about the fauna of the area.

Alfredo came to me to say that he would like me to be a co-author of his article. I said yes, but he is the one who deserves the credit for doing all the legwork. I just checked on some of the information regarding aquatic mammals. I am reproducing here the original article to make it readily available to a broader audience. I just corrected some typos from the published version. After toying with the idea of having translated into English, I realized that any translation from the passages by Gilij, written originally in Italian and later translated into Spanish, would lose a lot of the original writing's flavor. Hence, I left it in Spanish that includes all of the scientific names of the species mentioned. Our comments are in bold. Hopefully, this would be valuable to both historians and biologists interested in this topic.

SAGGIO
DI STORIA AMERICANA
O SIA
STORIA NATURALE, CIVILE, E SACRA

De' regni, e delle provincie Spagnuole di Terra-ferma
nell' America meridionale

DESCRITTA DALL' ABATE

FILIPPO SALVADORE GILIJ

E consecrata alla Santità di N. S.

PAPA PIO SESTO
FELICEMENTE REGNANTE.

TOMO IV.

Stato presente di Terra-ferma.

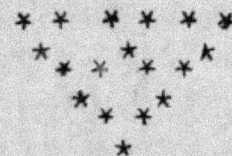

ROMA MDCCLXXXIV.
PER LUIGI PEREGO SALVIONI
Stampator Vaticano nella Sapienza

___

CON LICENZA DE' SUPERIORI.

Los relatos de la fauna orinoquense hechos por Felipe Salvador Gilij, evaluados con la óptica de la zoología del Siglo XX[436]

Alfredo Paolillo O.[437]
Aldemaro Romero Díaz[438]

Hemos revisado la traducción y el estudio preliminar hecho por Antonio Tovar, publicado en 1987 por la Academia Nacional de la Historia, de la obra publicada en 1782 por el misionero jesuita italiano Felipe Salvador Gilij titulada *Ensayo de Historia Americana o sea Historia Natural, Civil y Sacra de los Reinos y de las Provincias Españolas de Tierra Firme en la América Meridional; Tomo I: De la Historia Geográfica y Natural de la Provincia del Orinoco.*

Los autores analizan la visión propia del tiempo de Felipe Salvador Gilij, de la fauna de vertebrados del tramo central del rio Orinoco. Se han dejado al lado los fragmentos de su narración correspondientes a los invertebrados y a las plantas del mismo rio.

El aporte de Gilij al conocimiento de los vertebrados de dicha región no puede menos que considerarse variado y valioso. Al comparar la suya con las obras que conocemos escritas hasta la segunda mitad del siglo XVIII (e inclusive otras posteriores) y que de alguna manera clan información sobre la fauna de la actual Venezuela, no se puede menos que decir que el Torno I del *Ensayo de Historia Americana* de Gilij constituye un documento de primer orden para nuestra Historia Natural.

A los comentarios hechos por Gilij, aquí se presentan aclaratorias e identificaciones taxonómicas para un total de 46 especies de vertebrados, discriminados en dos de peces, 1 de anfibios, 8 de reptiles, 15 de aves y 20 de mamíferos silvestres, así como aclaratorias y aproximaciones de identificaciones taxonómicas para otras 19 descripciones zoológicas hechas por Gilij que involucran a mas de una especie posible para cada descripción.

Generalmente las descripciones hechas de la mayor parte de los animales mencionados por Gilij son tan precisas y ajustadas a la realidad que parecen establecer una diferencia apreciable con la mayor parte de los autores que le precedieron. No obstante, ello no impidió que en ocasiones el observador científico que parece Gilij se viera oscurecido por relatos de animales inexistentes, fabulosos y monstruosos, lo cual nos dice un poco del estado del conocimiento zoológico en la época que le tocó hacer su obra. La fauna americana, incluyendo la venezolana, apenas comenzaba a ser vista y estudiada por el continente europeo y con todas las conjeturas y aseveraciones hechas sobre supuestos animales y personas indescriptibles que poblaban el Nuevo Mundo, no es de extrañar que Gilij incurriera ocasionalmente en los mismos desaciertos.

Confiamos en que aquellos lectores poco familiarizados con la Zoología encuentren explicaciones y orientaciones precisas en los comentarios que hemos hecho a los relatos de Gilij, ya que estimamos que de otra manera algunas de las interpretaciones erróneas de Gilij sobre nuestra fauna orinoquense podrían tenerse como hechos reales. Los autores agradecen la cortesía de Angelina Pollak-Eltz (Universidad Católica Andrés Bello), quien nos solicitara preparar este

---

[436] Originally published in 1989 by *Revista Montalbán* (21):159-178.
[437] Director de Información Científica. Fundación Bioma, Apartado 1968, Caracas 1010-A, Venezuela.
[438] Director Ejecutivo. Fundación Bioma, Apartado 1968, Caracas 1010-A, Venezuela.

trabajo.

1. Refiriéndose a los peces del río Orinoco, Gilij dice: "No es, sin embargo notabilísima su variedad, y si son casi innumerables los individuos, no son demasiado numerosas las especies" (p. 90).

Los estudios modernos de la diversidad íctica de la cuenca del rio Orinoco demuestran lo contrario a lo afirmado entonces por Gilij. Se estima que allí habitan mas de 350 especies de peces.

2. "(...) es célebre la cachama (...) y las escamas son cenicientas (...) El morocoto, no muy distinto en el tamaño, pero mas piano que la cachama, es de escamas mas blancas (...)" (p. 90).

Se refiere a las especies del género *Colossoma*. En algunas regiones de la cuenca del Orinoco se usan indiferentemente ambos nombres comunes para las mismas especies.

3. "También de escamas, agradabilísimas, pero con un poco de sabor a lodo, en el que acaso se revuelcan, son las payaras" (p. 90).

Se refiere a la especie *Hydrolicus scomberoides*.

4. "Si nos atuviéramos al sabor, se preferiría la palometa a todo otro pescado (...) No se puede sin embargo hacer de ella sino un uso moderado, pues produce como efecto natural la fiebre" (p. 91).

Se refiere a algunas de las especies de los carácidos de los géneros *Mylossoma*, *Myleus* y *Metynnis*, de alto valor comercial en las pesquerías del Orinoco. La referencia al efecto febril de su carne no tiene fundamento.

5. "No es de mayor peso el pavón (...) nombre le fue dado por los españoles a causa de sus colores naturales, que parecen la cola de un pavo real" (p. 91).

Se refiere a cualquiera de las especies *Cichla ocellaris*, *C. temensis* y *C. nigrolineata*, altamente cotizadas entre los pescadores deportivos.

6. "El dorado (así lo llaman los españoles) es del tamaño de dos cachamas (...) Este pez habita en las cascadas" (p. 91).

Dorado *es el* nombre que actualmente se les da a algunos bagres del género *Brachyplatystoma*. Sin embargo, los bagres no son peces escamados, de manera que tal vez no se trate del mismo pez.

7. "Uno de los peces señalado del Orinoco es la curbinata (...) tiene en la cabeza dos huesecitos del tamaño de almendras sin cascara, trabajados bastante curiosamente por la naturaleza (...) pero que este pez sea confundido con (...) otros que (...) se in- dican como propios del solo Orinoco, lo <ludo mucho (...) lo creo común en otros muchos lugares, y fluvial y marino" (p. 91).

Gilij hace alusión a alguna especie de la familia *Sciaenidae*. Los "huesecitos" de la cabeza a los que se refiere son los otolitos, estructuras vinculadas al sistema auditivo. Como curbinata se conocen varias especies que no solamente viven en el Orinoco, sino también en ambientes estuarinos y marinas. Bajo el nombre de curbinata se conocen varias especies en Venezuela.

8. " (...) no debo callar el caribito. Llamase así por los españoles por el extraño amor que tiene a la carne humana" (p. 92).

Se refiere a varias especies posibles de los géneros *Serrasalmus*, *Pygocentrus* y *Pygopristis*. El pretendido amor de estos peces por la carne humana en realidad no es tal. Debido a que son carnívoros, en efecto muerden a los humanos, pero no por una predilección especial.

9. "el rey de los bagres (...) es el valentón, llamado por los indios laulau"

(p. 93).

Se refiere, probablemente, a las dos especies del género *Brachyplatystoma* conocidas hoy vulgarmente también como laulao (*B. vaillantii* y *B. filamentosum*).

10. "Del género de estos bagres con cuerno me parece que es el pez espada, que se dice tiene uno en la cabeza, largo y agudo y dentado por ambos lados (...) nunca los he visto" (p. 93).

Realmente sería especulativo decir a cuál pez se refería Gilij con una descripción tan vaga, pero es bastante improbable que se haya tratado de un bagre. Lo que si puede ser aclarado es que el lector no deberá confundir este pez con el pez espada marino, el cual no habita en el río Orinoco.

11. "No se si haya en el Orinoco, pero si en los lagos donde hay las palmeras muriche existe cierto pez llamado vulgarmente temblador (...) y según me es indicado por entendidos (...) tenemos, según les parece a los físicos, una máquina eléctrica en un ser acuático vivo (...) " (pp. 93-94).

El temblador si se halla en el Orinoco, aunque se le encuentra mas abundantemente en cuerpos de agua mas pequeños en toda su cuenca. Gilij hizo alusión a las facultades eléctricas de este pez, el cual es capaz de dar descargas hasta de 300 voltios. En condiciones naturales, el temblador se vale del campo eléctrico que se establece en su cuerpo para ubicar animales en su entorno. La descarga eléctrica también le sirve de mecanismo defensivo.

12. "Hay en abundancia una cierta especie de rayas (...) Se cree que el aguijón de la raya es de cualidad venenosa y frigidísima" (p. 95).

Las rayas del Orinoco a las que se refiere Gilij en realidad pertenecen a varias especies del género *Potamotrygon*. Efectivamente, tal como lo mencionó Gilij, el estilete óseo del cual esta provista la cola de estos peces produce heridas muy dolorosas e inyecta una potente toxina. Este veneno es aprovechado por algunos indígenas para proveer las puntas de sus flechas.

13. "Sabrosa también, y semejante en mucho a las nuestras, es una especie de anguila que los tamanacos llaman *camavá*. Están bajo los escollos en el agua (...) "(p. 95).

Se refiere a la especie *Synbranchus marmoratus*, conocida comúnmente como anguila de río.

14. "De los animales anfibios del Orinoco" (p. 96).

El lector debe tener en cuenta que el término "anfibio" utilizado por Gilij se refería a los hábitos tanto acuáticos como terrestres de algunas especies y no debe confundirlo con el significado que actualmente se le da al mismo término en el ámbito biológico (miembros de la Clase Amphibia, conformada por sapos, ranas, salamandras, tritones y cecilianos).

15. En muchas partes, pero especialmente al pie del monte Poco-pocori, llamado también el Capuchino, hay gran abundancia de una especie de gaviotas, llamadas por los españoles cotudas(...) se entienden muy bien en el agua, sin la cual no sobrevivirían quizás sino con esfuerzo (...) Son de color negro (...) "(p. 96).

Gilij se refiere a la cotúa zamurita (*Phalacrocorax olivacea*), ave ictiófaga y de hábitos gregarios.

16. "Están igualmente en las orillas, y se meten a menudo en el agua, los chigüiros(...) Su tamaño, pelaje y todo el resto se parece a un jabalí (...) son poquísimo sabrosos" (p. 96).

Los chigüires no están emparentados con los jabalíes y en realidad la semejanza que Gilij estableció entre ambos animales es bastante forzada. A diferencia del jabalí, el chigüire (*Hydrochaerus hydrochaeris*) es un roedor, siendo el roedor viviente mas grande del mundo. El desagrado de Gilij por la carne del chigüire debe verse como un gusto muy personal, pues lo cierto es que esta especie actualmente es manejada comercialmente en los Llanos venezolanos, principalmente para el consumo de su carne en el mercado nacional. Por otra parte, el uso de la denominación "chigüiro" a la que se refiere Gilij está restringido a Colombia.

17. "Pez sin duda es el manatí (...) ternero o vaca marina (...) cría a sus pequeñuelos, como los animales terrestres, a sus pechos (...) no es, como la imaginado raramente alguno, animal que ponga huevos, ya que es sin duda vivíparo (...) Quien tiene la suerte de matar al manatí, lo eleva a la aldea casi como en triunfo de su valor (...) y por el extraordinario gusto que todos tienen en el manatí (...) suelen llegar a aquellas aldeas guamos para matar y vendérselos a los orinoquenses. Suele hacerse, y la hacia yo también, cuando ya llega la cuaresma, provisión de uno o de varios manatíes para el ayuno cuaresmal" (pp. 97-98).

Debido a sus hábitos acuáticos y a la morfología de su cuerpo, el manatí era considerado en la época del Gilij y durante muchos años después como un pez. Esta especie del Orinoco (*Trichechus manatus*) es un mamífero y no un pez. No obstante, el mismo Gilij anotaba que sus crías se alimentan de sus pechos y descartaba la oviparidad de este animal. En su relato, se observa que la presión de cacería sobre el manatí era importante y actualmente esta considerado por la Unión Internacional para la Conservación de la Naturaleza y los Recursos Naturales como una especie vulnerable, en vista de sus reducidas poblaciones silvestres. Estudios recientes han evidenciado una preocupante ausencia de este animal en el Orinoco, donde antes fuera abundante. El uso del manatí como alimento en la Cuaresma refleja nuevamente la convicción propia de la época, aunque errónea, de que dicho animal era un pez, creyéndose por lo tanto propicio para el ayuno cuaresmal. La figura que acompaña la p. 82 muestra un indígena desollando un manatí, el cual, si bien tiene notables errores de dibujo, puede ser reconocido por el característico extremo caudal de los mamíferos sirenios.

18. "No es anfibio (...) un cierto animal que los españoles llaman tonina ... Pero no imagine nadie que la tonina del Orinoco es nuestro atún confundiéndose con el nombre italiano de *tonno* (...) sale del agua hasta la mitad de su cuerpo (...) Parece en la figura un puerco marino" (p. 99).

La tonina (*Inia geoffrensis*) es otro mamífero acuático que antiguamente fue con frecuencia considerado un pez. Es un cetáceo de agua dulce y la observación de Gilij en relación a su salida parcial del agua esta relacionada con la respiración pulmonar que la tonina efectúa. La comparación de su cuerpo con el de "un puerco marino" es forzada.

19. "Son de modo semejante anfibios, y reputadas también peces, las iguanas. Así son llamados en el Orinoco ciertos lagartos grandes o serpientes cuadrúpedas (...)" (p. 99).

Como se puede notar, no parece haber estado muy claro para Gilij si las iguanas eran peces, anfibios, lagartos o serpientes. Por el relato, parecía inclinarse mas bien por considerarlas serpientes, pues tal era el principal hecho. que a su juicio le producía rechazo a comer sus huevos ("Del todo malos no son. Pero son de serpientes, y ¿quien no dirá que son repugnantes para cualquiera que haya sido honradamente educado?"). (p. 100).

Las iguanas son lagartos y por lo tanto reptiles. Los hábitos anfibios mencionados por Gilij en realidad son bastante reducidos, siendo esencialmente un animal arborícola, consumidor de hojas y flores principalmente.

20. "En el Orinoco (...) hay una gran serpiente llamada por los españoles caimán (...) no es sino el cocodrilo, tan conocido en Egipto (...) Sus huevos (...) se comen (...) Quise, puesto que son comunísimos en el Orinoco, probarlos yo también" (pp. 100- 102).

Esta "gran serpiente" a la que se refería Gilij en realidad es un cocodrilo, no una serpiente como es entendida por la Zoología actual. Sin embargo, el caimán del Orinoco (*Crocodylus intermedius*) no es la misma especie de Egipto mencionada por Gilij. La referencia que el hace sobre lo común que era la especie en el Orinoco contrasta con la gravísima situación que este caimán enfrenta para su supervivencia. Es una especie en peligro de extinción, según lo definido por la Unión Internacional para la Conservación de la Naturaleza y los Recursos Naturales y en 1984 fue declarada una de las doce especies animales en mayor peligro a nivel mundial.

21. "No desemejantes en la figura de los caimanes, pero con mucho mas pequeñas (...) son las bavillas (...) Se vuelven locos por esta serpiente todos los indios (...) "(p. 103).

Estas "bavillas" son los reptiles conocidos en la actualidad como babos o babas en Venezuela. Babilla se le denomina principalmente en Colombia. Tampoco son serpientes en la interpretación moderna del término biológico.

22. "En las aguas del Orinoco (...) hay ciertos animales muy semejantes al perro (...) los españoles los llaman perros de agua (...) su pelaje (...) es suavísimo al tacto y estimado universalmente por todos" (p. 103).

Gilij se refiere en este párrafo al mamífero mustélido *Pteronura brasiliensis*, otra de las especies en peligro de extinción, según la Unión Internacional para la Conservación de la Naturaleza y los Recursos Naturales. La causa de su desaparición progresiva es la misma a la que se refiere Gilij: su cotizada piel. Aún sobreviven pequeños grupos de este mamífero acuático en algunos tributarios del Orinoco.

23. "El tiburón, animal o pez ferocísimo (...) Se encuentran poquísimos en el Orinoco (...) uno fue rescatado con anzuelo (...) en el escollo Aravacoto" (p. 104).

Según el mapa de la p. 31 (Carta del Fiume e Provincia dell' Orinoco Nell' America Merid.), la Encaramada aparece situada en la costa derecha del Orinoco, frente a la desembocadura del Apure. El escollo o roca a Aravacoto (p. 40) es ubicada por Gilij "bajo la Encaramada", de manera que ese tiburón fue pescado a unos 900 kilómetros del mar. Gilij al hacer su relato estaba señalando algo que la Ciencia moderna ha comprobado suficientemente, como lo es la incursión ocasional de tiburones oceánicos hacia ríos de gran caudal. No resulta posible indicar cual especie fue relatada por Gilij, debido a que varias especies de tiburones podrían remontar el Orinoco desde su desembocadura en el Océano Atlántico; sin embargo, el tiburón toro (*Carcharinus leucas*), que también habita en Venezuela, ha sido registrado en el río Amazonas en Perú.

24. "Las especies de tortugas son varias (...) dejando ahora las terrestres, divido las tortugas fluviales en dos clases. Unas se llaman terecayas (...) Otras se comprenden bajo el nombre de tortugas (...) "(p. 105).

Estas dos especies a las cuales se refiere Gilij son *Podocnemis unifilis* (terecay) y *Podocnemis expansa* (tortuga arrau). Sin embargo, en ese tramo del río Orinoco hoy día sabemos que existen al menos otras cinco especies acuáticas. La denominación "tortuga" en la región del Orinoco suele ser aplicada exclusivamente a la "arrau" *(Podocnemis expansa)*.

25. "El macho es muy pequeño (...) He visto en tantos años, pero son rarísimos, y se pretende allí que cada nido no tiene más que un solo macho. Es acaso diversa la cáscara del huevo de que nacen los machos (...) "(p. 105).

Efectivamente, la proporción natural de sexos de la tortuga *Podocnemis expansa* es de aproximadamente 30 hembras por cada macho. El reconocimiento de los pocos machos presentes en un nido no es posible tan solo observando la cascara de los huevos.

26. "Sin notarse ninguna disminución de ellas, se comen continuamente por los españoles y por los indios (...) "(p. 106).

El desmedido consumo de estas tortugas, basado casi exclusivamente en las hembras que salían a nidificar y en los huevos conseguidos en los nidos, finalmente permitió que si se notara una grave disminución de ellas. Actualmente la especie está en peligro de extinción, de acuerdo a lo establecido por la Unión Internacional para la Conservación de la Naturaleza y de los Recursos Naturales.

27. "En la playa amplísima que esta abajo de Uruana vi una tarde con mis propios ojos tanta multitud de tortugas, las unas cavando con sus patas la arena, otras poniendo huevos, que quedé sumamente maravillado" (p. 107).

La observación diurna de Gilij resulta importante a la luz de la abundancia actual de la tortuga. En el presente, su niclificaci6n es un proceso reproductivo que ocurre casi exclusivamente de noche. Esto es consecuencia de una disminución poblacional palpable y del incremento de los factores perturbadores de la tranquilidad que este animal requiere para salir a desovar.

28. "(...) y en parte de una ligera lluvia que sobreviene periódica- mente en aquella época y se llama por eso aguacero de las tortugas, saben los indios que es llegado el tiempo de satisfacer su golosina" (p. 108).

Durante aquella época, las indígenas consumían los tortuguillos recién eclosionados. Su nacimiento ocurre característicamente en los primeros días de mayo, cuando se inician las lluvias en el sector medio del rio Orinoco.

29. "Descargadas de los huevos, y vuelto al rio a su lecho natural, engordan (no si con peces pequeños o con frutas que caen de los árboles) las tortugas de maravilloso modo" (pp. 109- 110).

Los estudios biológicos realizados con la tortuga permiten afirmar hoy que su dieta es básicamente de origen vegetal, siendo preferidos los frutos. Sin embargo, también puede alimentarse de materia de origen animal (esponjas de rio, animales muertos).

30. "Del aceite que se extrae de los huevos de las tortugas" (pp. 110-114).

El relato que hace Gilij con respecto al proceso de extracción del aceite de los huevos de tortuga, así como su comercialización y usos es verdaderamente meticuloso. Puede decirse que dicho relate es inclusive mas precise que el hecho per Humboldt, quien hizo mas énfasis en los cálculos de los huevos explotados para la preparación del valioso aceite.

31. "En los tiempos lluviosos, no se ven mas que (...) las raras (...) Aprenden alguna palabra (...) los tamanacos, los caribes y otros muchos indios (...) las llaman *ara*. Los españoles (...) las llaman guacamayas (...) Existen estas, son de plumas unas rojas y otras turquí. La segunda especie es de tamaño

mas pequeño (...) son todas turquí La tercera finalmente son otras (...), son verdes" (pp. 114-115).

Las tres especies a las que se refería Gilij deben haber sido la guacamaya bandera (Ara *macao*), la guacamaya azul y amarilla (*Ara ararauna*) y la guacamaya verde (*Ara militaris*). Estas aves son muy solicitadas como mascotas y sus poblaciones en vida silvestre se encuentran amenazadas.

32. "Estimables (...) por su carne, son los paujíes (...) de color negro (...) los mas comunes, de los cuales abundan las selvas vecinas al Orinoco (...) son clcl color (...) que he dicho (...) El distintivo mas particular de Ios paujíes del Auvana son ciertas plumitas rizadas que a modo de mono o de cresta tienen en la cabeza" (pp. 115-116).

El sector del Orinoco en el cual estuvo Gilij tiene tres especies de paujíes. El que refiere con cresta rizada es bien el paují culo blanco (*Crax alector*) o bien el paují de copete (*Crax daubentoni*), mientras que la otra especie podría ser el paují culo colorado (*Mitu tornentosa*) o bien una de las antes indicadas. Todos los paujíes tienen actualmente problemas de sobrevivencia, debido principalmente a la destrucción de hábitat y a cacería excesiva.

33. "Seame permitido salir por poco tiempo de los límites que me he señalado, y hablar de una tercera especie de estos pájaros que vi en la Guaira. Esta en la cabeza, en vez de cresta o rizo; tiene un hueso de color castaiio (...) "(p. 116).

Gilij hace alusión al paují copete de piedra (*Pauxi pauxi*) el cual ya no es tan abundante como entonces y es el paují mas amenazado en Venezuela.

34. "Las orillas del Orinoco ... abundan de anades y de patos. Los mas hermosos y mas grandes son unos que los españoles llaman patos reales (...) "(p. 116).

Este pato es una valiosa especie de caceria. Se denomina científicamente *Cailina moschata*.

35. "Los anaclcs (...) tienen la carne mejor, y creo que puede ser causa de ello la larga morada que, en gran multitud, hacen en lugares húmedos que acaba de abandonar el rio (...)" (p. 116).

Estos patos gregarios son los conocidos como silbaclores (*Dendrocygna autumnalis*, D. *bicolor* y D. *11iduata*), también aves de cacería y que en algunas regiones llaneras actualmente representan cierto daño a los cultivos de arroz.

36. "En los tiempos secos (...) se ven muchos lugares llenos de pájaros blancos. Llamémoslos picazas (...) El rey, digámoslo así, es cierto ave (...) llamada por los españoles el soldado (...)" (p. 116).

Las "picazas" de Gilij son las aves zancudas conocidas actualmente como garzas, gabanes y garzones. El "soldado" al cual se refiere el misionero es el garzón soldado (*Jabiru mycteria*), cicónido de 1,30 m. de longitud corporal.

37. "Pero donde me dejo yo a los guanavares (...) Revolotean primero por el rio y por la playa, y sobre la cabeza misma de los navegantes (...) "(p. 117).

La denominación moderna de estas aves es "guanaguanare". Es conveniente hacer una corrección a la nota del traductor que aparece al pie de la pagina: "1 En español guananas o patos carreteros". El guanaguanare (*Phaetztsa simplex*) es una gaviota y no se le debe confundir con el pato carretero (*Neochen jubata*), especie que si bien vive en el Orinoco no corresponde a la descripción hecha por Gilij.

38. " (...) las guacharacas se reúnen en multitud en los montes vecinos a los

poblados (...) hacían ruido en tal abundancia (...) "(p. 208).

Esta ave que tantas veces sirvió a Gili j de sustento es la guacharaca común (*Ortalis rnficmula*).

39. "Se encuentran aves llamadas (...) por los españoles (...) predices" (p. 208).

Las perdices a las que se refiere Gilij son de la especie *Colinus cristatus*".

40. No vi nunca, porque es de matorrales mas espesos, el pájaro *nemi* (...) el huevo (...) es de cascara verde y de sabor bastante bueno" (p. 208).

Es muy probable que Gilij haya conocido los huevos de alguna especie de gallina de monte (*Tinamidae*), algunas de cuyas especies tienen la cascara del huevo verde azulado.

41. "El mas celebre entre los papagayos es el cori. Tiene en lo mas alto de la cabeza, un grupo de plumas rojas muy bonitas (...) aprenden muy bien a hablar y es un placer oírlos parlotear en las diversas lenguas de los indios" (p. 208).

Este "papagayo" debe haber sido el loro real (*Amazona ochrocephala*), el loro mas comúnmente utilizado como mascota en Venezuela, precisamente por su habilidad para repetir palabras y sonidos, así como por su colorido.

42. "No debe omitirse entre los volátiles singulares del Orinoco el pájaro vaco" (p. 210).

Como "pajaro vaco" se conocen actualmente dos especies de garzas, *Tigrisoma lineatum* y T. *fasciatum*.

43. "Es (...) bastante raro en las cercanías del Orinoco el célebre pájaro quiapocó. De este pájaro (...) se celebra sobre todo lo demás su pico (...) extraordinariamente grueso y no corresponde nada al resto del cuerpo" (p. 211).

Esta ave es el tucán, diostede o piapoco, termino este último que corresponde sin duda a la denominación "quiapocó" señalada por Gilij. En el rio Orinoco hay al menos unas cuatro especies del genero *Ramphastos*, al cual pertenece.

44. "El llamado cardenal, porque parece exactamente que lleva el capelo en la cabeza, es semejantemente de un canto muy agradable" (p. 211).

Posiblemente Gilij se haya referido a la cardenal bandera alemana (*Paroaria gularis*), de cabeza y copete rojo.

45. "El turpial merecería que su canto y por la singular belleza ser llevado a Italia" (p. 211).

El turpial (*Icterus icterus*) es el ave nacional de Venezuela, en reconocimiento a los atributos desde entonces ya apreciados por Gilij.

46. "Otro aborto de pájaro llamado en español pereza, es insufrible por los lamentos que de continuo emite por la noche. Este infeliz pájaro muchos creen que es una especie de fiera cuadrúpeda" (p. 212).

La pereza a la cual se refiere Gilij es en efecto un ave, denominada popularmente pereza de plumas. En la región central del Orinoco hay tres especies de esta ave nocturna, del genero *Nyctibius*.

47. "Pájaro también nocturno, o topo, o lo uno y lo otro, es el murciélago (...) en el Orinoco los murciélagos se mantienen, como de alimento natural, de sangre humana (...) Muerden mientras se duerme la extremidad de los dedos de los pies... yo los conocí de dos clases (...) v los dos son atraídos malamente por la sangre" (pp. 212-213).

En primer lugar, debe ser aclarado que los murciélagos son mamíferos voladores, no pájaros. Tampoco son ratones, termino no utilizado en la traducción pero que probablemente corresponda a lo que quiso decir Gilij. El lector deberá tomar en cuenta que en idioma italiano el termino para la

palabra española "ratón" es topo, de donde se sospecha la comparación indicada en la traducción. Por otra parte, solamente una ínfima parte de los murciélagos se alimenta de sangre.

> 48. "Pero hacen sus veces ciertos fetidísimos pájaros que se llaman gallinazos (...) suben tanto, que se pierden de vista en pocos momentos. Se dice que son de vista agudísima, y que por eso precisamente suben tan alto, para ver desde allí la presa y acudir en bandadas a devorarla (...) Su alimento son (...) animales cualesquiera, que yacen muertos por las campañas (...) las crías tiernas (¿quién lo creería?) son blancas" (pp. 214- 215).

El relato de Gilij sobre los gallinazos o zamuros, como comúnmente se les denomina en Venezuela, muy seguramente esta referido a la especie *Coragyps atratus*. Sin embargo, es posible que también haya observado algunos ejemplares de otras aves necrófagas similares a los zamuros, los oripopos (varias especies del genero *Cathartes*).

> 49. "Es cosa admirable en estos pájaros que siendo (...) todos negros, su rey (así se llama allá) es blanquísimo ... Yo me atendría al parecer de los que pretenden que este rey sea uno de los gallinazos viejísimos, y en efecto, cuanto mas envejecen, mas blancos se ponen" (p. 215).

Este animal al que hace alusión Gilij en realidad no es un ejemplar viejo del zamuro negro. Es una especie completamente diferente (*Sarcoramphus papa*). El respeto de sus "súbditos" no es otra cosa que una manifestación del comportamiento alimentario interespecífico ante el mismo alimento. El rey zamuro, como se le conoce a la especie indicada, tiene prioridad sobre los zamuros comunes al momento de consumir un cadáver.

> 50. "El araguato es del tamaño de un perro ordinario, pero de larga barba de color rojizo y de cola larga" (pp. 217-218).

Este mono es la especie *Alouatta seniculus*, muy común en los bosques del Orinoco.

> 51. "Pero el mas hermoso mono del Nuevo Mundo es, a lo que me parece, el caparro. Ha sido descubierto en estos ultimos tiempos, y que yo sepa no se encuentra mas que en el río Guaviare" (p. 218).

Gilij hizo referencia al mono caparro, conocido científicamente como *Lagothrix lagotricha*. Esta especie no ha sido registrada en territorio venezolano y todos los animales que se han mantenido cautivos en el país han procedido de la Amazonia de Colombia.

> 52. "El tigre (...) no se defiende sólo de quien le ataca, sino que ataca sin ser irritado, y busca cruelmente personas a quienes devorar. Existe en el Orinoco. . . en abundancia increíble (...) el tigre en aquellos lugares es un animal tan frecuente, que no creo haya en Italia país en que se vean tan frecuentemente los lobos (...) " (p. 219).

La aludida abundancia poblacional del tigre, o jaguar, como también se le conoce a la especie *Panthera onca* ha quedado como registro histórico en Venezuela. Considerado en peligro de extinción por la Unión Internacional para la Conservación de la Naturaleza y de los Recursos Naturales, se ha visto reducido progresivamente por la cacería excesiva y la pérdida de sus hábitats.

> 53. "He oído muchas veces decir que el tigre no mata con los dientes, sino con las uñas, las cuales son, a lo que se dice, muy venenosas. Su herida se encona al poco tiempo, y si no se cura prontamente, se gangrena" (p. 219).

El atributo tóxico de las uñas del tigre es una afirmación sin fundamento. Las gangrenas son producto de la contaminación posterior de las heridas.

54. "No sé si será creído, pero los habitantes del Orinoco son de parecer que el tigre, cuando viene de noche a alguna ranchería donde hay muchos durmiendo, escoge entre los durmientes para presa el mas débil. Si hay, pues, españoles, negros e indios, se lleva a estos ultimas, que son considerados de poco espíritu. Si no hay mas que los primeros, le toca al negro la fiesta. Al español, como al mas valiente, es el ultimo al que ataca el tigre" (p. 221).

Este es otro comentario sin ningún tipo de basamento. Es atribuible a la concepción social y humana de los tres grupos étnicos que caracterizaron el poblamiento de Venezuela, pro- ceso en el cual cl español fue el grupo dominante y que estableció las reglas de juego sobre los otros dos.

55. "El *uayapari* es otra especie de tigre. No es a pintas como los antedichos, sino de color entre el castaño y el rojizo (...) los españoles lo llaman león" (p. 221).

Este es cl león americano, o puma, conocido científicamente como *Felis concolor*.

56. "El año 1766 fue descubierta en el Cuchivero una nueva especie de tigres, y a lo que entendí, era toda negra, aunque del tamaño y hechuras de la otra" (p. 221).

Esta nueva especie en realidad no lo era. Considerados por mucho tiempo como animales diferentes al tigre común, se demostró posteriormente que los ejemplares negros tan sólo representan casos de melanismo (pigmentación oscura generalizada en todo el cuerpo).

57. "El tigrito debe contarse entre los animales de esta clase (...) Es con pintas (...) "(p. 221).

Este animal corresponde en realidad a otros dos felinos presentes en la región, conocidos actualmente como cunaguaros (*Felis vardalis* y *Felis wiedi*).

58. "El *avare* (...) es de pelo castaño y de un olor ingratísimo. La hembra, desde el pecho hasta el bajo vientre," tiene una abertura (...), dentro de la cual tiene agarradas a las mamas a las crías (...) tienen la cola (...) sin pelo alguno (...) y no salen a comer los frutos (...), sino de noche" (p. 222).

Esta descripción, muy precisa, corresponde al rabipelado (*Didelphis marsupialis*), mamífero marsupial muy común en todo el país.

59. "El *yuorocó* (...) Es de color rojizo, de orejas tiesas, y muy semejante al perro. Los españoles lo llaman zorra (...) "(p. 222).

Gilij está refiriéndose al zorro común (*Dusicyon thous*), mamífero cánido.

60. "Pero hablemos ya de un animal bípedo (...) Se encuentran en las grandes sabanas del Orinoco (...), ciertas fieras que, salvo pequeñas cosas, se parecen al hombre. Estos animales, que nosotros llamaremos el salvaje (...) De figura en todo lo restante humana, el salvaje no se diferencia mas que en los pies, cuyas puntas están naturalmente vueltas hacia atrás (...) Es todo peludo de cabeza a pies, sumamente libidinoso, y rapta si se le antoja a las mujeres (...) Sin embargo, no conocí a ningún indio que me dijese lo había visto con sus propios ojos. Aun- que esto mismo no es para mi argumento valedero para contradecir la voz de todas las naciones del Orinoco. Toda temen al salvaje, y como habita en lugares inaccesibles, nadie se atreve a acercarse a ellos por temer por su vida. Pero todos dicen las mismas cosas y narran de los hechos sucedidos a sus antepasados" (pp. 222-224).

Este pasaje del relato de Gilij no es otra cosa que una mas de las numerosas descripciones de seres monstruosos antropomórficos, mas digna de un bestiario que de una descripción zoológica propiamente dicha. El "salvaje" es

una creencia popular que en Venezuela actualmente aun persiste vivamente entre los habitantes de los Andes de Venezuela y de otros países andinos. En este ultimo caso, no se describe como un ser antropoide, sino que se le atribuyen los mismos detalles descritos por Gilij, pero al único oso suramericano viviente, el oso frontino o de anteojos (*Tremarctos ornatus*), conocido usualmente coma "el salvaje". El momento en que Gilij describe la fauna orinoquense, en cierta forma puede ser situado en un periodo transicional, entre la creencia de los mas variados mitos y leyendas sobre la fauna suramericana que alimentaron por mucho tiempo la imaginación europea y las descripciones precisas sobre la forma, hábitos y usos de dicha fauna. De hecho, cl mismo Gilij es uno de los primeros europeos que se refiere con relativa exactitud a muchos de los animales del Orinoco, pero como vemos no estuvo exento de creer y propagar la voz de ciertas bestias humanoides. Ya en la p. 104 del mismo libro que analizamos aquí, Gilij había asomado un detalle ilustrativo de esto último: " (...) entre las salivas y los negros no faltan personas que dicen haber visto cerca. La desembocadura del Paruasi sentados en los escollos ciertos animales semejantes a hombres. Si esto es verdad podríamos llamarlos sirenas (...) en el decir de muchos indios y de los españoles cabrutenses (...) una mañana al hacerse de día se vio (...) pasar por el Orinoco delante de Cabruta un animal de tan disforme mole, que parecía una pequeña casa. Dicen que estaba la mitad dentro y la mitad fuera del agua y que del alto Orinoco volvía al mar, de donde se creía venido".

61. "Oso, se llama en Orinoco, y en cierta manera se le asemeja, a un animal (...) de larga y hermosa cola, de color ceniciento; y de boca tan estrecha, que parece un pequeño agujero (...) se levanta con ligereza increíble, y con las patas y con las uñas les hace el daño que no puede con los dientes (...) El alimento mas grato del osito son las hormigas bachacos" (p. 225).

Este animal es el llamado oso palmero (*Myrmecophaga tridactyla*), un mamífero edentado. El termino oso probablemente se le haya adjudicado por la conducta que muestra de defenderse con las garras mientras se encuentra alzado sobre sus patas traseras.

62. "Pero he aquí un alimento mejor en los jabalíes. En Orinoco se hallan en tanta cantidad, que (...) puede parecer increíble (...) semejantes a estos pequeños cerdos, excepto en los pies, que son blancos, y alguna que otra mancha también blanca, son las báquiras o paquiras. Tienen en el lomo una pequeña prominencia, que algunos escritores creen que es su ombligo. M. Bomare la llama bolsita. Y no me meta a decidir en favor de ninguna de las partes" (pp. 225-226).

Gilij tuvo oportunidad de conocer a las dos especies de cerdos silvestres que viven en Venezuela, el báquiro cachete blanco (*Tayassu pecari*) y el báquiro de collar (*Tayassu tajacu*), ambos animales de hábitos gregarios muy importantes para la alimentación de las poblaciones indígenas y rurales de Venezuela. Con respecto al "ombligo" aludido por Gilij, en realidad no es otra cosa que una glándula de almizcle, utilizada para el reconocimiento individual y marcar sus territorios vitales.

63. "La danta, animal frecuente en el Orinoco (...) es del pelo y casi del tamaño de un asno. Babita en las selvas (...) "(pp. 226-227).

La danta, o danto o tapir, coma también se le conoce (*Tapirus terrestris*), es otra valiosa especie para el suministro de proteínas en el río Orinoco. Sus poblaciones al norte de este río se encuentran gravemente disminuidas y en algunas regiones han sido completamente exterminadas.

64. "Mas que las dantas abundan en el Orinoco los ciervos (...) Existen dos clases. Unos (...) tienen como los nuestros los cuernos ramificados. Otros (...) excepto los cuernos, que son de la longitud de medio dedo y peludo, son muy semejantes a las otros" (pp. 227-228).

La primera de las especies mencionadas por Gilij es el venado caramerudo (*Odocoileus virginianus*). En cuanto al segundo tipo de venado descrito, de cuernos cortos, pudo haberse tratado de cualquiera de estas dos especies de venado matacanes o lochos: *Mazama americana* y *M. gouazoubira*, ambas presentes en el Orinoco.

65. " (...) al *aruru*. Este animal es una especie de puercoespín. Sus espinas (...) son entreveradas de blanco y negro. Pesa cinco a seis libras (...)" (p. 228).

Este puercoespín corresponde a la especie *Coendu prehensilis*, muy común en los bosques orinoquenses.

66. "El cachicamo, cubierto todo de cabeza a pies de escamas apretadas y durísimas, no es (...) frecuente en el Orinoco. Pero en las llanuras del Meta se encuentran muchos (...) Vive comúnmente en los prados y en ellos hace agujeros en que habitar" (p. 229).

Parece ser que Gilij se refiere en realidad a al menos dos especies diferentes de cachicamos. En los bosques del Orinoco cl mas común suele ser *Dasypus noverncinctus*, aunque hay también otras especies de esta familia de edentados. La especie a la que se refiere de los llanos dcl Meta corresponde al cachicamo sabanero (*D. sabanicola*). Ambas especies son consumidas por los habitantes del Orinoco y actualmente la Medicina los está utilizando para el estudio de la lepra y su vacuna.

67. "En los prados de Cachichana (...) hay conejos salvajes. Son de color café, y de buen sabor, pero pequeños" (p. 229).

Este conejo silvestre es *Syfrilagus floridanus*. El lugar llamado por Gilij "Cachichana" hoy es conocido como Carichana.

68. "El *accuri*, que los españoles llaman picure, puede decirse también una especie de conejo, sino que es mas grande y acaso mas sabroso que estos mismos. Los hay por todas partes (...)" (p. 229).

Este fragmento de la narración de Gilij se presta a un poco de confusión al momento de asignarle identidad a los animales mencionados. Por una parte, en la región del Orinoco central hay un roedor silvestre conocido como acure (*Cavia porcellus*), demonización muy similar a la de "accuri". Por otra parte, también habita allí el picure (*Myoprocta pratti* y *Dasyprocta* spp.), nombre aparentemente dado por los españoles al mismo animal. En cualquiera de los dos casos, ninguno puede considerarse coma conejo.

69. "En los topos del Orinoco hallo una tercera especie de conejos (...) Los indios se vuelven locos por este topo, y van a buscarlo para dárselo como juguete a sus hijos" (p. 229).

Debido a que Gilij no describe rasgos morfológicos de este animal (aparentemente un roedor), no es posible su identificación. No obstante, la propensión a utilizarlo como mascota de los niños hace pensar en el picure pequeño (*Myoprocta pratti*), especie que actualmente es ampliamente usada como mascota por varias etnias indígenas del Orinoco.

70. "He visto la tercera especie de topos en los árboles de las selvas inundadas. Son de tamaño extraordinario (...)" (p. 229).

En las selvas inundadas del Orinoco habitan varias especies de roedores y marsupiales, estos últimos factiblemente confundidos por Gilij como roedores.

71. "La mas célebre entre todas las tortugas de tierra es el morrocoy (...) las escamas de sus patas son rojas (...) "(pp. 229- 230).

Sin duda se trata del morrocoy sabanero (*Geochelane carbollaria*), especie muy apetecida en todo el país.

72. "Mas grande que dos morrocoyes, pero de la misma forma, es el *timutú* (...)" (p. 230).

Esta es la otra especie de morrocoy presente en el Orinoco (*Geochelone denticulata*), mucho mas grande que *G. carbonaria* y de color amarillo.

73. "Son un poco diferentes los *tayelu*, que los españoles llaman terecayas. Pero los tayelos, aunque sean poco diferentes de las terecayas en la figura, son sin embargo distintos en su tamaño, que es notablemente pequeño (...) están buena parte del año fuera del agua (...)" (p. 230).

Gilij se refiere ahora al galápago llanero (*Podocnemis vogli*), efectivamente similar a la terecay (*P. unifilis*), pero de hábitos mas terrestres que esta. Es otra especie altamente consumida en la región central orinoquense.

74. "Pero matando al modo orinoqués una serpiente (...) Son muy cuidadosos de hacer una pequeña fosa donde meterla. Pero antes de enterrarla le parten la cabeza. Los huesos (...) son venenosos como sus dientes (...) "(p. 249).

En primer lugar, entre las serpientes del Orinoco Medio la menor proporción tiene facultades venenosas. Por otra parte, los huesos del cuerpo de las serpientes venenosas no tienen tales propiedades tóxicas. El veneno de estos animales se produce en una glándula situada junto al maxilar superior, la cual es drenada hasta el colmillo inoculador situado a cada lado de la cabeza.

75. "La maraca, llamada por los españoles la serpiente de cascabel, excede su potente veneno a toda otra raza de serpiente (...) Tiene en la extremidad de la cola varias sonajas, de donde tiene el nombre. Se dice que cada año echa una, y tantos años tiene la maraca como sonajas en la cola ... " (p. 250).

Esta serpiente es el cascabel común (*Crotalus durissus*). Modernamente se ha demostrado que los anillos de la sonaja no se corresponden con la edad del animal. Por el contrario, cada anillo representa una muda de piel de la serpiente y este hecho esta relacionado con el estado de salud y alimentación de cada individuo, no con su edad.

76. "Singular es también la serpiente *kiaucó-imu*, esto es, el padre de las hormigas bachacos, con las cuales convive. Los españoles la llaman la culebra de dos cabezas, pues a algunos les parece que ven otra en la cola del kiaucó-imu (...) me pareció mas bien que su extremidad *era* a manera de una cola cortada, entrando un poco hacia adentro, y no terminada en punta, coma otras serpientes (...) es de color ceniciento" (p. 250).

La llamada comúnmente "culebra de dos cabezas" ni es culebra ni tiene dos cabezas. Se trata de un reptil anfisbénido, cuyos extremos corporales son algo similares. Tal como le pareció a Gilij, el extremo posterior no es otra cosa que la cola del animal. En cuanto al significado de su nombre indígena, puede decirse que simboliza uno de sus hábitos de vida mas comunes, coma lo es vivir en bachaqueros y termiteros. Aunque en la región media del rio Orinoco hay al menos tres especies de estos animales, la especie sujeto de la descripción muy probablemente fue *Amphisbaena alba*.

77. "El *kiaucó-imu* se dice que es un remedio eficaz para las hernias y es muy buscado por los boticarios (...) El difunto (...) boticario del colegio de Santa Fe (...) pedía a menudo estas serpientes a los misioneros del Orinoco. Pero las quería no solo perfectamente matadas, sino bien

ahumadas y mantenidas por largo tiempo en un lugar muy seco (...) porque de otro modo, si no se hace así (...) vuelven enseguida a la vida con la lm- medad" (p. 250).

La facultad de volver a la vida de este animal es una fantasía mas de la época de Gilij. Sin embargo, debe ser ratificado el uso medicinal que se hace de *Amphisbaena alba* en el medio rural. Se le atribuyen propiedades curativas para fracturas de huesos y problemas respiratorios, entre otras. Es muy común observar actualmente que en las casas de los campesinos venezolanos hay una botella conteniendo una culebra de dos cabezas, o morrona, sumergida en aguardiente. Las aplicaciones son como fricción o bien ingiriendo el aguardiente añejado con el animal.

78. "(...) el buío (...) esta gran serpiente (...) por lo grueso, semejante a una viga. Es de color verde bastante oscuro, y habita en lugares húmedos en la proximidad de charcos. Los españoles la llaman tragavenados (...) Si, lo mis mismo que para las fieras, es también mortal para el hombre con el aliento que se dice exhala, no sabrían decirlo justamente, porque nunca he oído a los indios hablar de ello" (pp. 250-251).

La serpiente que describe Gilij es la culebra de agua o anaconda (*Eunectes murinus*), especie de hábitos semiacuáticos y común en el Orinoco. No es cierta la creencia, aun mantenida en nuestros tiempos por mucha gente, del poder letal de su aliento. También se dice que hipnotiza o atonta a los animales y personas con su "vaho". La denominación "tragavenados" no es aplicable a esta serpiente actualmente. Este es el nombre comúnmente aplicado a *Boa constrictor*.

79. " Bastante grande (...) son los sapos domésticos (...) Apenas venido el invierno, entran estos entran abundancia en las casas, y para no estar con huéspedes tan molestos, es preciso tener persona que continuamente los espante (...) "(p. 25 l).

Se trata del sapo común (*Bufo marinus*), conocidos por su afición a penetrar en medios antrópicos.

80. "El manatí no sólo se encuentra en el Orinoco (...) y otras partes de la América meridional mas caliente (...) Esto supuesto, no parecía que hubiera de vivir además en los mares fríos. Y, sin embargo, lo encontraron, y comieron en abundancia en él los rusos en la isla de Bering en cl mar glacial el año 1742" (p. 262).

Este fragmento citado en el capítulo "Notas y Aclaraciones" de la obra de Gilij, establece identidad entre el manatí del Caribe y norte de Sur América (*Trichechus* spp.) y cl también mamífero sirenio ártico conocido como "vaca marina de Steller" (*Hydrodamalis stelleri*). En realidad, estos animales pertenecen a familias distintas del Orden Sirenia. Resulta interesante mencionar que el año 1742, mencionado por Gilij como la fecha dcl hallazgo de la vaca marina de Steller, coincide on el año en el cual fue descubierto este animal por parte de la expedición del Capitán Vitus Bering. A consecuencia de la matanza irracional de este mamífero y a lo reducido de su distribución geogfica, *Hydrodamalis stelleri* fue extinguido por el hombre tan sólo 27 años después de su descubrimiento (Walker, E. et al. 1975. *Mammals of the World*, Third Edition, Volume II, The John Hopkins University Press; pp. 1334-1335).

81. "No se de qué manera creer que las tortugas (...) deban después, si se les impide ponerlos, esperar tranquilamente otro año para librarse de ellos. Yo (...) las creería capaces de ponerlos también en un espinar, si no queremos decir en el borde mismo del rio. Y sin embargo no es de esta

opinión cl P. Gumilla" (p. 263).

Gilij estaba en lo cierto en su suposición del capitulo "Notas y Aclaraciones". La tortuga al no encontrar el sitio ideal para nidificar o no poder contener el impulso de desovar, procede a expulsar los huevos en cualquier parte, inclusive dentro del agua.

82. "Se pretende que los huevos redondos contienen hembra, los alargados macho. Pero los de tortuga son todos redondos, los de las teracayas, alargados, según me parece" (p. 263).

En cl capitulo "Notas y Aclaraciones" Gilij insiste, acertadamente, en dudar sobre la correspondencia entre el sexo de las tortugas y la forma del huevo que las contienen. Efectivamente, también acierta al decir que los huevos de la tortuga son redondos, mientras que los de terecay son alargados.

83. "Tampoco vi nunca al mapurito. Pero no hay ninguno en el Orinoco, aunque contradiga lo que de el escribió Gumilla" (p. 276).

Hay que negar esta afirmación de Gilij aparecida en sus "Notas y Aclaraciones". En la región del Orinoco en la cual hizo sus anotaciones existe el mapurite o zorrillo (*Conepatus semistriatus*).

One of the drawings from Gilij's book (public domain image).

Section V. Letters to the Editor

This section reproduces three letters to the editor that have been published in major outlets. Two were to *The New York Times* and the other to *Science*, one of the world's top scientific journals. The letters were intended to correct significant mistakes about the history of science published in those media or clarify some statements made by the authors of those writings. I always believe that so much misinformation out there these days; we, in the academy, need to be more proactive in setting the record straight.

Humboldt's Travels[439]

To the Editor:

Candice Millard, in her review of Aaron Sachs's "Humboldt Current" (Aug. 13), mistakenly wrote that Alexander von Humboldt "felt a deep connection to the United States, where (...) he spent five years, from 1799 until 1804, traveling through the country, and 30 years writing about it." In fact. Humboldt and his companion Aimé Bonpland, spent most of those years in Latin America and only six weeks in the United States. And most of Humboldt's travel writing was about Latin America. Although he influenced American naturalists, they were not the exception but the rule: Humboldt had a tremendous impact on all naturalists of his and subsequent generations, from Charles Danwin down.

ALDEMARO ROMERO
Jonesboro, Arkansas
The writer is the Chairman of the Department of Biological Sciences at Arkansas State University.

Humboldt (left) and Bonpland (public domain illustration).

---

[439] Originally published as Romero, A. 2006f. Humboldt's travels. *The New York Times* **140**(53684): Section 7 (Book Reviews):5. 27 August 2006.

E- Letter responses to:
Review: Peter J. Bowler Darwin's Originality
Science 2009; **323**: 223-226 (Abstract)
The Erasmus Darwin's Influence on His Grandson Charles[440]
Aldemaro Romero (23 April 2009)
23 April 2009
Jonesboro, AR 72467, USA

I was surprised that Erasmus Darwin was not mentioned in P.J. Bowler's otherwise excellent article on Charles Darwin's ideas ("Darwin's originality," Review, 9 January 2009, p. 223).

Although Charles Darwin's synthetic and insightful mind led him to revolutionize the way we see nature, we should not forget the influence that his grandfather Erasmus had on him. When one reads Erasmus's *Zoonomia* (1794-1796) it is clear that he anticipated many of the main ideas that were later masterfully and abundantly substantiated by Charles. Charles was always very circumspect about the intellectual influence that his grandfather had on him; he said only that his grandfather had anticipated the erroneous ideas of Lamarck, despite the fact that Darwin followed mostly Lamarckian ideas when it came to explaining the rudimentation or loss of organs among many organisms, particularly cavernicolous ones *(1)*. This lack of acknowledgment may have been because, as Darlington put it, Charles wanted to distance himself from the most controversial writers on the topic that preceded him (2).

Aldemaro Romero
Department of Biological Sciences, Arkansas State University, Jonesboro, AR 72467, USA.

References
1. A. Romero, *Forum Public Pol.* 2, 867 (2007).
2. C.D. Darlington, *Sci. Amer.* 200, 60 (1959).

Erasmus Darwin (public domain illustration).

---

[440] Originally published as Romero, A. 2009. Erasmus Darwin's Influence on His Grandson Charles. *Science* E-Letter 23 April 2009 (http://www.sciencemag.org/cgi/eletters/323/5911/223).

Humboldt and Darwin

To the Editor:

Colin Thubron, in his review of Andrea Wulfs "The Invention of Nature: Alexander von Humboldt's New World" (Sept. 27), quotes Charles Darwin's reference to Humboldt as the "greatest scientific traveler who ever lived." However, Humboldt's influence on Darwin's work was much more than usually recognized. Humboldt's travel accounts about the American continent inspired Darwin's travel around the world on board the *Beagle*, and on that voyage, Darwin carried with him a copy of Humboldt's work.

Darwin's narrative, as well as those of many other 19th-century scientific explorers, was deeply influenced by Humboldt's style.

Further, Humboldt's observations on the distribution of plants and animals, as well as his belief that humans all around the world belong to the same species, had an important influence on Darwin's own views regarding natural selection and human evolution in general. No wonder Darwin cited Humboldt more than 400 times in his writings.

ALDEMARO ROMERO JR.
MARYVILLE, ILL.
The writer is *a professor of biology at SouthernIllinois University Edwardsville.*

A version of this letter appears in print on October 11, 2015, on page BR6 of the Sunday Book Review.

Alexander von Humboldt (left) and Charles Darwin (public domain images).

---

[441] Originally published as Romero, A. 2015. Humboldt and Darwin. *The New York Times* **155**(57,016): Section 7 (Book Reviews):6. 11 October 2015.

About the Author

Aldemaro Romero Jr. is a public intellectual, college professor, and author. He received his bachelor's degree in Biology from the University of Barcelona, Spain, and his Ph.D. in Biology from the University of Miami, Florida. He has published more than 1100 pieces including more than 30 books and monographs and hundreds of articles in both peer-reviewed and non-peer-reviewed publications. His experience in academia includes, but is not limited to, Director and Associate Professor of the Environmental Studies Program at Macalester College, MN (1998-2003), Chair and Professor of the Department of Biological Sciences at Arkansas State University (2003-2009), Dean and Professor of the College of Arts and Sciences at Southern Illinois University Edwardsville (2009-2014) and Dean of the Mildred and George Weissman School of Arts and Sciences at Baruch College-City University of New York (2016-2020). He has played an active role in fostering the need for promoting the value of a liberal arts and science education through the Council of the Colleges of Arts and Sciences (CCAS) for which he was the Chair of the Committee on Liberal Arts Institutions and as an active participant of the International Council of the Fine Arts Deans (ICFAD) for which he has been a board member. For more biographical information, visit http://aromerojr.net and http://en.wikipedia.org/wiki/Aldemaro_Romero_Jr.

Appendix 1. List of Published Contributions by Aldemaro Romero Jr., Ph.D., to the History and Philosophy of Science in Chronological Order

01. Romero, A. 1986. Charles Breder and the Mexican blind cave characid. *National Speleological Society News* **44**(1):16-18.
02. Romero, A. 1986. He wanted to know them all: Eigenmann and his blind vertebrates. *National Speleological Society News* **44**(11):379-381.
03. Paolillo, A. & A. Romero. 1989. Los relatos de la fauna orinoquense hechos por Felipe Salvador Gilij, evaluados con la óptica de la zoología del Siglo XX. *Revista Montalbán* (21):159-178.
04. Romero, A., A.I. Agudo & S.J. Blondell. 1997. The scientific discovery of the Amazon river dolphin *Inia geoffrensis*. *Marine Mammal Science* **13**(3):419-426.
05. Romero, A. 1997. Las Plantas del Mundo en la Historia. *Isis* **88**(4):699-700 (Book Review in English).
06. Romero, A. 1999. The blind cave fish that never was. *National Speleological Society News* **57**(6):180-181.
07. Romero, A. 1999. Myth and reality of the alleged blind cave fish from Pennsylvania. *Journal of Spelean History* **33**(4):67-75.
08. Romero, A. & Andrea Romero. 1999. Cope, Caves, and Skeletons in the Closet. *National Speleological Society News* **57**(11):341-343.
09. Romero, A. 2000. The speleologist who wrote too much. *National Speleological Society News* **58**(1):4-5.
10. Romero, A. 2000. Francisco Pelayo. Del diluvio al megaterio: Los orígenes de la paleontología en España (Book review in English). *Isis* **91**(1):134-135.
11. Romero, A. & K. Benz. 2000. The unsung heroes of speleology. *National Speleological Society News* **58**(4):106, 126.
12. Romero, A. & Z. Lomax. 2000. Jacques Besson, Cave eels and other alleged European fishes. *Journal of Spelean History* **34**(2):72-77.
13. Romero A. & K.M. Paulson. 2001. Humboldt's alleged subterranean fish from Ecuador. *Journal of Spelean History* **36**(2):56-59.
14. Romero, A. 2001. Scientists prefer them blind: The history of hypogean fish research. *Environmental Biology of Fishes* **62**(1-3):43-71
15. Romero, A. 2002. Between the first blind cave fish and the Last of the Mohicans: the scientific romanticism of James E. DeKay. *Journal of Spelean History* **36**(1): 19-29.
16. Romero, A. 2002. The life and work of a little known biospeleologist: Theodor Tellkampf. *Journal of Spelean History* **36**(2):68-76.
17. Romero, A. & J.S. Woodward. 2005. On white fish and black men: did Stephen Bishop really discover the blind cave fish of Mammoth Cave? *Journal of Spelean History* **39**(1):23-32.
18. Romero, A. 2006a. Steven Palmer. *From Popular Medicine to Medical Populism: Doctors, Healers, and Public Power in Costa Rica, 1800-1940*. *Isis* **97**(2):369-370.
19. Romero, A. 2006b. Humboldt's travels. *The New York Times* **140**(53684): Section 7 (Book Reviews):5. 27 August 2006.
20. Romero, A. 2006d. The big issue between science and religion: purpose vs. uncertainty. *Forum on Public Policy* **2**(4):867-881.
21. Romero, A. 2007. The discovery of the first Cuban blind cave fish: the untold story. *Journal of Spelean History* **41**(131):16-22.
22. Romero, A. & S.D. Kannada. 2008. Columbus, Christopher. Pp. 163-164, *In*: *Encyclopedia of Tourism and Recreation in Marine Environments* (M. Lück, Ed.). Oxfordshire, UK: CAB International.
23. Romero, A. & S.D. Kannada. 2008. Vespucci, Amerigo. P.170, *In*: *Encyclopedia of Tourism and Recreation in Marine Environments* (M. Lück, Ed.). Oxfordshire, UK: CAB International.

24. Romero, A. 2009. Erasmus Darwin's Influence on His Grandson Charles. *Science* E-Letter 23 April 2009 (http://www.sciencemag.org/cgi/eletters/323/5911/223).
25. Romero, A. 2009. *Cave Biology: Life in Darkness*. 291 pp. Cambridge: Cambridge University Press.
26. Romero, A. 2012. When whales became mammals: the scientific journey of cetaceans from fish to mammals in the history of science. Pp. 4-30. *In*: Romero, A. & E.O. Keith (Eds.). 2012. *New Approaches to the Study of Marine Mammals*. Rijeka, Croatia: InTech.
27. Romero, A. 2013. Floating Gold: A Natural (and Unnatural) History of Ambergris. Christopher Kemp. Chicago and London: The University of Chicago Press, 2012. 187 pp. $22.50 cloth (ISBN – 13: 978-0-226-43036-2). *Polymath* **3**(2):56. (Book review).
28. Romero, A. 2014. Felipe Poey, hyperbole, and the myth of the "isolated genius" among Spanish and Latin American scientists. *Polymath* **4**(4):85-95.
29. Romero, A. 2014. Constituciones De La Real Y Pontificia Universidad De San Gerónimo: Fundada En El Convento De San Juan De Letrán, Orden De Predicadores, De La Ciudad De San Cristobal De La Habana, En La Isla De Cuba – Primary Source Edition. (Facsimile Reproduction) Charleston, NC: Nabu Press 2013. 136 pp. $16.66 paperback (ISBN 978-1289778668). *Polymath* **4**(4):96 (book review).
30. Romero, A. 2015. Humboldt and Darwin. *The New York Times* **155**(57,016): Section 7 (Book Reviews):6. 11 October 2015.
31. Romero, A. 2016. The influence of religion on science: the case of the idea of predestination in biospeleology. *Research Ideas and Outcomes* **2**: e9015. DOI: 10.3897/rio.2.e9015. 20 pp.
32. Romero, A. 2016. *The New Celebrity Scientists. Out of the Lab and into the Limelight.* Fahy, Declan. Lanham: Rowman & Littlefield, 2015. 287 pp. $38.00 hardcover (ISBN 978-1-4422-3342-3). *Polymath* **6**(1):11-15.
33. Romero, A. 2016. Siddhartha Mukherjee. *The Gene. An Intimate History*. New York: Scribner, 2016. 593 pp. $32.00 cloth (ISBN 978-1-4767-3350-0). *Polymath* **6**(2):79-83.
34. Romero, A. 2017. Newman, Eric A., Alfonso Araque, Janet M. Dubinsky, Larry W. Swanson, Lyndel Saunders King, and Eric Himmel. *The Beautiful Brain: The Drawings of Santiago Ramón y Cajal*. New York: Abrams, 2017. 207 pp. $40.00 hardcover (ISBN: 9781419722271). *Polymath* **7**(1):32-34. (Book Review).
35. Romero, A. 2017. Sobel, Dava. *The Glass Universe: How the Ladies of the Harvard Observatory Took the Measure of the Stars*. New York: Viking, 2016. 324 pp. $30.00 hardcover (ISBN 9780670016952). *Polymath* **7**(1):35-38. (Book Review).
36. Romero, A. 2018. In Memoriam: Herbert L. Needleman (1927-2017). *Environmental Research* **165**:507-509. (DOI: https://doi.org/10.1016/j.envres.2017.11.048)

www.ingramcontent.com/pod-product-compliance
Lightning Source LLC
Chambersburg PA
CBHW081423220526
45466CB00008B/2255